MATH-
STAT.
LIBRARY

Recent Titles in This Series

153 **Richard S. Elman, Murray M. Schacher, and V. S. Varadarajan, Editors,** Linear algebraic groups and their representations, 1993

152 **Christopher K. McCord, Editor,** Nielsen theory and dynamical systems, 1993

151 **Matatyahu Rubin,** The reconstruction of trees from their automorphism groups, 1993

150 **Carl-Friedrich Bödigheimer and Richard M. Hain, Editors,** Mapping class groups and moduli spaces of Riemann surfaces, 1993

149 **Harry Cohn, Editor,** Doeblin and modern probability, 1993

148 **Jeffrey Fox and Peter Haskell, Editors,** Index theory and operator algebras, 1993

147 **Neil Robertson and Paul Seymour, Editors,** Graph structure theory, 1993

146 **Martin C. Tangora, Editor,** Algebraic topology, 1993

145 **Jeffrey Adams, Rebecca Herb, Stephen Kudla, Jian-Shu Li, Ron Lipsman, Jonathan Rosenberg, Editors,** Representation theory of groups and algebras, 1993

144 **Bor-Luh Lin and William B. Johnson, Editors,** Banach spaces, 1993

143 **Marvin Knopp and Mark Sheingorn, Editors,** A tribute to Emil Grosswald: Number theory and related analysis, 1993

142 **Chung-Chun Yang and Sheng Gong, Editors,** Several complex variables in China, 1993

141 **A. Y. Cheer and C. P. van Dam, Editors,** Fluid dynamics in biology, 1993

140 **Eric L. Grinberg, Editor,** Geometric analysis, 1992

139 **Vinay Deodhar, Editor,** Kazhdan-Lusztig theory and related topics, 1992

138 **Donald St. P. Richards, Editor,** Hypergeometric functions on domains of positivity, Jack polynomials, and applications, 1992

137 **Alexander Nagel and Edgar Lee Stout, Editors,** The Madison symposium on complex analysis, 1992

136 **Ron Donagi, Editor,** Curves, Jacobians, and Abelian varieties, 1992

135 **Peter Walters, Editor,** Symbolic dynamics and its applications, 1992

134 **Murray Gerstenhaber and Jim Stasheff, Editors,** Deformation theory and quantum groups with applications to mathematical physics, 1992

133 **Alan Adolphson, Steven Sperber, and Marvin Tretkoff, Editors,** p-Adic methods in number theory and algebraic geometry, 1992

132 **Mark Gotay, Jerrold Marsden, and Vincent Moncrief, Editors,** Mathematical aspects of classical field theory, 1992

131 **L. A. Bokut′, Yu. L. Ershov, and A. I. Kostrikin, Editors,** Proceedings of the International Conference on Algebra Dedicated to the Memory of A. I. Mal′cev, Parts 1, 2, and 3, 1992

130 **L. Fuchs, K. R. Goodearl, J. T. Stafford, and C. Vinsonhaler, Editors,** Abelian groups and noncommutative rings, 1992

129 **John R. Graef and Jack K. Hale, Editors,** Oscillation and dynamics in delay equations, 1992

128 **Ridgley Lange and Shengwang Wang,** New approaches in spectral decomposition, 1992

127 **Vladimir Oliker and Andrejs Treibergs, Editors,** Geometry and nonlinear partial differential equations, 1992

126 **R. Keith Dennis, Claudio Pedrini, and Michael R. Stein, Editors,** Algebraic K-theory, commutative algebra, and algebraic geometry, 1992

125 **F. Thomas Bruss, Thomas S. Ferguson, and Stephen M. Samuels, Editors,** Strategies for sequential search and selection in real time, 1992

124 **Darrell Haile and James Osterburg, Editors,** Azumaya algebras, actions, and modules, 1992

123 **Steven L. Kleiman and Anders Thorup, Editors,** Enumerative algebraic geometry, 1991

(*Continued in the back of this publication*)

Linear Algebraic Groups and Their Representations

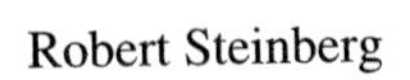

Robert Steinberg

CONTEMPORARY
MATHEMATICS

153

Linear Algebraic Groups and Their Representations

Conference on Linear Algebraic
Groups and their Representations
March 25–28, 1992
Los Angeles, California

Richard S. Elman
Murray M. Schacher
V. S. Varadarajan
Editors

American Mathematical Society
Providence, Rhode Island

The Conference on Linear Algebraic Groups and Their Representations was held at the University of California, Los Angeles, on March 25–28, 1992, with support from the National Security Agency Grant No. NSA MDA904-92-H-3002.

1991 *Mathematics Subject Classification.* Primary 20G05;
Secondary 14L30, 14D25.

Library of Congress Cataloging-in-Publication Data

Linear algebraic groups and their representations/Richard S. Elman, Murray M. Schacher, V. S. Varadarajan, editors.
p. cm.—(Contemporary mathematics; v. 153)
"Proceedings...[of] the Conference on Linear Algebraic Groups and Their Representations, held at UCLA, March 25–28, 1992"—Introd.
Includes bibliographical references.
ISBN 0-8218-5161-6
1. Linear algebraic groups—Congresses. 2. Representations of groups—Congresses. I. Elman, Richard S., 1945– . II. Schacher, Murray M., 1941– . III. Varadarajan, V. S. IV. Conference on Linear Algebraic Groups and Their Representations (1992:UCLA) V. Series: Contemporary mathematics (American Mathematical Society); v. 153.
QA179.L56 1993   93-5642
512′.5—dc20   CIP

Printed in the United States of America.
The paper used in this book is acid-free and falls within the guidelines established to ensure permanence and durability. ♾
Printed on recycled paper. ♻

All articles in this volume were printed from copy prepared by the authors.
Some articles were typeset using AMS-TEX or AMS-LATEX, the American Mathematical Society's TEX macro system.

10 9 8 7 6 5 4 3 2 1   98 97 96 95 94 93

Contents

Introduction

आ नो॑ भ॒द्राः क्रत॑वो यन्तु वि॒श्वत : ।

May the good things come to us from everywhere

These Proceedings contain papers requested by the organizers or presented in person at the Conference on Linear Algebraic Groups and Their Representations held at UCLA March 25–28, 1992, on the occasion of the retirement of Professor Robert Steinberg. Support was generously provided by the National Security Agency[1], the Mathematical Sciences Research Institute, the Chancellor's Office of UCLA, and the Mathematics Department of UCLA. Mathematicians came to UCLA from Asia, Australia, Europe, and North America to participate in this conference honoring Robert Steinberg. They ranged from senior mathematicians to recently graduated mathematicians and graduate students. In addition to the many visitors, mathematicians throughout Southern and Northern California participated in the conference on a daily or occasional basis.

The idea of the conference was to discuss the current status of the field and to suggest directions for future research in those areas of linear algebraic groups and their representations that have been decisively influenced by Robert Steinberg throughout his distinguished career. The success of this goal is indicated by the papers contained here. All papers appearing in this volume are final and will

[1]NSA MDA904-92-H-3002

not be submitted for publication elsewhere in more detailed form.

The Organizing Committee for the conference consisted of Walter Feit, T.A. Springer, Murray Schacher and V.S. Varadarajan. Special assistance was rendered by David Cantor in the preparation of grant proposals. Richard Elman joined us to edit this volume. We were ably assisted by Goody Dhillon, Julie Speckart, Alice Ramirez, Elaine Barth, and Michael Olsson, all of the UCLA Mathematics Department. Without their patient aid, this conference could never have happened. Editorial assistance for these Proceedings has been graciously rendered by Christine Thivierge and Donna Harmon of the American Mathematical Society, and external referees whom we thank while respecting their anonymity. We also wish to thank the American Mathematical Society for publishing the proceedings of this conference in their Contemporary Mathematics series.

Some of the papers bear dedications setting forth the authors' own affection and scientific indebtedness. In the other cases, contributors requested to be included in a general expression of these sentiments. On behalf of all the participants, we present Bob with these efforts. We hope they are worthy of him. We thank him for the standards that he has set for us. We rejoice that we continue to enjoy and to cherish his advice and his friendship.

We truly feel that the great diversity and range of the talks at the conference were genuinely reflective of the spirit of the Vedic invocation given on the previous page[2] with which the conference began.

Richard S. Elman
Murray M. Schacher
V. S. Varadarajan

[2] *Rg Veda, 1, 89*

Speakers and Titles of Talks

R. Boltje	Extending constructible sets on 1-dimensional characters
V. Deodhar	A survey of recent results in Kazhdan-Lustig theory and related topics
M. Dyer	Dihedral groups and projective functors
W. Feit	Extending Steinberg Characters
S. Gelfand	Examples of tensor categories from representation theory
R. Guralnick	Primitive Permutation Representations
W. Haboush	Parabolic schemes on algebraic groups
G. Lusztig	The quantum Frobenius map and Steinberg's tensor product theorem
K. Misra	Quantum affine Lie algebras and affine crystals
B. Parshall	Ext groups and duality
M.S. Raghunathan	The first Betti Number of Locally Symmetric Spaces
N. Reshetikhin	Representations of the double of Hopf algebras
R. Richardson	Parabolic subgroups with abelian unipotent radical
Z. Richstein	Matrix invariants
J. Rickard	Finite group actions and l-adic cohomology
M. Rosso	*-balanced categories and subfunctors
D. Saltman	Rationality of invariant fields
G. Schwarz	The linearization of equivariant Serre problems for reductive groups
L. Scott	Kazhdan-Lustig theory and some problems on automorphisms

G. Seelinger	A completion of the space of matrix invariants
C.S. Seshadri	Moduli of vector bundles on curves
T.A. Springer	Graded Root Data
B. Srinivasan	Variations on themes of Steinberg

Contemporary Mathematics
Volume **153**, 1993

EXTENDING STEINBERG CHARACTERS

WALTER FEIT

Dedicated to Bob Steinberg on the occasion of his 70th birthday May 25, 1992

ABSTRACT. let G be a semi simple group of Lie type over a finite field of characteristic p. Let θ be the Steinberg character of G. If $G \triangleleft H$ it is known that θ extends to a character χ of H. The purpose of this paper is to evaluate χ on p'-elements. The problem is considered in a more general context and various properties of Steinberg characters are discussed.

§1 Introduction.

Let $G(q)$ be a simply connected semi-simple group of Lie type over $\mathbf{F}_q$, where q is a power of the prime p. Let $St_{G(q)} = St$ be the Steinberg character of $G(q)$. See [St1, 15.5]. Suppose that $G(q) \triangleleft H$. Then it is known that St extends to a character of H. See [Sc1], [Sc2], [M]. The main object of this paper is to get some information about the values of this extended character. This information is contained in Theorem C. We first need some definitions and preliminary results.

Let p be a prime and let G be a finite group. A *Steinberg character with respect to* p or a p*-Steinberg character* of G is an irreducible character Θ of G such that $\Theta(x) = \pm|\mathbf{C}_G(x)|_p$ for every p'-element x in G. Here $\mathbf{C}_G(x)$ denotes the centralizer of x in G and n_p is the p-part of the integer n.

Since $\Theta(1) = |G|_p$ for a p-Steinberg character Θ, it has p-defect 0 and so $\Theta(y) = 0$ for every p-singular element y in G.

Suppose that p is a prime and $G \triangleleft H$ are finite groups. A *relative* (G, H) p*-Steinberg character* is an irreducible rational valued character Θ of H such that $\Theta(x) = \pm|\mathbf{C}_G(x)|_p$ for every p'-element x in H.

Observe that if $p \nmid |H : G|$, then a relative (G, H) p-Steinberg character is a p-Steinberg character of H. However, in the general case the values of a

---

1991 *Mathematics Subject Classification.* Primary 20C15, Secondary 20C25, 20G40.
This paper is in final form and no version of it will be submitted for publication elsewhere.

relative (G, H) p-Steinberg character of H at p-singular elements of H are not determined.

Let χ be a character of G which is afforded by a complex representation M. If $y \in G$ let $d_\chi(y) = det(M(y))$. Then d_χ is a linear character of G. If $d_\chi = 1_G$, χ is said to be a *unimodular* character. The next theorem follows from a more general result proved in Section 2.

Theorem A. *Suppose that G has a p-Steinberg character Θ.*

(i) If $p = 2$, then Θ is the unique p-Steinberg character of G.

(ii) If $p \neq 2$, the map $\lambda \to \Theta\lambda$ is a bijection from the set of all rational valued linear characters λ of G to the set of all p-Steinberg characters of G.

(iii) If $p \neq 2$, then G has a unique unimodular p-Steinberg character.

If $p = 2$, Statement (iii) of Theorem A need not be true. Consider for instance the irreducible character of degree 2 of Σ_3. However it is usually true, it is only false if the S_2-group is cyclic.

Suppose that G has a p-Steinberg character. Define a *basic p-Steinberg character* Θ of G as follows.

If $p = 2$, Θ is a 2-Steinberg character of G.

If $p \neq 2$, Θ is a unimodular p-Steinberg character of G.

Theorem A implies that if G has a p-Steinberg character then it has a *unique* basic p-Steinberg character.

Thus the problem of uniqueness is settled by Theorem A. The question of the existence of p-Steinberg characters is more difficult. If $p \nmid |G|$ then 1_G is the basic p-Steinberg character of G. Hence this question is only of interest in case $p \mid |G|$.

Of course, the most important class of p-Steinberg characters are those constructed by Steinberg [St1, 15.5] for the simply connected semi-simple groups $G(q)$ of Lie type over $\mathbf{F}_q$, where q is a power of p.

It is not known which groups have Steinberg characters.

Question 1. *Let G be a finite simple group whose order is divisible by the prime p. Suppose that G has a p-Steinberg character. Does it follow that G is a semi-simple group of Lie type in characteristic p?*

Since the finite simple groups have been classified, it should be possible to answer Question 1. If the hypothesis of simplicity is dropped, other examples exist. Section 6 contains a list of all finite groups G with $|G|_p = p$ which have a faithful p-Steinberg character.

Question 2. *Suppose that $G \triangleleft H$ and G has a p-Steinberg character. Does there exist a relative (G, H) p-Steinberg character of H?*

In this connection the following can be proved.

Theorem B. *Suppose that $G \triangleleft H$ and Θ is the basic p-Steinberg character of G. Then there exists a rational representation M of H which affords χ with $\chi_G = \Theta$.*

As mentioned above, Theorem B is known in case $G = G(q)$ is a simply connected semi-simple group of Lie type in characteristic p. In fact, the proof of Theorem B is quite similar to the argument in [Sc1]. The next result gives a partial answer to Question 2.

Theorem C. *Let $G = G(q)$ be a finite simply connected semi-simple group of Lie type in characteristic p. Suppose that $G \triangleleft H$.*

(i) If χ is a rational valued character of H such that $\chi_G = St$, then χ is a relative (G, H) p-Steinberg character.

(ii) H has a relative (G, H) p-Steinberg character.

(iii) If $p \neq 2$, H has a unique unimodular relative (G, H) p-Steinberg character.

This immediately implies

Corollary D. *Let G, H, p be as in Theorem C. Then H has a relative (G, H) p-Steinberg character χ such that if x is any p'-element in H and $L = G\langle x\rangle$, then χ_L is the basic p-Steinberg character of L.*

The signs of the values of St can be described in terms of properties of the Weyl group of $G(q)$ [Sr, p.8]. It should similarly be possible to find the precise values of the character χ of Corollary D on p'-elements of H, though this problem is not addressed here. That of course would still leave the question of the values of χ on p-singular elements[1]

The proof of Theorem C depends on properties of automorphisms of $G(q)$. These are stated in Theorem 5.1. I am indebted to G. Seitz and R. Steinberg who independently provided proofs of this result [Se], [St2]. It is rather surprising that it does not appear to be in the literature.

§2 Characters of p-Steinberg type.

A generalized character Θ of G is of *Steinberg type with respect to p* or of *p-Steinberg type* if the following are satisfied.

(I) If x is a p'-element in G then $\Theta(x) = \pm|\mathbf{C}_G(x)|_p$.

(II) If x is a p-singular element in G, then $\Theta(x) = 0$.

Thus a p-Steinberg character of G is of p-Steinberg type. Furthermore, a generalized character of p-Steinberg type is an integral linear combination of indecomposable projective characters in characteristic p.

Suppose that G has a normal p-group P. Let $\bar{G} = G/P$. If $\bar{G}$ has a p-Steinberg character Θ then the projective cover of Θ is a character of p-Steinberg type of G. See [L] or [F4, (IV.4.26)]. Thus in particular a parabolic subgroup of a simply connected semi-simple group of Lie type in characteristic p has a character of p-Steinberg type. If $|P| \neq 1$ this is not a p-Steinberg character since an irreducible Brauer character of G has P in its kernel. Hence the class of characters of p-Steinberg type is strictly larger than the class of p-Steinberg characters.

We begin with an elementary Lemma.

Lemma 2.1. *Let η be a rational valued character of a group G. Then d_η is rational valued. If furthermore $\eta(1)$ is odd then $d_\eta\eta$ is unimodular.*

Proof. Let M be a representation of G over an algebraic number field which affords η. Let σ be an automorphism of $\mathbf{Q}(d_\eta)$. Then $d_\eta^\sigma(y) = det(M^\sigma(y))$ for all $y \in G$. Since η is rational valued, M is equivalent to M^σ over the complex numbers and so $d_\eta^\sigma = d_\eta$. Hence d_η is rational valued.

[1]This question will be considered in a forthcoming paper.

If $\eta(1)$ is odd and $y \in G$ then $d_\eta(y) = \pm 1$ and

$$det(d_\eta(y)M(y)) = d_\eta(y)^{\eta(1)}d_\eta(y) = d_\eta(y)^2 = 1.$$

Therefore $d_\eta \eta$ is afforded by the unimodular representation $d_\eta M$. □

Theorem 2.2. *Suppose that G has a generalized character Θ of p-Steinberg type. Replacing Θ by $-\Theta$ if necessary it may be assumed that $\Theta(1) > 0$.*

(i) If $p = 2$, Θ is the unique generalized character of 2-Steinberg type with $\Theta(1) > 0$.

(ii) If $p \neq 2$, the map $\lambda \to \Theta\lambda = \Theta_1$ is a bijection from the set of all rational valued linear characters λ of G to the set of all generalized characters Θ_1 of p-Steinberg type with $\Theta_1 > 0$.

Proof. Let R be the ring of all integral linear combinations of irreducible Brauer characters in characteristic p. Let I be the ideal of R consisting of all integral linear combinations of projective Brauer characters. By [F2, Theorem 1] Θ is a generator of I. Thus if Θ_1 is another character of p-Steinberg type with $\Theta_1(1) > 0$ then $\Theta_1 = \alpha\Theta$ for some $\alpha \in R$. Hence $\alpha(x) = \pm 1$ for every p'-element x in G. Define the class function λ on G by $\lambda(y) = \alpha(x)$, where x is the p'-part of y. Brauer's characterization of characters implies that λ is a generalized character of G. Since $\lambda(y) = \pm 1$ for all $y \in G$ it follows that $\Sigma_{y \in G}|\lambda(y)|^2 = |G|$. Hence $\pm\lambda$ is an irreducible character. As $\lambda(1) = 1, \lambda$ is a linear character of G.

(i) Θ and Θ_1 vanish on 2-singular elements while $\lambda(y) = 1$ for y of odd order. Hence $\Theta_1 = \lambda\Theta = \Theta$.

(ii) Suppose that λ_1 and λ_2 are rational valued linear characters such that $\lambda_1\Theta = \lambda_2\Theta$. Then $\lambda_1(x) = \lambda_2(x)$ for all p'-elements x in G. As $p \neq 2$ this implies that $\lambda_1\lambda_2(x) = 1$ for every 2-element x in G. Hence the subgroup G_0 of G generated by all 2-elements is in the kernel of $\lambda_1\lambda_2$. As $|G/G_0|$ is odd and $(\lambda_1\lambda_2)^2 = 1_G$ this yields that $\lambda_1\lambda_2 = 1_G$ and so $\lambda_1 = \lambda_2$. □

Theorem 2.3. *Suppose that $p \neq 2$ and G has a character Θ of p-Steinberg type. Then $d_\Theta\Theta$ is the unique unimodular character of G of p-Steinberg type.*

Proof. By Lemma 2.1, it may be assumed that Θ is unimodular. Let Θ_1 be another unimodular character of p-Steinberg type. By Theorem 2.2 $\Theta_1 = \lambda\Theta$ for some rational valued linear character λ of G. If $\Theta_1 \neq \Theta$ there exists a p'-element x with $\lambda(x) = -1$. Let M be a representation which affords Θ. Hence

$$det(\lambda(x)M(x)) = (-1)^{\Theta(1)} = -1,$$

contrary to the fact that Θ_1 is unimodular. □

Theorem A is a direct consequence of Theorems 2.2 and 2.3.

§3 The Proof of Theorem B.

Proof of Theorem B. Let $m = m(\Theta)$ be the Schur index of Θ. Then $m \mid \Theta(1)$ and so $m = 1$ by [F3, Theorem 2]. Thus there exists a rational representation M_0 of G which affords Θ.

By Theorem A, Θ is unique. Thus $\Theta(x^{-1}yx) = \Theta(y)$ for all $y \in G, x \in H$ and the kernel of Θ is normal in H. Hence by factoring out this kernel it may be assumed that Θ is faithful.

Let Z be the center of G. Then $|Z| \mid 2$ as Θ is rational valued. Suppose that $|Z| \neq 1$. Let z be an involution in Z. Then $d_\Theta(z) = (-1)^{\Theta(1)}$. If $p = 2$, then $\Theta(z) \neq 0$ contrary to the definition of Θ. If $p \neq 2$, then $(-1)^{\Theta(1)} = -1$ contrary to the fact that Θ is unimodular. Therefore $|Z| = 1$ in all cases.

Hence $G\mathbf{C}_H(G) = G \times \mathbf{C}_H(G)$. Thus Θ extends to $G\mathbf{C}_H(G)$ by defining $\Theta(y) = \Theta(1)$ for $y \in \mathbf{C}_H(G)$. Replacing H by $H/\mathbf{C}_H(G)$ it may be assumed that $|\mathbf{C}_H(G)| = 1$. The main theorem of [Sc1] now yields the existence of M and χ. □

§4 Steinberg Characters.

Lemma 4.1. *Let $G \lhd H$ with $|H : G| = q$, where q is a prime, $q \neq p$. Let ζ be a rational valued character of H such that ζ_G is a p-Steinberg character. Then $\zeta(x) = 0$ for p-singular elements in G and if x is a p'-element then*

$$\zeta(x) = |\mathbf{C}_H(x)|_p \; a(x)$$

for $a(x)$ a nonzero rational integer.

Proof. If x is p-singular, then $\zeta(x) = 0$ since ζ is of defect 0 for p. Suppose now that x is a p'-element. Let ω be the central character corresponding to ζ. Then

$$\omega(x) = \frac{|H|}{|\mathbf{C}_H(x)|} \frac{\zeta(x)}{\zeta(1)}$$

is an algebraic (and hence a rational) integer. Thus $\zeta(x)/|\mathbf{C}_H(x)|$ is a local integer at p and so $\zeta(x) = |\mathbf{C}_H(x)|_p a(x)$ for an integer $a(x)$. It remains to show that $\zeta(x) \neq 0$.

Let $x = x_q x' = x' x_q$, where x_q is the q-part of x. Thus $x' \in G$. Therefore

$$\zeta(x) \equiv \zeta(x') \equiv \pm|\mathbf{C}_G(x')|_p \not\equiv 0 (mod\ q) \quad \square$$

We will be concerned with the following condition.

(*) *$G \lhd H$ and p is a prime. If x is a p'-element of H, then $\mathbf{C}_G(x)$ contains exactly $|\mathbf{C}_G(x)|_p^2$ p-elements.*

Lemma 4.2. *Suppose that (*) is satisfied and $|H : G| = q \neq p$, for q a prime. Let ζ be a rational valued character of H such that ζ_G is a p-Steinberg character of G. Then ζ is a p-Steinberg character of H.*

Proof. If x is a p'-element in H, let $S(x)$ be the p'-section of H which contains x. In other words $S(x)$ consists of all elements in H whose p'-part is x. Since $|\mathbf{C}_H(x)|_p = |\mathbf{C}_G(x)|_p$, (*) implies that

$$|S(x)| = |\mathbf{C}_H(x)|_p^2.$$

Hence Lemma 4.1 yields that

$$\sum_{S(x)} |\zeta(y)|^2 = \zeta(x)^2 = |S(x)|a(x)^2. \tag{4.3}$$

Let T be the set of all p'-elements in H. Then $\{S(x) \mid x \in T\}$ is a partition of H. Hence

$$|H| = \sum_{x \in T} |S(x)|.$$

As ζ is irreducible, (4.3) implies that

$$|H| = \sum_{x \in T} |S(x)|\ a(x)^2.$$

Consequently

$$0 = \sum_{x \in T} |S(x)|(a(x)^2 - 1).$$

By Lemma 4.1, $a(x) \neq 0$ and so $a(x)^2 \geq 1$. Thus $a(x)^2 = 1$ for all $x \in T$ as required. □

Lemma 4.4. *Suppose that (*) is satisfied and H/G is a cyclic p'-group. If G has a p-Steinberg character then so does H.*

Proof. Induction on $|H : G|$. If $|H : G| = 1$ there is nothing to prove. Suppose that $|H : G| > 1$.

Let $G < G_1 < H$ with $|H : G_1| = q$ a prime. By induction, G_1 has a p-Steinberg character. By changing notation, it may be assumed that $G = G_1$ and $|H : G| = q$. By Theorems A and B, H has a character ζ such that ζ_G is the basic p-Steinberg character of G. The result follows from Lemma 4.2. □

Before proceeding, we need the following result of Gallagher [I, (6.17)].

Lemma 4.5. *Suppose that $G \triangleleft H$. Let ζ be an irreducible character of H such that ζ_G is irreducible. If χ is an irreducible character of H such that $\zeta_G = \chi_G$, then $\chi = \lambda\zeta$ for some linear character λ of H.*

Theorem 4.6. *Suppose that (*) is satisfied and G has a p-Steinberg character.*

(i) If χ is a rational valued character of H such that χ_G is a p-Steinberg character of G then χ is a relative (G, H) p-Steinberg character.

(ii) H has a relative (G, H) p-Steinberg character.

(iii) If $p \neq 2$, H has a unique unimodular relative (G, H) p-Steinberg character.

Proof. (i) By Lemma 2.1 $d_\chi \chi$ is unimodular and rational valued if $p \neq 2$. Thus by changing notation, it may be assumed that χ is unimodular if $p \neq 2$.

Let x be a p'-element in H and let $L = G < x >$. By Theorem A and Lemma 4.4 there exists a basic p-Steinberg character ζ of L. Then $\zeta_G = \Theta = \chi_G$, where Θ is the basic p-Steinberg character of G.

By Lemma 4.5, $\chi_L = \lambda\zeta$ for a suitable linear character λ. Hence $\chi(y) = \lambda(y)\zeta(y)$ for all $y \in L$. As χ is rational valued $\lambda(x) = \pm 1$ for any p'-element x in L and so

$$\chi(x) = \pm\zeta(x) = \pm|\mathbf{C}_L(x)|_p = \pm|\mathbf{C}_G(x)|_p.$$

(ii) This follows from (i) and Theorems A and B.

(iii) By Lemma 2.1 and (ii), H has a unimodular relative (G, H) p-Steinberg character χ. Suppose that χ_1 is another such character. By Lemma 4.5, $\chi_1 = \lambda\chi$ for some linear character λ of H. Let M be a complex representation which affords χ. If y is an element of H with $\chi(y) \neq 0$, then $\lambda(y)$ is rational valued and so $\lambda(y) = \pm 1$. Furthermore

$$1 = det(\lambda(y)M(y)) = \lambda(y)^{\chi(1)} = \lambda(y).$$

Thus $\lambda(y) = 1$ if $\chi(y) \neq 0$. Hence $\chi_1 = \lambda\chi = \chi$. □

§5 The proof of Theorem C.

The proof of Theorem C depends on the following result. As mentioned in the introduction it is due to G. Seitz and R. Steinberg.

Theorem 5.1. *Let $G = G(q)$ be a connected semi-simple group of Lie type over $\mathbf{F}_q$, where q is a power of the prime p. Let τ be an automorphism of G whose order is not divisible by p and let G_0 be the group of fixed points of τ. Let K be the subgroup of G_0 which is generated by all the p-elements in G_0. Then K is a connected semi-simple group of Lie type over a finite extension of $\mathbf{F}_q$.*

For the purpose of this paper, we need the following consequence of Theorem 5.1.

Corollary 5.2. *Let $G = G(q)$ be a connected semi-simple group of Lie type over $\mathbf{F}_q$, where q is a power of p. Let $G \lhd H$ and let x be a p'-element in H. Then $\mathbf{C}_G(x)$ contains exactly $|\mathbf{C}_G(x)|_p^2$ p-elements.*

Proof. The element x induces an inner automorphism of H and so an automorphism τ of G whose order is prime to p. Let K be the subgroup of H generated by all p elements in $C_G(x)$. Then

$$|K|_p = |\mathbf{C}_G(x)|_p.$$

Thus it suffices to show that K contains exactly $|K|_p^2$ p-elements.

By Theorem 5.1, K is a connected semi-simple group of Lie type over an extension field of $\mathbf{F}_q$. Thus the result follows from [St1, 15.1]. □

Theorem C follows directly from Theorem 4.6 and Corollary 5.2.

§6 The Case $|G|_p = p$.

Let q be a natural number. Let A_q denote the group of permutations of $F = \mathbf{F}_{2^q}$ of the form $\alpha \rightarrow a\alpha^\tau + b$, where $a, b \in F, a \neq 0$ and τ is an automorphism of F. Thus $|A_q| = 2^q(2^q - 1)q$.

Theorem 6.1. *Let p be a prime and let G be a finite group with $|G|_p = p$. Suppose that G has a faithful character of p-Steinberg type. Then $G = Z \times G_0$, where $|Z| \mid 2$ and $|Z| = 1$ if $p = 2$. Furthermore one of the following holds.*

(I) $|G_0| = p$ or G_0 is a Frobenius group of order pe for some divisor e of $p-1$.

(II) $p = 2^q - 1$ for some prime q and G_0 is either a Frobenius goup of order $p(p+1)$ with Frobenius kernel of order $p + 1 = 2^q$ or $G \simeq A_q$.

(III) G_0 is isomorphic to $PSL_2(p)$ or $PGL_2(p)$.

Before proving Theorem 6.1, we deduce the following consequence.

Theorem 6.2. *Let p be a prime and let G be a finite group with $|G|_p = p$. The following are equivalent.*

(i) G has a faithful p-Steinberg character.

(ii) Condition (II) or (III) of Theorem 6.1 holds.

Proof. (i)$\Rightarrow$ (ii). If Condition (I) of Theorem 6.1 holds, then G has no irreducible character of degree p. The result follows from Theorem 6.1.

(ii)$\Rightarrow$ (i). In either Case (II) or (III), G_0 is a doubly transitive group on $p+1$ letters. Thus the permutation character is of the form $1_G + \Theta$ for some irreducible character Θ of G_0 with $\Theta(1) = p$. The subgroup G_1 of G_0 which fixes a given letter is the normalizer of some S_p-group P. Thus either $G_1 = P$ or G_1 is a Frobenius group with Frobenius kernel P. Hence, G_0 is either a Frobenius group or a Zassenhaus group and Θ is a p-Steinberg character of G. Thus if λ is a faithful linear character of Z then $\lambda\Theta$ is a faithful p-Steinberg character of G. □

Proof of Theorem 6.1. Let P be a S_p-group of G. If $P \subseteq \mathbf{Z}(G)$, the center of G, then $G = P \times G_1$ for a p'-group G_1. Furthermore, $\Theta = \rho\lambda$, where ρ is the character afforded by the regular representation of $G/G_1 \simeq P$ and λ is a faithful linear character of G_1. Thus $|G_1| \mid 2$, and $|G_1| = 1$ if $p = 2$. Thus (I) holds. Suppose now that $\mathbf{Z}(G)$ is a p'-group.

Let x be a p'-element in G such that $p \mid |\mathbf{C}_G(x)|$. Then $\Theta(x) = \pm\Theta(1)$ and so $x \in \mathbf{Z}(G)$ with $x^2 = 1$ as Θ is faithful. Hence $|\mathbf{Z}(G)| \mid 2$. If $|\mathbf{Z}(G)| \neq 1$, then $p \neq 2$. Furthermore, $d_\Theta(x) = -1$ for $x \in \mathbf{Z}(G), x \neq 1$. Hence $G = G_0 \times \mathbf{Z}(G)$. Thus $\mathbf{Z}(G)$ is the kernel of $d_\Theta\Theta$. Therefore by changing notations, it may be assumed that $|\mathbf{Z}(G)| = 1$ and $\Theta(x) = \pm 1$ for every p'-element $x \neq 1$ of G.

Let $N = \mathbf{N}_G(P)$. Then either $N = P$ or N is a Frobenius group with Frobenius kernel P. Let $|N| = ep$. Then $|G| = ep(1 + kp)$ for some integer $k \geq 0$. If $k = 0$, (I) holds, and so it may be assumed that $k > 0$.

The number of p-elements in G is $1 + (p-1)(1 + kp)$. Therefore

$$|G| - (p-1)(1+kp) - 1 + p^2 = \sum_{x \in G} |\Theta(x)|^2 \geq |G|.$$

Hence $p^2 - 1 \geq (1 + kp)(p - 1)$ and so $1 + p \geq 1 + kp$. Hence $k = 1$ and $|G| = ep(1 + p)$.

If $e = 1$, then G is a Frobenius group with Frobenius kernel of order $1 + p$. Hence $1 + p = r^n$ for some prime r as $p \mid |G|$. If $p = 2$, then $G \simeq PSL_2(2)$. If $p \neq 2$ then $r = 2$ and $2^n - 1 = p$. Thus $n = q$ is a prime and (II) holds.

Suppose finally that $e \neq 1$. Then N is a Frobenius group with Frobenius kernel P. Since N acts faithfully on the p S_p-groups other than P, it follows that G acts as a Zassenhaus group on $p+1$ letters. Hence either $G \simeq PSL_2(p)$ or $PGL_2(p)$ or G contains a normal subgroup of order $p+1$. In the latter case, (II) holds by [F1, Lemma 4.1]. □

References

[F1] W. Feit, *On a class of doubly transitive permutation groups*, Ill. J. Math. **4** (1960)), 170–188.

[F2] W. Feit, *Divisibility of projective modules of finite groups*, J. Pure and Applied Algebra **8** (1976), 183–185.

[F3] W. Feit, *Some properties of characters of finite groups*, Bull. L.M.S. **14** (1982), 129–132.
[F4] W. Feit, *The representation theory of finite groups*, North Holland, Amsterdam, New York, Oxford, 1982.
[I] M. Isaacs, *Character theory of finite groups*, Academic Press, N.Y. San Francisco, London, 1976.
[L] G. Lusztig, *Divisibility of projective modules of finite Chevalley groups by the Steinberg module*, Bull. L.M.S. **8** (1976), 130–134.
[M] G. Malle, *Generalized Deligne-Lusztig characters*, to appear.
[Sc1] P. Schmid, *Rational matrix groups of a special type*, Linear Algebra and its Applications **71** (1985), 289–293.
[Sc2] P. Schmid, *Extending the Steinberg representation*, Journal of Algebra **150** (1992), 254–256.
[Se] G. Seitz, *Private Communication.*
[Sr] B. Srinivasan, *On the Steinberg character for a finite simple group of Lie type*, J. Australian Math. Soc. **12** (1971), 1–14.
[St1] R. Steinberg, *Endomorphisms of linear algebraic groups*, Memoirs A.M.S. **80** (1968).
[St2] R. Steinberg, *Private communication.*

DEPARTMENT OF MATHEMATICS, YALE UNIVERSITY, NEW HAVEN, CT 06520
E-mail address: feit@pascal.math.yale.edu

Contemporary Mathematics
Volume **153**, 1993

A comparison of bases of quantized enveloping algebras

I. GROJNOWSKI AND G. LUSZTIG

To Robert Steinberg on the occasion of his seventieth birthday

Let $\mathbf{U}^-$ be the $-$part of the quantized enveloping algebra associated by Drinfeld and Jimbo to a symmetric generalized Cartan matrix. For type A, D, E a canonical basis of $\mathbf{U}^-$ has been introduced in [**L3**] both by an algebraic and by a topological method.

This has been extended to the general case by Kashiwara [**K**] (by an algebraic method) and by the second author [**L5**] (by a topological method). In [**L4**] it has been shown that for type A, D, E, Kashiwara's approach leads to the same basis as the one in [**L3**]. The main result of this paper is that, in the general case, the basis in [**K**] coincides with the basis in [**L5**].

1. Let X be an algebraic variety over $\mathbf{C}$ with an action of a connected algebraic group G. A complex on X is by definition an object of the bounded derived category of constructible sheaves on X. As in [**BBD**], we say that a complex A on X is semisimple if it is isomorphic to $\oplus_i {}^pH^iA[-i]$ and each ${}^pH^iA$ is a semisimple perverse sheaf. If, in addition, each ${}^pH^iA$ is G-equivariant, we say that A is G-equivariant.

2. Let A, B be two G-equivariant semisimple complexes on X. We will define, for any $j \in \mathbf{Z}$, an integer $d_j(X, G; A, B)$ as follows. We choose an integer $m \geq 1$ and a smooth irreducible algebraic variety Γ with a free action of G such that $H^i(\Gamma, \mathbf{C}) = 0$ for $i = 1, \ldots, m$. Then G acts diagonally on $\Gamma \times X$ and we form ${}_\Gamma X = G \backslash (\Gamma \times X)$. In the diagram

$$X \xleftarrow{a} \Gamma \times X \xrightarrow{b} {}_\Gamma X$$

(with the obvious maps a, b) we have $a^*A = b^*{}_\Gamma A$ and $a^*B = b^*{}_\Gamma B$ for well defined semisimple complexes ${}_\Gamma A, {}_\Gamma B$ on ${}_\Gamma X$.

1991 *Mathematics Subject Classification.* Primary 20G99.

G. L. is supported in part by the National Science Foundation.
This paper is in final form and will not be submitted for publication elsewhere

By the argument in [**L2, 1.1, 1.2**], we see that, if m is large enough, then

$$\begin{aligned}&\dim H_c^{j+2\dim\Gamma-2\dim G}({}_\Gamma X, {}_\Gamma A\otimes {}_\Gamma B)\\&=\dim H_c^j({}_\Gamma X, {}_\Gamma A[\dim G\backslash\Gamma]\otimes {}_\Gamma B[\dim G\backslash\Gamma])\end{aligned}$$

(hypercohomology with compact support) is independent of m,Γ; we denote it by $d_j(X,G;A,B)$.

It is clear that

(a) $$d_j(X,G;A,B)=d_j(X,G;B,A),$$

(b) $$d_j(X,G;A[n],B[m])=d_{j+n+m}(X,G;A,B),$$

(c) $$d_j(X,G;A\oplus A',B)=d_j(X,G;A,B)+d_j(X,G;A',B),$$

if A,A',B are semisimple G-equivariant complexes on X and n,m are integers.

3. If A,B in §2 are known to be perverse sheaves, then so are ${}_\Gamma A[\dim G\backslash\Gamma]$ and ${}_\Gamma B[\dim G\backslash\Gamma]$ and by [**L1, 7.4**] we have

$$d_j(X,G;A,B)=0 \text{ for all } j>0;$$

if in addition A,B are simple, then $d_0(X,G;A,B)$ is 1 if B is isomorphic to the Verdier dual of A and is zero, otherwise.

4. We consider two finite sets I and Ω with I non-empty, a map $\Omega\to I$ denoted $h\mapsto h'$ and a map $\Omega\to I$ denoted $h\mapsto h''$, such that $h'\neq h''$ for all $h\in\Omega$. This datum is the same as an oriented graph.

Let $\mathbf{V}$ be a finite dimensional I-graded vector space over $\mathbf{C}$. Let $\mathbf{E_V}=\oplus_{h\in\Omega}\mathrm{Hom}(V_{h'},V_{h''})$. The algebraic group $G_\mathbf{V}=\prod_i\mathrm{Aut}(\mathbf{V}_i)$ acts on $\mathbf{E_V}$ by $(g,x)\to gx=x'$ where $x'_h=g_{h''}x_hg_{h'}^{-1}$ for all $h\in\Omega$.

Let $\mathcal{P}_\mathbf{V}$ be the subset of the set of isomorphism classes of simple $G_\mathbf{V}$-equivariant perverse sheaves on $\mathbf{E_V}$ defined in [**L5, 2.2**] (where it is denoted $\mathcal{P}_{\mathbf{V},\Omega}$). Let $\mathcal{Q}_\mathbf{V}$ be the set of isomorphism classes of complexes on $\mathbf{E_V}$ which are isomorphic to a direct sum of finitely many complexes in $\mathcal{P}_\mathbf{V}$, with shifts. The complexes in $\mathcal{Q}_\mathbf{V}$ are semisimple, $G_\mathbf{V}$-equivariant.

As in [**L5, 10.1**], we define $\mathcal{K}_\mathbf{V}$ to be the abelian group with one generator (L) for each element of $\mathcal{Q}_\mathbf{V}$ and with relations $(L)+(L')=(L'')$ whenever L'' is isomorphic to $L\oplus L'$. We regard $\mathcal{K}_\mathbf{V}$ as a module over $\mathbf{Z}[v,v^{-1}]$ (where v is an indeterminate) by $v^n(L)=(L[n])$. Clearly $\mathcal{K}_\mathbf{V}$ is a free abelian group with basis (L) where L runs over $\mathcal{P}_\mathbf{V}$.

For A,B in $\mathcal{Q}_\mathbf{V}$, we define

(a) $$(A,B)=\sum_{j\in\mathbf{Z}}d_j(\mathbf{E_V},G_\mathbf{V};A,B)v^{-j}\in\mathbf{Z}((v))$$

(the last inclusion follows from §3).

By 2(a),(b),(c), this defines a symmetric $\mathbf{Z}[v,v^{-1}]$-bilinear form $(,)$ on $\mathcal{K}_\mathbf{V}$ with values in $\mathbf{Z}((v))$.

5. Now let $\mathbf{V}', \mathbf{V}''$ be two finite dimensional I-graded vector spaces. Then $\mathbf{E}_{\mathbf{V}'}, \mathbf{E}_{\mathbf{V}''}, G_{\mathbf{V}'}, G_{\mathbf{V}''}$ are defined as in §4. The actions of $G_{\mathbf{V}'}$ on $\mathbf{E}_{\mathbf{V}'}$ and of $G_{\mathbf{V}''}$ on $\mathbf{E}_{\mathbf{V}''}$ give a product action of $G_{\mathbf{V}'} \times G_{\mathbf{V}''}$ on $\mathbf{E}_{\mathbf{V}'} \times \mathbf{E}_{\mathbf{V}''}$.

As in [**L5, 4.1**], let $\mathcal{Q}_{\mathbf{V}',\mathbf{V}''}$ be the set of isomorphism classes of complexes on $\mathbf{E}_{\mathbf{V}'} \times \mathbf{E}_{\mathbf{V}''}$ which are direct sums of finitely many complexes of form $A \otimes B$ where A is in $\mathcal{Q}_{\mathbf{V}'}$ and B is in $\mathcal{Q}_{\mathbf{V}''}$. As in [**L5, 10.4**], let $\mathcal{K}_{\mathbf{V}',\mathbf{V}''}$ be the abelian group with one generator (L) for each element of $\mathcal{Q}_{\mathbf{V}',\mathbf{V}''}$ and with relations $(L) + (L') = (L'')$ whenever L'' is isomorphic to $L \oplus L'$. We regard $\mathcal{K}_{\mathbf{V}',\mathbf{V}''}$ as a module over $\mathbf{Z}[v, v^{-1}]$ by $v^n(L) = (L[n])$. Clearly $\mathcal{K}_{\mathbf{V}',\mathbf{V}''}$ is a free $\mathbf{Z}[v, v^{-1}]$-module with basis $(L' \otimes L'')$ where L' (resp. L'') runs over $\mathcal{P}_{\mathbf{V}'}$ (resp. $\mathcal{P}_{\mathbf{V}''}$). Thus, we have canonically

(a) $$\mathcal{K}_{\mathbf{V}',\mathbf{V}''} = \mathcal{K}_{\mathbf{V}'} \otimes_{\mathbf{Z}[v,v^{-1}]} \mathcal{K}_{\mathbf{V}''}.$$

If A', B' are in $\mathcal{Q}_{\mathbf{V}'}$ and A'', B'' are in $\mathcal{Q}_{\mathbf{V}''}$, then

(b) $$\begin{aligned} &d_j(\mathbf{E}_{\mathbf{V}'} \times \mathbf{E}_{\mathbf{V}''}, G_{\mathbf{V}'} \times G_{\mathbf{V}''}; A' \otimes A'', B' \otimes B'') \\ &= \sum_{j'+j''=j} d_{j'}(\mathbf{E}_{\mathbf{V}'}, G_{\mathbf{V}'}; A', B') d_{j''}(\mathbf{E}_{\mathbf{V}''}, G_{\mathbf{V}''}; A'', B''). \end{aligned}$$

This follows easily from the definitions and the Künneth formula. (The sum is finite by 4(a).)

For A, B in $\mathcal{Q}_{\mathbf{V}',\mathbf{V}''}$, we define

$$(A, B) = \sum_{j \in \mathbf{Z}} d_j(\mathbf{E}_{\mathbf{V}'} \times \mathbf{E}_{\mathbf{V}''}, G_{\mathbf{V}'} \times G_{\mathbf{V}''}; A, B) v^{-j} \in \mathbf{Z}((v))$$

(again, the last inclusion is by §3). As in §4, this defines a symmetric $\mathbf{Z}[v, v^{-1}]$-bilinear form $(,)$ on $\mathcal{K}_{\mathbf{V}',\mathbf{V}''}$ with values in $\mathbf{Z}((v))$.

With this notation, we can rewrite (b) as follows:

(c) $$(A' \otimes A'', B' \otimes B'') = (A', B')(A'', B'')$$

(product of power series).

6. Let $\mathbf{V}$ be as in §4 and let $\mathbf{V}', \mathbf{V}''$ be as in §5. We now assume that $\mathbf{V}''$ is an I-graded subspace of $\mathbf{V}$ and that $\mathbf{V}' = \mathbf{V}/\mathbf{V}''$ as an I-graded vector space.

Let P be the stabilizer of $\mathbf{V}''$ in $G_{\mathbf{V}}$. This is a parabolic subgroup of $G_{\mathbf{V}}$. Its unipotent radical is denoted U. We have canonically, $P/U = G_{\mathbf{V}'} \times G_{\mathbf{V}''}$.

Let F be the closed subvariety of $\mathbf{E}_{\mathbf{V}}$ consisting of all $x \in \mathbf{E}_{\mathbf{V}}$ such that $x_h(\mathbf{V}''_{h'}) \subset \mathbf{V}''_{h''}$ for all $h \in \Omega$. We denote $i : F \to \mathbf{E}_{\mathbf{V}}$ the inclusion. Note that P acts on F (restriction of the $G_{\mathbf{V}}$-action on $\mathbf{E}_{\mathbf{V}}$).

If $x \in F$, then x induces elements $x' \in \mathbf{E}_{\mathbf{V}'}$ and $x'' \in \mathbf{E}_{\mathbf{V}''}$; the map $x \mapsto (x', x'')$ is a vector bundle $\phi : F \to \mathbf{E}_{\mathbf{V}'} \times \mathbf{E}_{\mathbf{V}''}$. Now P acts on $\mathbf{E}_{\mathbf{V}'} \times \mathbf{E}_{\mathbf{V}''}$ through its quotient $P/U = G_{\mathbf{V}'} \times G_{\mathbf{V}''}$, as in §5. The map ϕ is compatible with the P-actions.

We set $G_{\mathbf{V}} = G, P/U = \bar{G}$, $\mathbf{E}_{\mathbf{V}} = E$, $\mathbf{E}_{\mathbf{V}'} \times \mathbf{E}_{\mathbf{V}''} = \bar{E}$. We have a diagram

$$\bar{E} \xleftarrow{\phi} F \xrightarrow{i} E.$$

Let $E'' = G \times_P F, E' = G \times_U F$. We have a diagram (see [**L5, 3.1**])

$$\bar{E} \xleftarrow{p_1} E' \xrightarrow{p_2} E'' \xrightarrow{p_3} E$$

where $p_1(g, f) = \phi(f)$; $p_2(g, f) = (g, f)$; $p_3(g, f) = g(i(f))$.

Note that p_1 is smooth with connected fibres, p_2 is a $\bar{G}$-principal bundle and p_3 is proper.

Let A', A'', B be complexes on $\mathbf{E_{V'}}, \mathbf{E_{V''}}, \mathbf{E_V}$ respectively which belong to $\mathcal{Q}_{\mathbf{V'}}, \mathcal{Q}_{\mathbf{V''}}, \mathcal{Q}_{\mathbf{V}}$ (respectively).

Then $A = A' \otimes A''$ (a complex on $\bar{E}$) belongs to $\mathcal{Q}_{\mathbf{V'},\mathbf{V''}}$.

We can form $\phi_!(i^*B)$; this complex on $\bar{E}$ belongs to $\mathcal{Q}_{\mathbf{V'},\mathbf{V''}}$, by [**L5, 4.2(a)**].

There is a well defined semisimple G-equivariant complex $\tilde{A}$ on E'' such that $p_1^*A \cong p_2^*\tilde{A}$; then $(p_3)_!\tilde{A}$ is an object in $\mathcal{Q}_{\mathbf{V}}$ (see [**L5, 3.2(a)**]).

LEMMA 7. *We have for any $j \in \mathbf{Z}$:*

$$d_j(\bar{E}, \bar{G}; A, \phi_!(i^*B)) = d_{j'}(E, G; (p_3)_!\tilde{A}, B)$$

where $j' = j + 2\dim G/P$.

Let $m \geq 1$ be large (with respect to j) and let Γ be a smooth irreducible variety with a free action of G, such that $H^i(\Gamma, \mathbf{C}) = 0$ for $i = 1, \ldots, m$. Let $\bar{\Gamma} = U\backslash\Gamma$. Then $\bar{\Gamma}$ has a free action of $\bar{G}$ (induced by that of G) and $H^i(\bar{\Gamma}, \mathbf{C}) = 0$ for $i = 1, \ldots, m$.

To the G-variety E and to the complexes $(p_3)_!\tilde{A}$ and B on it, we can associate the complexes ${}_\Gamma((p_3)_!\tilde{A})$ and ${}_\Gamma B$ on ${}_\Gamma E$ as in §2.

To the $\bar{G}$-variety $\bar{E}$ and to the complexes A and $\phi_!(i^*B)$ on it, we can associate the complexes ${}_{\bar{\Gamma}}A$ and ${}_{\bar{\Gamma}}(\phi_!(i^*B))$ on ${}_{\bar{\Gamma}}\bar{E}$ as in §2.

It is then enough to prove that

$$\begin{aligned} &H_c^{j+2\dim\bar{\Gamma}-2\dim\bar{G}}({}_{\bar{\Gamma}}\bar{E}, {}_{\bar{\Gamma}}A \otimes {}_{\bar{\Gamma}}(\phi_!(i^*B))) \\ &\cong H_c^{j'+2\dim\Gamma-2\dim G}({}_\Gamma E, {}_\Gamma((p_3)_!\tilde{A}) \otimes {}_\Gamma B). \end{aligned}$$

We will in fact show the stronger statement that

(a) $$H_c^i({}_{\bar{\Gamma}}\bar{E}, {}_{\bar{\Gamma}}A \otimes {}_{\bar{\Gamma}}(\phi_!(i^*B))) \cong H_c^i({}_\Gamma E, {}_\Gamma((p_3)_!\tilde{A}) \otimes {}_\Gamma B)$$

for any $i \in \mathbf{Z}$.

We now introduce some further notation.

To the G-variety E'' and to the complex $\tilde{A}$ on it, we can associate the complex ${}_\Gamma\tilde{A}$ on ${}_\Gamma E''$ as in §2.

Let

$$\tilde{p}_3 : {}_\Gamma E'' \to {}_\Gamma E$$

be the proper map induced by p_3.

Let

$$\psi : P\backslash(\Gamma \times F) \to P\backslash(\Gamma \times \bar{E})$$

be the map induced by ϕ; this is a vector bundle.

Consider the diagram

$$E \xleftarrow{a'} \Gamma \times E \xrightarrow{b'} P\backslash(\Gamma \times E)$$

as in §2 (with P instead of G). Let $\tilde{B}$ be the unique semisimple complex on $P\backslash(\Gamma \times E)$ such that $a'^*B = b'^*\tilde{B}$.

We have a diagram

$$P\backslash(\Gamma \times \bar{E}) \xleftarrow{\tilde{\phi}} P\backslash(\Gamma \times F) \xrightarrow{\tilde{i}} P\backslash(\Gamma \times E)$$

with morphisms $\tilde{\phi}, \tilde{i}$ induced by ϕ, i.

By the decomposition theorem [**BBD**], the complex $(\tilde{p}_3)_!({}_\Gamma\tilde{A})$ on ${}_\Gamma E$ is semisimple; it satisfies the definition of ${}_\Gamma((p_3)_!\tilde{A})$ hence it is isomorphic to it:

(b) $$_\Gamma((p_3)_!\tilde{A}) = (\tilde{p}_3)_!({}_\Gamma\tilde{A}).$$

From the definitions, we see that

(c) $${}_\Gamma\tilde{A} = \psi^*({}_{\bar{\Gamma}}A).$$

The proof in [**L5,§4**] showing that $\phi_!(i^*B)$ is a semisimple complex can be repeated essentially word by word and gives that $\tilde{\phi}_!(\tilde{i}^*\tilde{B})$ is semisimple. It then satisfies the definition of ${}_{\bar{\Gamma}}(\phi_!(i^*B))$, so it is isomorphic to it:

(d) $$\tilde{\phi}_!(\tilde{i}^*\tilde{B}) = {}_{\bar{\Gamma}}(\phi_!(i^*B)).$$

From (b), we have

$${}_\Gamma((p_3)_!\tilde{A}) \otimes {}_\Gamma B = (\tilde{p}_3)_!({}_\Gamma\tilde{A}) \otimes {}_\Gamma B = (\tilde{p}_3)_!({}_\Gamma\tilde{A} \otimes (\tilde{p}_3)^*{}_\Gamma B).$$

Thus the right hand side of (a) is equal to

(e) $$H^i_c({}_\Gamma E'', {}_\Gamma\tilde{A} \otimes (\tilde{p}_3)^*{}_\Gamma B).$$

Now ${}_\Gamma E'' = P\backslash(\Gamma \times F)$ is a vector bundle (via ψ above) over $P\backslash(\Gamma \times \bar{E}) = \bar{G}\backslash(\bar{\Gamma} \times \bar{E}) = {}_{\bar{\Gamma}}\bar{E}$.

Using (c), the vector space (e) becomes

$$\begin{aligned} H^i_c({}_\Gamma E'', \psi^*({}_{\bar{\Gamma}}A) \otimes (\tilde{p}_3)^*{}_\Gamma B) &= H^i_c({}_{\bar{\Gamma}}\bar{E}, \psi_!(\psi^*({}_{\bar{\Gamma}}A) \otimes (\tilde{p}_3)^*{}_\Gamma B)) \\ &= H^i_c({}_{\bar{\Gamma}}\bar{E}, {}_{\bar{\Gamma}}A \otimes \psi_!(\tilde{p}_3)^*{}_\Gamma B)). \end{aligned}$$

It is clear that $\psi_!(\tilde{p}_3)^*{}_\Gamma B = \tilde{\phi}_!(\tilde{i}^*\tilde{B})$ and this is the same as ${}_{\bar{\Gamma}}(\phi_!(i^*B))$ by (d). Hence the last H^i_c is the same as the left hand side of (a). Thus, the lemma is proved.

8. With the notations of [**L5, 3.1, 4,1**], we have

$$(p_3)_!\tilde{A} = A' * A'' \text{ and } \phi_!(i^*B) = \mathrm{res}_{\mathbf{V}',\mathbf{V}''}B.$$

Hence the previous lemma can be restated as follows.

(a) $$(A' \otimes A'', \mathrm{res}_{\mathbf{V}',\mathbf{V}''}B) = v^{2\dim G/P}(A' * A'', B).$$

Now $A', A'' \mapsto A' * A''$ defines a linear map $\mathcal{K}_{\mathbf{V}'} \otimes \mathcal{K}_{\mathbf{V}''} \to \mathcal{K}_{\mathbf{V}}$. As in [**L5, 10.2**], we modify it to a linear map $A' \circ A'' = v^m A' * A''$ where

$$m = \sum_{h\in\Omega} \dim \mathbf{V}'_{h'} \dim \mathbf{V}''_{h''} + \sum_{i\in I} \dim \mathbf{V}'_i \dim \mathbf{V}''_i.$$

Similarly, $B \mapsto \mathrm{res}_{\mathbf{V}',\mathbf{V}''}B$ defines a linear map $\mathcal{K}_{\mathbf{V}} \to \mathcal{K}_{\mathbf{V}'} \otimes \mathcal{K}_{\mathbf{V}''}$; as in [**L5, 10.10**], we modify it to a linear map $B \mapsto \widetilde{\mathrm{res}}_{\mathbf{V}',\mathbf{V}''}B = v^s \mathrm{res}_{\mathbf{V}',\mathbf{V}''}B$ where

$$s = \sum_{h\in\Omega} \dim \mathbf{V}'_{h'} \dim \mathbf{V}''_{h''} - \sum_{i\in I} \dim \mathbf{V}'_i \dim \mathbf{V}''_i.$$

(The second author would like to use this opportunity to point out that in the formula for $s(\tau,\omega)$ given on line -3 of [**L5, p.397**] there should be a minus sign in front of $s(\tau,\omega)$; this is what the computations in [**L5, p.398**] actually give.)

Using the equality $\dim G/P = \sum_{i\in I} \dim \mathbf{V}'_i \dim \mathbf{V}''_i$, we can rewrite (a) as follows:

(b) $$(A' \otimes A'', \tilde{\mathrm{res}}_{\mathbf{V}',\mathbf{V}''}B) = (A' \circ A'', B).$$

9. Let $\mathbf{U}^-$ be the $-$part of the quantized enveloping algebra $\mathbf{U}$ over $\mathbf{Q}(v)$ attached by Drinfeld and Jimbo to the generalized Cartan matrix $(a_{ij})_{i,j\in I}$ where $a_{ii} = 2$ and, for $i \neq j$, $-a_{ij}$ equals the number of $h \in \Omega$ such that $\{h', h''\} = \{i, j\}$ (as an unordered set).

We have the standard decomposition $\mathbf{U}^- = \oplus_\nu \mathbf{U}^-_\nu$ with ν running over $\mathbf{N}^I$ (as in [**L5, 10.15**]).

If $\nu \in \mathbf{N}^I$, we shall write K_ν instead of $K_{\mathbf{V}}$ where $\mathbf{V}$ is any I-graded vector space with $\dim \mathbf{V}_i = \nu(i)$ for all i. Similarly, if $\mathbf{V}, \mathbf{V}', \mathbf{V}''$ are as in §6, we shall write $\tilde{\mathrm{res}}_{\tau,\omega}$ instead of $\tilde{\mathrm{res}}_{\mathbf{V}',\mathbf{V}''}$ where $\tau, \omega \in \mathbf{N}^I$ are such that $\tau(i) = \dim \mathbf{V}'_i$ and $\omega(i) = \dim \mathbf{V}''_i$ for all i.

As in [**L5, 10.17**], we may identify the algebra $\mathbf{U}^-$ with the $\mathbf{Q}(v)$-vector space $\oplus_\nu \mathbf{Q}(v) \otimes_{\mathbf{Z}[v,v^{-1}]} \mathcal{K}_\nu$ where ν runs over $\mathbf{N}^I$; the multiplication on the direct sum is given by the operation $A', A'' \to A' \circ A''$. The summands $\mathbf{U}^-_\nu$ and $\mathbf{Q}(v) \otimes_{\mathbf{Z}[v,v^{-1}]} \mathcal{K}_\nu$ correspond.

Let $\mathbf{U}^0$ be the zero-part of $\mathbf{U}$, that is the group algebra of $\mathbf{Z}^I$ over $\mathbf{Q}(v)$ with standard basis $K_\sigma (\sigma \in \mathbf{Z}^I)$.

For the usual comultiplication Δ in $\mathbf{U}^0 \otimes \mathbf{U}^-$ we have

$$\Delta(u) = \sum (1 \otimes K_{-\tau}) \Delta_{\tau,\omega} u$$

for all $u \in \mathbf{U}_\nu^-$; the sum is taken over all τ and ω in $\mathbf{N}^I$ such that $\nu = \tau + \omega$ and $\Delta_{\tau,\omega} u \in \mathbf{U}_\tau^- \otimes \mathbf{U}_\omega^-$. In our identification we have $\Delta_{\tau,\omega} u = \tilde{\text{res}}_{\tau,\omega} u$ for all $u \in \mathbf{U}_\nu^-$. (See [**L5, 10.10**].)

For any $i \in I$, the standard generator F_i of $\mathbf{U}^-$ corresponds to the standard basis element of $\mathcal{K}_\nu$ (with $\nu(j) = \delta_{ij}$) given by the unique simple perverse sheaf over a point.

The form $(,)$ defined in §4, with scalars extended to $\mathbf{Q}(v)$, becomes a symmetric $\mathbf{Q}(v)$-bilinear form $(,) : \mathbf{U}^- \times \mathbf{U}^- \to \mathbf{Q}((v))$ such that

(a) $$(\mathbf{U}_\nu^-, \mathbf{U}_{\nu'}^-) = 0 \text{ for } \nu \neq \nu',$$

(b) $$(u' \otimes u'', \Delta_{\tau,\omega} u) = (u'u'', u)$$

for all $u' \in \mathbf{U}_\tau^-, u'' \in \mathbf{U}_\omega^-, u \in \mathbf{U}_{\tau+\omega}^-$; in the left hand side we have the natural extension of $(,)$ to a bilinear form on $(\mathbf{U}^- \otimes \mathbf{U}^-) \times (\mathbf{U}^- \otimes \mathbf{U}^-)$. Note also that

(c) $$(1,1) = 1 \text{ and } (F_i, F_i) = 1 + v^2 + v^4 + \cdots = (1 - v^2)^{-1}$$

for all i. (The last formula comes from the equivariant cohomology of a point with respect to the group $\mathbf{C}^*$.) Hence $(,)$ on $\mathbf{U}^-$ coincides up to factors in $1 + v\mathbf{Z}[[v]]$ with the form in [**K, 3.4.4**] (which is a variant of a result of Drinfeld).

10. Let $\mathbf{B}$ be the basis of $\mathbf{U}^-$ corresponding to the union of the bases of the various $\mathcal{K}_\mathbf{V}$ given by the objects in $\mathcal{P}_\mathbf{V}$. By definition, we have $\mathbf{B} = \cup_\nu \mathbf{B}_\nu$ (disjoint union) where $\mathbf{B}_\nu = \mathbf{B} \cap \mathbf{U}_\nu^-$.

Let U^- be the $\mathbf{Z}[v, v^{-1}]$-subalgebra of $\mathbf{U}^-$ generated by the elements $F_i^{(a)} = F_i^a/[a]^!$ for $i \in I$ and $a \in \mathbf{N}$ ($[a]^!$ as in [**L5, 10.12**]). According to [**L5, 11.3**], U^- coincides with the $\mathbf{Z}[v, v^{-1}]$-submodule of $\mathbf{U}^-$ generated by $\mathbf{B}$.

11. For any $i \in I$, let $r_i : \mathbf{B} \to \mathbf{N}$ be the function defined by $b \in F_i^{(r_i(b))}\mathbf{U}^-$, $b \notin F_i^{(r_i(b)+1)}\mathbf{U}^-$. For any ν and any $t \geq 0$, let $\mathbf{B}_{\nu;i,t} = \{b \in B_\nu | r_i(b) = t\}$.

12. The following property of $\mathbf{B}$ is proved in [**L5, §6**].

Let $b \in \mathbf{B}_{\nu;i,t}$ and let $\nu' \in \mathbf{N}^I$ be defined by $\nu'(i) = \nu(i) - t$ and $\nu'(j) = \nu(j)$ for $j \neq i$. Then there exists $b' \in \mathbf{B}_{\nu';i,0}$ such that $F_i^{(t)} b'$ is equal to b plus an element in $F_i^{(t+1)}\mathbf{U}^-$.

13. The following property of $\mathbf{B}$ is proved in [**L5, 7.2**].

Let $b \in \mathbf{B}_\nu$ where $\nu \neq 0$. Then there exists $i \in I$ and $t > 0$ such that $b \in \mathbf{B}_{\nu;i,t}$.

14. Now Verdier duality induces a $\mathbf{Z}$-linear involution D of $\mathcal{K}_\nu$ which maps the canonical basis into itself. Under the identification in §9, this becomes an involution $^- : U^- \to U^-$ which preserves the ring structure (see [**L5, 3.8**]), and satisfies $\overline{vu} = v^{-1}\bar{u}$ for all $u \in U^-$. Moreover, it clearly leaves fixed the elements $F_i^{(a)}$. These properties determine the involution $^-$ uniquely.

It is clear that $^-$ carries $\mathbf{B}_\nu$ into itself for any ν. It is also clear that the subspace $F_i^{(t)}\mathbf{U}^-$ is stable under $^-$ for any $t \geq 0$.

If b, b' are as in §12 and if $\bar{b}' = b'$, then applying $^-$ to the equality in §12, we see that $\bar{b}$ is equal to b plus a linear combination of elements of $\mathbf{B}$ other than

b. Since $\bar{b} \in \mathbf{B}$ and $\mathbf{B}$ is a basis, it follows that $\bar{b} = b$. This argument shows by induction on $\sum_i \nu(i)$ that

(a) $$\bar{b} = b \text{ for all } b \in \mathbf{B}.$$

(The argument above is always applicable when $\nu \neq 0$, in view of §13.) An equivalent statement is that D keeps fixed any simple perverse sheaf in $\mathcal{P}_{\mathbf{V}}$.

The following result is an analogue of [**K, 5.1.2**].

LEMMA 15. *Let $u = \sum_{j,b} n_{j,b} v^j b$ be an element of U^-. (Here b runs over $\mathbf{B}$, j runs over $\mathbf{Z}$ and $n_{j,b}$ are integers, zero for all but finitely many indices.) The following conditions for u are equivalent:*

(a) $(u,u) \in \mathbf{Z}[[v]]$;

(b) $n_{j,b} = 0$ for all b and all $j < 0$.

On the other hand the following conditions on u are equivalent:

(a1) $(u,u) \in 1 + v\mathbf{Z}[[v]]$;

(b1) there exists $b_0 \in \mathbf{B}$ such that $n_{0,b_0} = \pm 1$ and $n_{j,b} = 0$ if $j < 0$ or if $j = 0, b \neq b_0$.

Using §3 and the last sentence in §14, we see that

(c) $$(b, b') \in \delta_{b,b'} + v\mathbf{N}[[v]]$$

for any b, b' in $\mathbf{B}$. If $u = 0$ there is nothing to prove. We assume that $u \neq 0$. Let j_0 be the minimum value of j such that $n_{j,b} \neq 0$ for some b. From (a) we see that $(u,u) \in \sum_b n_{j_0,b}^2 v^{2j_0} + v^{2j_0+1}\mathbf{Z}[[v]]$. The lemma follows.

16. Let $\mathcal{B}$ be the basis of $\mathbf{U}^-$ defined in [**K**]. It is known that $\mathcal{B}$ satisfies the analogue of 15(c):

(a) $$(\beta, \beta') = \delta_{\beta,\beta'} + v\mathbf{Z}[[v]]$$

for any $\beta, \beta' \in \mathcal{B}$ (see [**K**] and the last sentence in §9). Moreover, $\mathcal{B} \subset U^-$. Using Lemma 15, it follows that for any $\beta \in \mathcal{B}$, there exists $b \in \mathbf{B}$ and $\epsilon = \pm 1$ such that $\beta - \epsilon b$ is a linear combination of elements in $\mathbf{B}$ with coefficients in $v\mathbf{Z}[v]$. Now b is fixed by $^-$ by 14(a) and β is fixed by $^-$ by definition. Applying $^-$ to $\overline{\beta - \epsilon b}$ we get on the one hand $\beta - \epsilon b$ and on the other hand, an element of $\overline{\sum_{b'} v\mathbf{Z}[v]b'} = \sum_{b'} v^{-1}\mathbf{Z}[v^{-1}]b'$ where b' runs over $\mathbf{B}$. Since $\sum_{b'} v^{-1}\mathbf{Z}[v^{-1}]b' \cap \sum_{b'} v\mathbf{Z}[v]b' = 0$, it follows that $\beta - \epsilon b = 0$ and $\beta = \epsilon b$. Thus, there is a map $\epsilon : \mathbf{B} \to \{1, -1\}$ such that $\mathcal{B} = \{\epsilon(b)b | b \in \mathbf{B}\}$.

LEMMA 17. *We have $\epsilon(b) = 1$ for all $b \in \mathbf{B}$.*

This is clear if $b \in \mathbf{B}_\nu$ with $\nu = 0$. Hence we may assume that $b \in \mathbf{B}_\nu$ with $\nu \neq 0$ and that the result is already proved when ν is replaced by ν' with $\sum_j \nu'(j) < \sum_j \nu(j)$. By §13 we can find $i \in I$ and $t > 0$ such that $b \in \mathbf{B}_{\nu;i,t}$. We associate ν', b' to b as in §12. By the induction hypothesis, we have $\epsilon(b') = 1$, hence $b' \in \mathcal{B}$. By definition, we have $b' \notin F_i\mathbf{U}^-$. By a known property of $\mathcal{B}$, the product $F_i^{(t)}b'$ is equal to an element $\beta_1 \in \mathcal{B}$ plus an element of $F_i^{(t+1)}\mathbf{U}^-$; moreover, we must have $\beta_1 = \epsilon(b_1)b_1$ for some $b_1 \in \mathbf{B}$. On the other hand, by §12, $F_i^{(t)}b'$ is equal to b plus an element of $F_i^{(t+1)}\mathbf{U}^-$. Thus, $b-\epsilon(b_1)b_1 \in F_i^{(t+1)}\mathbf{U}^-$. Since $F_i^{(t+1)}\mathbf{U}^-$ is spanned by elements of $\mathbf{B}$ other than b, it follows that $b = \epsilon(b_1)b_1$. Since both b, b_1 belong to the basis $\mathbf{B}$, we must have $\epsilon(b_1) = 1$ and $b = b_1$. Thus, $\epsilon(b) = 1$. The lemma is proved.

We have proved the following result.

THEOREM 18. $\mathbf{B} = \mathcal{B}$.

REFERENCES

[BBD] A. Beilinson, J. Bernstein and P. Deligne, *Faisceaux pervers*, Astérisque **100** (1982).

[K] M. Kashiwara, *On crystal bases of the q-analogue of universal enveloping algebras*, Duke Math. J. **63** (1991), 465-516.

[L1] G. Lusztig, *Character sheaves, II*, Adv. in Math. **57** (1985), 226-265.

[L2] G. Lusztig, *Cuspidal local systems and graded Hecke algebras*, Publications Mathématiques **67** (1988), 145-202.

[L3] G. Lusztig, *Canonical bases arising from quantized enveloping algebras*, J. Amer. Math. Soc. **3** (1990), 447-498.

[L4] G. Lusztig, *Canonical bases arising from quantized enveloping algebras, II*, Progr. Theor. Phys. Suppl. **102** (1990), 175-201.

[L5] G. Lusztig, *Quivers, perverse sheaves and quantized enveloping algebras*, J. Amer. Math. Soc. **4** (1991), 365-421.

MASSACHUSETTS INSTITUTE OF TECHNOLOGY, CAMBRIDGE, MA 02139

Contemporary Mathematics
Volume **153**, 1993

Transitive Permutation Lattices in the Same Genus and Embeddings of Groups

ROBERT M. GURALNICK AND AL WEISS

ABSTRACT. We consider the problem of finding nonconjugate subgroups of a finite group such that the permutation lattices corresponding to these subgroups are closely related. In particular, we show that examples with the lattices locally isomorphic can occur with most isomorphism types of subgroups. We consider some related problems for finite subgroups of infinite groups.

1. Introduction

Let G be a finite group. If X is a G-set and R is a commutative ring, we will let RX denote the corresponding RG-permutation module. In particular, if H is a subgroup of G, we let G/H denote the G-set consisting of the left cosets of H in G. In this note, we wish to study the condition $RX \cong RY$ for transitive G-sets X and Y, in particular when R is a semilocalization of the integers or the integers. This problem (for various choices of R) has applications to many areas including geometry, number theory, and invariant theory (cf [J], [Su]).

If R is a field of characteristic zero (or more generally, $|G|$ is invertible in a semilocal ring R), then $R[G/H] \cong R[G/K]$ is equivalent to the equality of the permutation characters 1_H^G and 1_K^G. It is well known that this is equivalent to:

$$(1.1) \qquad |x^G \cap H| = |x^G \cap K| \quad \text{for all} \quad x \in G,$$

where x^G is the conjugacy class of x in G.

1991 *Mathematics Subject Classification.* Primary 20C10; Secondary 20C11, 20B15, 19A22.

Key words and phrases. permutation lattice, genus, permutation character, conjugacy class.

The first author was supported in part by an NSF Grant. He would also like to thank the University of Alberta for its warm hospitality. The second author was partially supported by the N.S.E.R.C.

This paper is in final form and no version of it will be submitted for publication elsewhere.

If R is a field or local ring, there is an analog of (1.1) essentially due to Conlon. We will use this criterion in the next section to construct examples of a finite group G (with G a symmetric group) and nonconjugate subgroups H, K such that $R[G/H] \cong R[G/K]$ for $R = \mathbb{Z}_p$, the completion of $\mathbb{Z}$ at the prime p, for all primes p (i.e. $\mathbb{Z}[G/H]$ and $\mathbb{Z}[G/K]$ are in the same genus). L. Scott produced the first such examples (with $G = L_2(q), q \equiv \pm 1 mod\ 30$ and H, K each isomorphic to A_5). When $q = 29$, these integral permutation lattices are in fact isomorphic (see [S]). In our construction, $H \cong K$ can be any finite group such that $H/O_p(H)$ is not cyclic for any prime p.

Condition (1.1) can be studied for finite subgroups of any group G (finite or infinite). In section 3, we make some remarks regarding this condition when G is a complex semisimple Lie group. If $H \cong K$ and $G = GL_n(\mathbb{C})$, then condition (1.1) implies that the corresponding characters are equivalent and so in particular H and K are conjugate. The same is true for $G = O_n(\mathbb{C})$ or $Sp_n(\mathbb{C})$. It is easy to construct examples where this fails for $PGL_n(\mathbb{C})$. We will discuss an example of Borovik which shows that the same is not true for $E_8(\mathbb{C})$ (and so in particular, there is no such result in general for simply connected semisimple Lie groups).

Borovik's example also works over finite fields. His technique was used in [GS] to produce infinitely many examples of finite groups G with subgroups H, K satisfying (1.1) with H maximal and K nonmaximal, thus answering a question of Wielandt. Forster and Kovacs [FK] had already proved that a minimal counterexample to Wielandt's question would be an almost simple group. In the final section, we consider isomorphic permutation lattices where one of the G-sets is primitive and obtain some positive results.

2. Permutation Latttices in the Same Genus

Let G be a finite group. Let $\mathcal{C}_p(G)$ be the collection of subgroups C of G such that $C/O_p(C)$ is cyclic. Let $\mathcal{C}(G) = \cup_p \mathcal{C}_p(G)$. If X is a G-set, let X^G denote the set of fixed points of G on X.

The following theorem which is essentially due to Conlon will be quite useful. See [W, Theorem 1.2] or [S, Prop. 3.1].

THEOREM 2.1. *Let G be a finite group and p a prime. Let F be a field of characteristic p. Let X, Y be two finite G-sets. Let $V = FX$ and $W = FY$ be the corresponding FG-modules. Then $V \cong W$ if and only X^C and Y^C have the same cardinality for all subgroups $C \in \mathcal{C}_p(G)$.*

Since $FX \cong FY$ (for permutation modules) implies that $\mathbb{Z}_p X \cong \mathbb{Z}_p Y$, this yields:

COROLLARY 2.2. *Let G be a finite group. Let X, Y be two finite G-sets. Let $V = \mathbb{Z}X$ and $W = \mathbb{Z}Y$ be the corresponding $\mathbb{Z}G$-permutation lattices. The following are equivalent:*

(i) *V and W are in the same genus.*
(ii) *X^C and Y^C have the same cardinality for all subgroups $C \in \mathcal{C}(G)$.*

It will be useful to restate the two previous results in the case that X and Y are transitive G-sets. If V is a $\mathbb{Z}G$-module, let V_p denote the p-adic completion of V at the prime p. We will make use of the easy fact that if $X = G/H$ and $C \leq G$, then

$$|X^C| = |H|^{-1}|\{g \in G : C^g \subseteq H\}| = \frac{|N_G(C)|}{|H|} m_{G,H}(C),$$

where $m_{G,H}(C)$ is the number of G-conjugates of C contained in H.

COROLLARY 2.3. *Let G be a finite group with subgroups K and L. Let $X = G/K$ and $Y = G/L$. Let $V = \mathbb{Z}X$ and $W = \mathbb{Z}Y$. The following are equivalent:*

(i) $V_p \cong W_p$.
(ii) X^C *and* Y^C *have the same cardinality for all subgroups* $C \in \mathcal{C}_p(G)$.
(iii) $m_{G,H}(C) = m_{G,K}(C)$ *for all* $C \in \mathcal{C}_p(G)$.

Proof. The equivalence of (i) and (ii) follows from Theorem 2.1. The equivalence of (ii) and (iii) follows from the observation prior to the statement of the Corollary (noting that either condition implies that $|H| = |K|$).

The previous result extends the well known result that the two permutation characters 1_K^G and 1_L^G are equal if and only if $|K \cap x^G| = |L \cap x^G|$ for all $x \in G$. (See (1.1).)

In particular, Corollary 2.3 asserts that V and W are in the same genus if and only if (iii) holds for all $C \in \mathcal{C}(G)$.

We now apply the previous result to the case where K and L are isomorphic. There is a similar version for p-adic isomorphism for a single prime p.

COROLLARY 2.4. *Let G, H be finite groups. Let $\phi_i : H \to G$ for $i = 1, 2$ be embeddings. If C is a subgroup of H, let $C_i = \phi_i(C)$. Assume that if $C \in \mathcal{C}(H)$, then C_1 and C_2 are conjugate in G. Then $\mathbb{Z}(G/H_1)$ and $\mathbb{Z}(G/H_2)$ are in the same genus.*

Proof. If $D \subset G$, set $m_i(D) = m_{G,H_i}(D)$. By the previous result, it suffices to prove that $m_1(D) = m_2(D)$ for all $D \in \mathcal{C}(G)$.

Let $D \in \mathcal{C}(G)$. By hypothesis, D is conjugate to a subgroup of H_1 if and only if D is conjugate to a subgroup of H_2. Thus $m_1(D) = 0$ if and only if $m_2(D) = 0$. So it suffices to consider subgroups of H_1. The equality $m_1(C_1) = m_2(C_2)$ for all $C \in \mathcal{C}(H)$ follows immediately from the hypothesis.

Of course, in the previous result, we want H_1 and H_2 to be nonconjugate. This is impossible if $H/O_p(H)$ is cyclic for some p.

Let $\Omega(H)$ be the Burnside ring of H. Let $H_i, 1 \leq i \leq m$, be a set of representatives for the conjugacy classes of subgroups of H. If X is a finite H-set, let x_i be the cardinality of X^{H_i}. Then it is well known that the map $\phi : \Omega(H) \to \prod_{i=1}^m \mathbb{Z}$ defined by $\phi(X) = (x_1, \dots, x_m)$ is an injection with finite cokernel (if we order the transitive G-sets and the H_i appropriately, the matrix of ϕ will be upper triangular with nonzero entries on the diagonal).

Thus, if $H \notin \mathcal{C}(H)$ (respectively, $\mathcal{C}_p(H)$), we can find two nonisomorphic H-sets X and Y such that $x_i = y_i$ for all i with $H_i \in \mathcal{C}(H)$ (respectively $\mathcal{C}_p(H)$). We can assume that H acts faithfully on X (and so necessarily on Y). This implies that $|X| = |Y| = n$. Indeed, this implies more generally that X and Y are isomorphic as C-sets for any subgroup C of H such that $C \in \mathcal{C}(H)$ (respectively $\mathcal{C}_p(H)$) (if T is any subgroup of C, then $|X^T| = |Y^T|$ and so X and Y are equal in $\Omega(C)$). Applying Theorem 2.1 yields:

COROLLARY 2.5. *Let H be a finite group. Let $\mathcal{P}$ be the set of primes p such that $H/O_p(H)$ is cyclic. Then there exist (not necessarily transitive) nonisomorphic H-sets X and Y such that $\mathbb{Z}_q X \cong \mathbb{Z}_q Y$ for all primes $q \notin \mathcal{P}$. In particular, if $H/O_p(H)$ is not cyclic for any prime p, $\mathbb{Z}X$ and $\mathbb{Z}Y$ are in the same genus.*

Since the genus class group of any lattice is finite, it follows that if $\mathbb{Z}X$ and $\mathbb{Z}Y$ are in the same genus, then $\mathbb{Z}X^m \cong \mathbb{Z}Y^m$ for some m, where X^m is the disjoint union of m H-sets each isomorphic to X. If X and Y are nonisomorphic H-sets, then so are X^m and Y^m. Thus:

COROLLARY 2.6. *Let H be a finite group with $H \notin \mathcal{C}(H)$. Then there exist nonisomorphic H-sets X and Y such that $\mathbb{Z}X \cong \mathbb{Z}Y$.*

We now want to produce examples of *transitive* permutation lattices in the same genus. Let H be a finite group with $H \notin \mathcal{C}(H)$. As above, we can find nonisomorphic H-sets X and Y of some cardinality n with $|X^C| = |Y^C|$ for all $C \in \mathcal{C}(H)$. Thus we have two embeddings α_X and α_Y of H into $G = S_n$ (by the actions of H on X and on Y). Denote the images by K and L respectively. Clearly K and L are not conjugate in G (since X and Y are nonisomorphic H-sets). On the other hand, if $C \in \mathcal{C}(H)$, then $\alpha_X(C)$ and $\alpha_Y(C)$ are conjugate (since X and Y are isomorphic as C-sets). We can apply Corollary 2.4 and obtain:

THEOREM 2.7. *Let H be any finite group which is not cyclic modulo $O_p(H)$ for any p. Then there exists a positive integer n and two nonconjugate subgroups K and L of $G = S_n$, each isomorphic to H, such that $\mathbb{Z}(G/K)$ and $\mathbb{Z}(G/L)$ are in the same genus.*

One can improve the theorem a bit by using a result of Oliver. Keep notation as above. By the previous result, we can find G-sets X and Y with the corresponding permutation lattices in the same genus. Let $\Gamma \supseteq \mathbb{Z}G$ be a maximal order in $\mathbb{Q}G$. Since $\mathbb{Q}$ is a splitting field for any symmetric group, Γ has trivial class group and so $\Gamma \cdot \mathbb{Z}X \cong \Gamma \cdot \mathbb{Z}Y$. By [O, Theorem 6], there exists a G-set T such that $\mathbb{Z}X \oplus \mathbb{Z}T \cong \mathbb{Z}Y \oplus \mathbb{Z}T$. Thus:

COROLLARY 2.8. *Let H be any finite group which is not cyclic modulo $O_p(H)$ for any p. Then there exists a positive integer n and two nonconjugate subgroups K and L of $G = S_n$, each isomorphic to H, such that $\mathbb{Z}(G/K)$ and $\mathbb{Z}(G/L)$ are stably isomorphic in the category of permutation lattices.*

The next lemma shows that the reduction to symmetric groups is not a severe one.

LEMMA 2.9. *Let K be a subgroup of G. Assume that G acts faithfully on G/K. View G as a transitive subgroup of $Sym(G/K) = \hat{G}$. Suppose that $R(G/K) \cong R(G/L)$ with K and L nonconjugate in G. Then K and L are not conjugate in $\hat{G}$ and $R(\hat{G}/K) \cong R(\hat{G}/L)$.*

Proof. Since K has an orbit of size 1 on G/K and L does not, K and L are not conjugate in $\hat{G}$. The isomorphism of the modules is obvious.

Let $H = A_5$ and consider the two nonconjugate embeddings into $L_2(29) = G$. Then G embeds into $\hat{G} = S_{203}$ and the two classes of A_5 are nonconjugate in $\hat{G}$. The two permutation lattices are isomorphic by Scott [S]. In fact, one can find two nonconjugate embeddings of A_5 into S_{81} so that the two permutation lattices are in the same genus.

D. Hahn has constructed some examples of solvable groups G with nonisomorphic G-sets which give rise to permutation lattices in the same genus. Apparently some of these examples give rise to nonisomorphic lattices.

One way to proceed to try to produce examples with groups other than symmetric groups would be to take H to be a group with $H \notin \mathcal{C}(H)$. Embed H in a symmetric group in two ways and try to find a smaller group containing the two embeddings which satisfy Corollary 2.4.

For example, if we take $H \cong D_{12}$, the dihedral group of order 12, then the choice of X and Y is essentially unique (because every proper subgroup is in $\mathcal{C}(H)$). It is straightforward to compute that $|X| = |Y| = 24$ is the smallest possibility. We can assume that the two embeddings agree on the element of order 12. We then need to adjoin elements in S_{24} which fuse the conjugacy classes of proper subgroups of the two images.

We close this section by mentioning some open questions.

(2.10) If $\mathbb{Z}(G/K)$ and $\mathbb{Z}(G/L)$ are in the same genus, must the modules be isomorphic?

The previous question can be asked for any coefficient ring R. An interesting special case is $R = \mathbb{Z}[1/|G|]$. In this case, the genus condition is equivalent to the equality of the two permutation characters.

(2.11) If $\mathbb{Z}(G/K)$ and $\mathbb{Z}(G/L)$ are in the same genus, must K and L be isomorphic?

(2.12) If $\mathbb{Q}(G/K) \cong \mathbb{Q}(G/L)$ with K solvable, must L be solvable?

The last question has been posed by Thompson. There are examples which show that not only is it possible for K and L to be nonisomorphic but they need not have the same composition factors (eg, take $G = M_{23}$, $H = M_{21}.2$, and $K = 2^4A_7$). Let $a_n(H)$ be the number of elements of order n in a group H. It is easy to see that $R(G/K) \cong R(G/L)$ implies that $a_n(K) = a_n(L)$ (for any choice of R). Conversely, if $a_n(K) = a_n(L)$ for all n, then $\mathbb{Q}(S_n/K) \cong \mathbb{Q}(S_n/L)$, where $n = |K|$ and K, L are embedded in S_n via the regular representation. Thus question (2.12) can be rephrased as:

(2.13) If K and L are finite groups with K solvable and $a_n(K) = a_n(L)$ for all n, is L solvable?

The previous two questions have an affirmative answer if solvable is replaced by nilpotent. Let p be a prime and $m = |K| = |L|$. Let m_p be the largest power of p dividing m. Since K is nilpotent, $\sum a_i(K) = m_p$, where i ranges over all powers of p. The same is true for L and it follows that the Sylow p-subgroup of L is normal. Since this is true for every prime, L is nilpotent.

The simplest example of two nonisomorphic subgroups K and L which induce the same permutation character in a group G is to take K and L to be nonisomorphic groups of order $m = p^3$ and exponent p (for an odd prime p) embedded in S_m via the regular representations.

However, it is even possible for this to occur with G a p-group. We will construct an example of a p-group P with two nonisomorphic subgroups A and B such that $1_A^P = 1_B^P$. Our example will not be corefree (i.e. $A_P = B_P \neq 1$). However, if Q is any p-group for which P is a subgroup of $\mathrm{Aut}(Q)$, then the group $G = QP$ is still a p-group with $1_A^G = 1_B^G$ (this is clear by first inducing up to P) and $A_G = B_G = 1$ since P acts faithfully on Q. We note that a similar example has been independently constructed in [CGM]. Their example is based on an example of the first author.

First consider the case $p \geq 3$. Let V be the 4-dimensional vector space of column vectors over $\mathbb{F}_p$. Consider the two matrices,

$$a = \begin{pmatrix} 1 & 1 & 0 & 0 \\ 0 & 1 & 1 & 0 \\ 0 & 0 & 1 & 1 \\ 0 & 0 & 0 & 1 \end{pmatrix}$$

$$b = \begin{pmatrix} 1 & 0 & 1 & 0 \\ 0 & 1 & 0 & 1 \\ 0 & 0 & 1 & 0 \\ 0 & 0 & 0 & 1 \end{pmatrix}.$$

Let $A_0 = \langle a, b \rangle \leq GL_4(p)$. Note that A_0 is abelian. Let V_0 be the one dimensional subspace of V fixed by A_0 and choose a nontrivial element v in V_0. Let $P = VA_0$, $A = \langle a, b, v \rangle$ and $B = \langle a, bw, v \rangle$, where $w \in V - V_0$ with w fixed by A in V/V_0. If $K = A_P$, it is straightforward to compute that $K = B_P$ as well and that A/K and B/K induce the same permutation character in P/K (this example was exhibited by Jehne [J] in this form). Here A/K and B/K are both elementary abelian groups of order p^2. It therefore follows that $1_A^P = 1_B^P$. Since A_0 is abelian, so is A. On the other hand, since a does not fix w, B is not abelian and in particular is not isomorphic to A. (Note that for $p > 3$, A is elementary abelian of order p^3 while for $p = 3$, $A \cong C_9 \times C_3 \times C_3$).

A similar (and in fact easier) example can be constructed for $p = 2$. Let V be the cyclic group of order 16 generated by x. Let $A_0 = Aut(V)$. Let $a, b \in A_0$ with $x^a = x^{-1}$ and $x^b = x^3$. Note that $A_0 = \langle a, b\rangle$. Let $A = \langle a, b, x^8\rangle$ and $B = \langle a, bx^4, x^8\rangle$. Then A is abelian while B is not. Clearly, $K = A_P = B_P = \langle x^8, b^2\rangle$ and it is straightforward to see that A/K and B/K induce the same permutation character in the group of order P/K of order 32. Thus, the same is true for A and B as promised. In this case, P has order 128.

Question (2.11) can also be rephrased similarly. If A and B are finite groups, let $a_A(B)$ be the number of subgroups of B which are isomorphic to A.

PROPOSITION 2.14. *Let H and K be two nonisomorphic finite groups such that $a_C(H) = a_C(K)$ for all $C \in \mathcal{C}(H) \cup \mathcal{C}(K)$. Then*

(i) *$|H| = |K|$, and*

(ii) *if we view H and K as subgroups of $G = S_n$, $n = |H|$, where each subgroup is embedded via the regular representation, then $\mathbb{Z}(G/H)$ and $\mathbb{Z}(G/K)$ are in the same genus.*

Proof. (i) follows by considering cyclic subgroups (there are the same number of elements of each order). Now (ii) follows by Corollary 2.3 (since $a_C(H) = m_{S_n,H}(C)$ when H is embedded in S_n via the regular representation).

Thus question (2.11) can be rephrased as:

(2.15) If $G_i, i = 1, 2$ are finite groups with $a_C(G_1) = a_C(G_2)$ for all $C \in \mathcal{C}(G_1) \cup \mathcal{C}(G_2)$, is $G_1 \cong G_2$?

We suspect the answer is no.

3. Embeddings into Semisimple Lie Groups

In this section, we consider a stronger version of (1.1) for G a semisimple complex Lie group. We will also assume that the subgroups are isomorphic. So assume H is a finite group and $\phi_i, i = 1, 2$ are homomorphisms into G such that

$$\phi_1(h)^G = \phi_2(h)^G \quad \text{for all} \quad h \in H. \tag{3.1}$$

If $G = GL_n(\mathbb{C}), SL_n(\mathbb{C}), Sp_n(\mathbb{C})$, or $O_n(\mathbb{C})$, then (3.1) implies that the two representations have equal characters and so the representations are equivalent. There are two questions one could ask.

(3.2) Does (3.1) imply that there exists $g \in G$ such that $\phi_1(h) = g\phi_2(h)g^{-1}$ for all $h \in H$?

(3.3) Does (3.1) imply that $\phi_1(H) = g\phi_2(H)g^{-1}$ for some $g \in G$?

Of course an affirmative answer to (3.2) implies the same for (3.3). As we remarked above, (3.2) does have an affirmative answer for G one of the classical groups described above. Now consider the situation for $G = PGL_n(\mathbb{C})$. Essentially, we are asking whether projective representations are determined by their 'projective' characters. The answer is easily is seen to be no. For (3.2), we take

H to be $C_p \times C_p$, for p an odd prime. Let P be the nonabelian group of order p^3 and exponent p. If Z is the center of P, then $P/Z = H$. Now P has $p-1$ distinct irreducible representations $\phi_i, 1 \leq i \leq p-1$, of degree p. The character of each representation is zero outside of Z and so the ϕ_i are obviously distinct on Z. These representations give rise to embeddings of H into $PGL_p(\mathbb{C})$ satisfying (3.1) (since the character values are 0 outside of Z). On the other hand, no two of the embeddings are equivalent, since if they were, it would follow that the corresponding representations, say α and β would satisfy

$$\alpha(h) = g\beta(h)g^{-1}\gamma(h),$$

for some $g \in GL_p(\mathbb{C})$ where $\gamma(h)$ is some scalar. It follows that γ is a one dimensional character of H and so is trivial on $Z = P'$. This implies α and β agree on Z, a contradiction.

This shows (3.2) has a negative answer for $PGL_n(\mathbb{C}), n \geq 3$. See also [Bl]. (Since $PGL_2(\mathbb{C})$ is isomorphic to $SO_3(\mathbb{C})$, (3.2) has an affirmative answer in this case). However, for these examples, $\phi_i = \phi_1 \circ \gamma$ for some automorphism γ of P. Thus, this does not answer (3.3) in the negative. A minor modification will provide an example. Consider the two embeddings of H into $PGL_{p^2}(\mathbb{C})$, where the first is p copies of the representation described above and the second is via the regular representation of H. It is straightforward to verify condition (3.1) is valid. However, their images cannot be conjugate, for if they were their inverse images would also be conjugate in $GL_{p^2}(\mathbb{C})$ and their inverses images are not even isomorphic (one is abelian and the other is not).

After seeing these examples, one might ask whether the problem is that $PGL_n(\mathbb{C})$ is not simply connected. This seems quite plausible. However, Borovik [B] found a counterexample with the simply connected group $E_8(\mathbb{C})$. See also [L] for examples with Spin groups. We will now describe a slightly modified version of his example.

As we have seen above, one way to obtain counterexamples to (3.2) is to consider projective representations. Let J be the triple cover of A_6. Then there exist $\phi_i, i = 1, 2$, inequivalent nine dimensional irreducible representations of J such that the characters agree on all elements of minimal order in a given coset of the center of J. (Indeed, we can take ϕ_1 to be faithful and ϕ_2 to have a kernel of order 3.) This implies that (3.1) holds for the corresponding embeddings of A_6 into $G_0 = SL_9(\mathbb{C})/W$, where W is a central subgroup of order 3 in $SL_9(\mathbb{C})$. Note that G_0 has center Z of order 3. Thus, (3.1) holds for $H = C_3 \times A_6$, where C_3 is mapped onto the center. The images of the two embeddings are not conjugate (even under $\mathrm{Aut}(G_0)$) since their inverse images in $SL_9(\mathbb{C})$ are not isomorphic (one is a central product of J with C_9 and the other is $A_6 \times C_9$). Since G_0 embeds into $G = E_8(\mathbb{C})$, we obtain two embeddings of A_6 into G. We recall that $G_0 = C_G(Z)$. Let $H_i, i = 1, 2$ denote the images of the two embeddings. If H_1 and H_2 are conjugate, it follows that they are conjugate in $N_G(Z)$ and so are conjugate via an element of $\mathrm{Aut}(G_0)$, a contradiction.

Borovik's example also provides a nice counterexample to a problem in invariant theory. Let G be a reductive group. Then G acts on $V = G \times G$ by simultaneous conjugation. One very nice problem is to describe the invariants, $\mathbb{C}[V]^G$. If $G = GL_n(\mathbb{C})$ (or $SL_n(\mathbb{C})$), then the invariants are generated by the functions $\mathrm{Tr}(w(X,Y))$, where w is some monomial in the noncommuting variables X and Y and Tr is the usual trace of a matrix. X and Y are the projections of V onto G. It was open whether the same was true for other simply connected groups (replace Tr by any (all) character for a faithful representation of G). Borovik [B] shows that this fails for $G = E_8(\mathbb{C})$.

This can be seen as follows. Choose a pair of generators a, b for H with a of order 15 (so a^5 generates the center of H). Let $a_i = \phi_i(a)$ and $b_i = \phi_i(b)$. Consider the G-orbits, $U_i = (a_i, b_i)^G$. We claim that each U_i is a closed orbit. Since the centralizer of $\phi_i(H)$ is finite, U_i has dimension 248. Suppose (α, β) is in $\bar{U_i}$. Since any word w in a_i, b_i has finite order (and so is semisimple), it follows that the orbit w^G is closed and so the corresponding word in α, β is in the same G-conjugacy class. This shows that first of all the group generated by α, β is also isomorphic to H and moreover that the adjoint representation of H on the Lie algebra L of G determined by α, β is equivalent to that of H_i (indeed, the same is true for any representation of G). In particular, since H_i has finite centralizer, H_i has no fixed points on L. The same is thus true for the group generated by α, β. Thus $(\alpha, \beta)^G$ has dimension 248 and so $(\alpha, \beta)^G = U_i$. Thus U_i is closed for each i. As we observed above if ρ is the character of any finite dimensional representation of G and $w(X, Y)$ is any monomial, then $\rho(w(a_1, b_1)) = \rho(w(a_2, b_2))$. Thus, these functions do not separate closed orbits and so cannot generate the invariants. See also [B, Proposition 3.1].

Recently, Springer has proved a result which in particular implies that there are only finitely many nonconjugate embeddings of H into G. Springer proved:

THEOREM 3.4. [Sp] *Let G be a connected algebraic group over an algebraically closed field k of characteristic $p \geq 0$. Let H be a finite group with $|H| \neq 0$ in k. Let X be the variety of homomorphisms of H into G.*

(i) *Each irreducible component of X is a G-orbit.*
(ii) *The number of inequivalent homomorphisms of H into G is finite.*
(iii) *If $H \leq G$, then H is conjugate in G to a subgroup of $G(K)$ where K is a finite extension of the field of definition of G.*

In particular, Theorem 3.4 implies that if G is a simple algebraic group and H is a finite subgroup of G whose order is not divisible by the characteristic of the field, then H is conjugate to a subgroup of $G(K)$ where K is finite (for positive characteristic) or a number field (for characteristic zero). Theorem 3.4 is not true if $|H|$ is divisible by the characteristic p. It is true if H is cyclic, but it fails already for H elementary abelian of order p^2 and $G = SL_2(k)$.

4. Primitive Permutation Lattices

Borovik's example will also work over finite fields as long as the representations are defined over the field. So for example, one gets two embeddings of A_6 into $PSL_9(7) = G$ satisfying (3.1) (and so (1.1)). Indeed one can extend these representations to M_{10}. However, neither image is maximal. In [GS], it was found that there were two embeddings of M_{22} into $G = PSL_{45}(43)$ satisfying (3.1). Moreover, one image was maximal and the other was not. This was used to construct an infinite family of examples of simple groups G with subgroups H and K inducing the same permutation character with H maximal and K nonmaximal. Breuer [Br] also found two embeddings of $M_{22}.2$ into J_4 with the same fusion. Again, one image is maximal and the other is not.

Let H be a maximal subgroup of a finite group G of index n. Suppose K is a subgroup of G and

$$R(G/H) \cong R(G/K) \tag{4.1}$$

for some commutative ring R. Since $H_G := \cap_{g \in G} H^g = K_G$, we may assume that $H_G = K_G = 1$. We would like to characterize the possible K (not just determine whether K is necessarily maximal) and extend the results of [FK]. In [FK], the problem of the maximality of K was reduced to the case of almost simple groups.

Recall (cf [AS]) that if G is a finite group and H is a maximal subgroup of G such that $H_G = 1$ (this is equivalent to saying that G is a primitive faithful group of permutations on the set of cosets of H), then the following holds:

Let Q be a minimal normal subgroup of G, L a minimal normal subgroup of Q, and let $\Delta = \{L = L_1, \ldots, L_t\}$ be the set of G conjugates of L. Then $G = HQ$ and precisely one of the following holds:

(A) L is of prime order p, Q is an elementary abelian p-group, and $H \cap Q = 1$.
(B) $F^*(G) = Q \times S$, where $Q \cong S$ and $H \cap Q = 1$.
(C1) $F^*(G) = Q$ is nonabelian, $H \cap Q = 1$, and $t > 1$.
(C2) $F^*(G) = Q$ is nonabelian, $H \cap Q \neq 1 = H \cap L$.
(C3) $F^*(G) = Q$ is nonabelian, $H \cap L_i = H_i \neq 1$, and $H \cap Q = H_1 \times \ldots \times H_t$.

We investigate the cases separately. We still are assuming (4.1). We also note that (4.1) valid for any R implies that (4.1) holds for $\mathbb{Q}$ (or any field of characteristic zero).

LEMMA 4.2. *Let F be a nontrivial normal subgroup of G.*

(a) *$G = HF = KF$ and $R(F/(F \cap H)) \cong R(F/(F \cap K))$.*
(b) *If $F \cap H = 1$, then $F \cap K = 1$ and so there exists $\delta : H \to F$ such that $K = \{h\delta(h) : h \in H\}$. Moreover, $h\delta(h)$ is conjugate to h.*

Proof. (a) We know that $G = HF$ since H is maximal and $H_G = 1$. Since F has a one dimensional fixed space on $\mathbb{Q}(G/H)$, the same is true on $\mathbb{Q}(G/K)$, whence $G = KF$. The last statement holds by restricting the modules to F.

(b) The first statement holds by (a). Since $1_K^G = 1_H^G$, it follows that $h\delta(h) = s^g$ for some $s \in H$ and $g \in G$. Since $G = HF$, we can assume that $g \in F$ which forces $s = h$ and completes the proof.

PROPOSITION 4.3. *Assume* (A) *holds. Then* K *is maximal and corresponds to some* $\alpha \in H^1(H, Q)$ *in the bijection between conjugacy classes of complements to* Q *and elements of* $H^1(H, Q)$.

(i) (4.1) *holds with* R *a field of characteristic* q *if and only if* $\alpha \in \ker \mathrm{res} : H^1(H, Q) \to H^1(C, Q)$ *for all* $C \in \mathcal{C}_q(H)$.
(ii) *Assume that* $q \neq p$. *Then* (4.1) *holds for* R *a field of characteristic* q *if and only if* (4.1) *holds for* $R = \mathbb{Q}$.
(iii) (4.1) *holds with* R *of characteristic* p *if and only if* H *and* K *are conjugate in* G.

Proof. Let δ be a derivation from H to Q which gives rise to the class α. Thus $K = \{h\delta(h) : h \in H\}$. Let $C \in \mathcal{C}_q(H)$. Then C is conjugate to a subgroup of K if and only if δ is inner when to restricted to C. By Corollary 2.3, this is a necessary condition for (4.1) to hold with R a field of characteristic q. By the q-adic version of Corollary 2.4, this is also sufficient. This proves (i).

The forward implication of (ii) is certainly true as we have observed above. Now assume that $p \neq q$. Let $C \in \mathcal{C}_q(H)$ and let $D = O_q(C)$. Let E be a cyclic complement to D in C. Since we are assuming that (4.1) holds for $R = \mathbb{Q}$, it follows by (i) that δ is inner when restricted to E. Since $q \neq p$, $\mathrm{res}\, H^1(C, Q) \to H^1(E, Q)$ is injective, it follows that δ is inner on C, and the result follows by (i).

Now consider the case $p = q$. Since δ restricted to P is inner for a Sylow p-subgroup of H, it follows as above that δ is inner on H and so H and K are conjugate.

We note that for G solvable, all complements are conjugate (see [AG]).

COROLLARY 4.4. *Assume the hypotheses of Proposition* 4.3. *The number of conjugacy classes of subgroups satisfying* (4.1) *is at most* $n^{1/2}$.

Proof. First observe that $|Q| = n$. The result follows from the Proposition and the fact that $|H^1(H, Q)| \leq |Q|^{1/2}$ (the bound with $1/2$ replaced by $2/3$ is proved in [G] - the better bound is an unpublished result of the first author).

The inequality in Corollary 4.4 can be sharp. For example, if $G = Q.SL_2(q)$ with Q the natural module for $SL_2(q), q = 2^e, e > 1$, then there are exactly q conjugacy classes of complements to Q in G and they all induce the same permutation character.

PROPOSITION 4.5. *If* (B) *holds, then* H *and* K *are conjugate.*

Proof. We use the notation of (B). We can identify S with Q and (by conjugating if necessary in $\mathrm{Aut}(Q \times Q)$) assume that $H \cap (Q \times Q) = \{(x, x) : x \in Q\}$.

By Lemma 4.2, $K \cap (Q \times Q)$ has the same order as H and intersects both Q and S trivially. It follows that $K \cap (Q \times Q) = \{(x, \sigma(x)) : x \in Q\}$, where $\sigma \in \mathrm{Aut}(Q)$. If (4.1) holds, then, by Lemma 4.2(b), σ must preserve conjugacy classes. It follows from [FS] that σ is inner and so by conjugation, we can assume that $H \cap (Q \times Q) = K \cap (Q \times Q) = D$. Since H is maximal, it follows that $K \subseteq H = N_G(D)$. Since (4.1) implies that H and K have the same order, this implies that H and K are conjugate.

Essentially, the same proof yields:

PROPOSTION 4.6. *If* (C2) *holds, then H and K are conjugate.*

Finally, we consider (C1).

PROPOSITION 4.7. *If* (C1) *holds, then H and K are conjugate.*

Proof. In this case, H and K are both complements to Q (K is a complement by Lemma 4.2.) Set $L = L_1$, $N = N_G(L)$ and $T = L_2 \times \ldots \times L_t$. By [AS, Theorem 2], there is a conjugacy class preserving bijection between complements of Q in G and complements of L in M where $M = N/T$. The bijection is given by $W \to \mu(W) = N_W(L)$ (note that $N_W(L) \cap T = 1$ and so we identify this group with its image in the quotient group).

Since N contains Q, $G = NW$, and so arguing as in Lemma 4.2, it follows that $1^N_{(N\cap H)} = 1^N_{(N\cap K)}$. This implies that $1^M_{\mu(H)} = 1^M_{\mu(K)}$. In particular, it follows that $C := C_{\mu(H)}(L) = \mu(H)_M = \mu(K)_M = C_{\mu(K)}(L)$.

Let $I = L \times C_M(L)$ be the subgroup of M which induces inner automorphisms on L and let $I(W) = I \cap \mu(W)$. Since $M = L\mu(W)$, $I = LI(W)$ and $N_M(I(W)) = \mu(W)C_L(I(W))$.

Since H is maximal in G, it follows by [AS, Theorem 1] that $I(H)/C \cong L$ (i.e. all inner automorphisms of L are induced by elements of $N_H(L)$). Thus $\mu(H)$ is the normalizer of $I(H)$. Since $|\mu(K)| = |\mu(H)|$, the conjugacy of $I(H)$ and $I(K)$ will imply the conjugacy of $\mu(H)$ and $\mu(K)$.

Consider $J = I/C = L(I(H)/C)$. Since $M = I\mu(H) = I\mu(K)$ and $C \subseteq I(H) \cap I(K)$, it follows that $1^J_{(I(H)/C)} = 1^J_{(I(K)/C)}$. As we observed above, $I(H)/C \cong L$, and so $J \cong L \times L$. Thus $I(H)/C$ and $I(K)/C$ are diagonal subgroups of J inducing the same permutation character in J. Again, it follows as above (using [FS]) that these subgroups are conjugate in J. Hence $I(H)$ and $I(K)$ are conjugate. This implies that $\mu(H)$ and $\mu(K)$ are conjugate and so by [AS], H and K are conjugate.

In case (C3), one reduces the problem from G to $N_G(L)/C_G(L)$ which is almost simple. As we remarked above, there are examples of (4.1) holding for $R = \mathbb{Q}$ when H is maximal and K is not in that case (indeed infinitely many examples). We close with two questions.

(4.8) If H is maximal in G and $\mathbb{Z}[G/H]$ and $\mathbb{Z}[G/K]$ are in the same genus, must K be maximal ?

The example of Scott [S] shows that we cannot expect conjugacy.

The examples constructed in [GS] have the property that the lattices are p-adically isomorphic for every prime $p \neq 3$.

(4.9) Find a bound on the number of conjugacy classes of subgroups which induce the same permutation character as the maximal subgroup H. In particular, is $n^{1/2}$ a bound?

By the results above and [FK], this problem reduces to the study of almost simple groups.

REFERENCES

[AG] M. Aschbacher and R. Guralnick, *Solvable generation of groups and sylow subgroups of the lower central series*, J. Algebra **77** (1982), 189-201.

[AS] M. Aschbacher and L. Scott, *Maximal subgroups of finite groups*, J. Algebra **92** (1985), 44–80.

[Bl] D. Blasius, *On multiplicities for SL_n*, preprint.

[B] A. Borovik, *The structure of finite subgroups of simple algebraic groups*, Algebra and Logic (1989), no. 3, 249–279. (Russian)

[Br] T. Breuer, *Potenzabbildungen, Untergruppenfusionen, Tafel-Automorphismen*, Diplomarbeit, Lehrstuhl D für Mathematik, Aachen, Germany, 1991.

[CGM] A. Caranti, N. Gavioli, and S. Mattarei, *Subgroups of finite p-groups inducing the same permutation character*, preprint.

[FK] P. Förster and L. G. Kovács, *A Problem of Wielandt on finite permutation groups*, J. London Math. Soc. **41** (1990), 231–243.

[FS] W. Feit and G. Seitz, *On finite rational groups and related topics*, Illinois J. Math. **33** (1989), 103–131.

[G] R. Guralnick, *Generation of simple groups*, J. Algebra **103** (1986), 381–401.

[GS] R. Guralnick and J. Saxl, *Primitive permutation characters*, Groups, Combinatorics and Geometry, LNS #165 (M. Liebeck and J. Saxl, eds.), Cambridge University Press, Cambridge, 1992, pp. 361–364.

[J] W. Jehne, *Kronecker classes of algebraic number fields*, J. Number Theory **9** (1977), 279–320.

[L] M. Larsen, *On the conjugacy of element-conjugate homomorphisms*, preprint.

[O] R. Oliver, *G-Actions on disks and permutation representations*, J. Algebra **50** (1978), 44-62.

[S] L. L. Scott, *Integral equivalence of permutation representations*, preprint, 1975.

[Sp] T. Springer, *Representations of finite groups in algebraic groups*, unpublished.

[Su] T. Sunada, *Riemannian coverings and isospectral manifolds*, Ann. Math. **121** (1985), 169–186.

[W] P. J. Webb, *Complexes, group cohomology, and an induction theorem for the Green ring*, J. of Algebra **104** (1986), 351–357.

DEPARTMENT OF MATHEMATICS, UNIVERSITY OF SOUTHERN CALIFORNIA, LOS ANGELES, CA 90089-1113, USA

E-mail address: guralnic@mtha.usc.edu

DEPARTMENT OF MATHEMATICS, UNIVERSITY OF ALBERTA, EDMONTON, ALBERTA T6G 2G1, CANADA

E-mail address: userwyss@mts.ucs.ualberta.ca

Contemporary Mathematics
Volume **153**, 1993

Varieties of Unseparated Flags

WILLIAM HABOUSH AND NIELS LAURITZEN

Introduction

Classically the main method employed in the study of representations of semisimple algebraic groups has been the application of the geometry of homogeneous spaces to the analysis of the structure of algebraically induced representations. The main objects of study have been the generalized flag variety, the cells of the Bruhat decomposition and line bundles induced by dominant weights. While this may be a fairly complete list of fundamental objects of study over fields of characteristic zero, in positive characteristic it omits certain basic objects of fundamental importance. That is, in focusing on these objects it is assumed that the only complete homogeneous spaces accessible to meaningful research are the generalized flag and incomplete flag varieties. These are the homogeneous spaces with reduced parabolic stabilizers.

There are, in positive characteristic, also non-reduced parabolic subgroup schemes of semisimple algebraic groups. They have been classified and described by C. Wenzel [**14**]. Their structure can be said to be very well understood. We call the homogeneous spaces with non-reduced parabolic stabilizers *unseparated* flag varieties. In this paper we will describe them, review what has been proven about them and give several new results including a vanishing theorem for the cohomology of ample line bundles.

The most basic result about these spaces is their rationality. This has been established by Wenzel [**15**] who has given a T-stable parametrization of a "big cell" in these varieties. (Here T is a maximal torus in the corresponding semisimple group.) After this, there are a host of questions about these spaces many of which remain unanswered. One of the authors has shown that the canonical line bundles on these spaces are in general neither ample nor negative ample [**10**]

1991 *Mathematics Subject Classification.* Primary 20G05; Secondary 14B10, 14F17.
The first author was funded in part by NSF grant #DMS 90-06321.
This paper is in final form and no version of it will be submitted for publication elsewhere.

0271-4132/93 $1.00 + $.25 per page

and that they in general violate the Kodaira vanishing theorem [**9**]. All of these results will be discussed.

We begin with an account of the classification. We use a new method in part because it suggests the possibility of inductive arguments of a certain type. This is the content of the first section. In section 2 we compute the characters of parabolic subgroup schemes. It is proven that a line bundle has global sections if and only if it is dominant. Then we describe the canonical line bundle and give a character formula for spaces of sections in homogeneous line bundles which coincides with the Weyl character formula for usual flag varieties. An unfortunate fact is that varieties of unseparated flags only are Frobenius split in the trivial cases, when they are isomorphic to ordinary flag varieties. The reason for this is that the canonical line bundle in general fails to be negative. In section 4 we give a systematic method for computing the weight defining the canonical line bundle. Section 5 gives a vanishing theorem for sufficiently dominant line bundles. Finally section 6 deals with the simplest non-trivial case of non-reduced parabolic stabilizers. In this section we give examples showing that the vanishing theorem for ample line bundles and Kodaira's vanishing theorem break down for small dominant weights for varieties of unseparated flags.

Much of this work was done while the authors were visiting the Tata Institute of Fundamental Research in Bombay and the Southern Petroleum Industries Corporation Science Foundation in Madras, India. We wish to express special thanks to all at those two institutions who facilitated our work. We also wish to thank several mathematicians at those institutions, particularly Vikram Mehta, A. Ramanathan and C. S. Seshadri, for their fellowship conversation and inspiration. Thanks are also due to H. H. Andersen for observing that some horrendous computer calculations in the end of section 6 could be carried out by human beings via Jantzen's sum formula.

1. Basic Notation; the Classification

Throughout this work, k will denote an algebraically closed field of characteristic $p \geq 5$, G, B and T will denote a semisimple,simply connected algebraic group, a Borel subgroup and a torus contained in it respectively. Let Φ denote the roots of G with respect to T and Φ^+ the roots of the opposite Borel subgroup B^0. Also, Δ will be a set of simple positive roots. The module of weights will be written, $\mathbf{P}$ and the root lattice will be written $\mathbf{R}$. For $\alpha \in \Phi$, U_α will denote the root subgroup corresponding to α and u_α will denote the parametrizing isomorphism, $u_\alpha : U_\alpha \to G$.

If A is a commutative k-algebra then, A_ν denotes the k-algebra which is equal to A as a set but with k-structure, $a \cdot f = a^{p^\nu} f$. The ν'th order Frobenius map, $\phi_\nu(x) = x^{p^\nu}$ is a k-algebra map from A to A_ν. The corresponding map of schemes $\phi_\nu : \mathrm{Spec}(A_\nu) \to \mathrm{Spec}(A)$ is what we call the Frobenius covering. When X is a k-scheme and $\mathcal{O}$ is a sheaf of k-algebras, let $\mathcal{O}_\nu$ be defined by $\mathcal{O}_\nu(U) = (\mathcal{O}(U))_\nu$. When $\mathcal{O} = \mathcal{O}_X$, the pair $(X, \mathcal{O}_\nu)$ is the Frobenius cover. We denote it X_ν and we write $\phi_\nu : X_\nu \to X$ for the map. It is a proper morphism (even a homeomorphism) for the schemes we shall discuss as they are all of finite type over a field.

If $I \subseteq A$ is an ideal, then let:

$$\sqrt[p^\nu]{I} = \{a : a^{p^\nu} \in I\} \tag{1}$$

We shall simply refer to this ideal as the p^ν-radical of I. The ideals, $\sqrt[p^\nu]{I}$ are an ascending sequence of ideals whose union is the radical of I. If A is Noetherian then $\sqrt[p^\nu]{I} = \sqrt{I}$ for all sufficiently large ν. The ideal I is set theoretically equal to a certain ideal in A_ν which we denote I_ν. Clearly $\phi_\nu^{-1}(I_\nu) = \sqrt[p^\nu]{I}$. This means that if Y is the subscheme of X defined by I, then $\phi_\nu(Y_\nu)$ is the subscheme defined by $\sqrt[p^\nu]{I}$.

If A is the coordinate ring of an affine group scheme and $G = \mathrm{Spec}(A)$, then G_ν is a group and ϕ_ν is a morphism of group schemes. If P is a closed subgroup scheme of G, then P_ν is a subgroup scheme of G_ν and so $\phi_\nu(P_\nu)$ is a closed subgroup of G as well. Thus, if I is the ideal defining P, $\sqrt[p^\nu]{I}$ is the ideal defining $\phi_\nu(P_\nu)$ and so if I is a Hopf ideal then so is $\sqrt[p^\nu]{I}$. Let ${}^\nu P = \phi_\nu(P_\nu)$. Then, since G is semisimple and hence Noetherian, the sequence ${}^\nu P$ is a decreasing sequence of closed subgroups with lower bound ${}^\nu P = P_{red}$, the reduced subgroup scheme of P for all sufficiently large ν.

For any algebraic group denoted by a Roman letter, the corresponding fraktur letter will denote its Lie algebra. Thus $\mathfrak{g}$, $\mathfrak{t}$, $\mathfrak{b}$ and $\mathfrak{p}$ will denote the Lie algebras of G, T, B and P respectively. In addition, ${}^\nu\mathfrak{p}$ will denote the Lie algebra of ${}^\nu P$. Since the group schemes, ${}^\nu P$, are a decreasing sequence of parabolic subgroup schemes each containing B, the Lie algebras, ${}^\nu\mathfrak{p}$, constitute a decreasing sequence of restricted Lie subalgebras of $\mathfrak{g}$ each containing $\mathfrak{b}$.

We shall write $\{X_\alpha\}_{\alpha\in\Phi}, \{H_\alpha\}_{\alpha\in\Delta}$ for a Chevalley basis of $\mathfrak{g}$. We will write $N_{\alpha,\beta}$ for the integer such that $[X_\alpha, X_\beta] = N_{\alpha,\beta}X_{\alpha+\beta}$ and we shall call it the Chevalley integer associated to (α, β). Let $X_\alpha^{[\nu]}$ be the ν'th divided power of the X_α.Since $N_{\alpha,\beta}$ is a product of the lengths of complementary segments of a root string, it only admits two and three as prime factors. Write D_B, D_T, D_G and D_P for the rings of invariant differential operators (*i.e.* the hyperalgebra) of B, T, G and P respectively but ${}^\nu D_P$ for the hyperalgebra of ${}^\nu P$. Write D_α for the hyperalgebra of U_α. Since G_ν is simply a base extension of G we may identify their rings of differential operators. We recall, that D_G is the continuous dual of $k[G]$ for the $\mathfrak{m}$-adic topology, where $\mathfrak{m}$ denotes the ideal of functions vanishing at the identity of G. Thus we can write $J^\perp$ for J a subspace of D_G or $k[G]$ and note that its properties are determined by the duality. For properties of the hyperalgebra consult [**4**].

Now ϕ_ν induces a corresponding endomorphism of D_G. Its action on divided powers of root vectors is:

$$(2)\qquad \begin{aligned} \phi_\nu(X_\alpha^{[q]}) &= X^{[q/p^\nu]} \quad \text{when} \quad p^\nu | q \\ \phi_\nu(X_\alpha^{[q]}) &= 0 \quad \text{when} \quad p^\nu \not| q \end{aligned}$$

For any algebraic group, G, $G^{(\nu)}$ denotes the kernel of ϕ^ν, the ν'th order Frobenius kernel and $D_G^{(\nu)}$ denotes its hyperalgebra. Notice that $D_G^{(\nu)} \cap D_\alpha = D_\alpha^{(\nu)} = \sum_{j<p^\nu} kX_\alpha^{[j]}$.

PROPOSITION 3. *Let $\mathfrak{p}$ be a restricted Lie algebra containing $\mathfrak{b}$ and B-stable under the adjoint action. Then there is a reduced parabolic subgroup of G, P, such that $\mathfrak{p}$ is the Lie algebra of P. That is, there is a subset of Δ, S so that if Ψ^+ is the set of positive roots which are integral linear combinations of elements of S, then,*

$$\mathfrak{p} = \mathfrak{b} \oplus \coprod_{\beta \in \Psi^+} kX_\beta$$

PROOF. Since $\mathfrak{p}$ contains $\mathfrak{b}$ and is T-stable, we may write, $\mathfrak{p} = \mathfrak{b} \oplus \coprod_{\beta \in Q} kX_\beta$. All we must show is that Q is the set of positive roots in the subroot system spanned by some subset of Δ,S. Let $S = Q \cap \Delta$.

Let β be any element of Q. If $\beta \notin \Delta$ then $\beta = \sum_{\alpha\in\Delta} n_\alpha \alpha$ with at least two of the $n_\alpha \neq 0$.If $n_\alpha \neq 0$, then $-\alpha + \beta \in \Phi$ and so the Chevalley integer, $N_{-\alpha,\beta} \neq 0$ in k (recall that $p \geq 5$) and, moreover, $[X_{-\alpha}, X_\beta] = N_{-\alpha,\beta} X_{-\alpha+\beta} \in \mathfrak{p}$. Thus, if $\beta \in Q$ then $\alpha \in Q$ for each α such that $n_\alpha \neq 0$. It immediately follows that Q is the set of positive roots in the root system spanned by S. $\square$

If $\mathfrak{p}$ is a restricted Lie subalgebra of $\mathfrak{g}$ satisfying the hypotheses of 3 we shall simply refer to it as parabolic. We will write $\Phi^+(\mathfrak{p})$ for the set of roots $\beta \in \Phi^+$ such that $X_\beta \in \mathfrak{p}$ and we will write $\Delta(\mathfrak{p}) = \Delta \cap \Phi^+(\mathfrak{p})$

LEMMA 4. $\Phi^+({}^\nu\mathfrak{p}) = \{\beta | \beta \in \Phi^+, X_\beta^{[p^\nu]} \in {}^\nu D_P\}$

PROOF. This just follows from the fact that ${}^\nu P = \phi_\nu(P)$, formula (2) and the definition of $\Phi^+({}^\nu\mathfrak{p})$. $\square$

For each α, the subalgebra of $D_{U_\alpha/k}$ generated by $X_\alpha, X_\alpha^{[p]}, \ldots, X_\alpha^{[p^{\nu-1}]}$ has as a linear basis the divided powers $X_\alpha^{[r]}$, for $0 \leq r < p^\nu$.

DEFINITION 5. *Write $\mathbb{N}^*$ to signify the set of non-negative natural numbers together with ∞. A W-function on Φ^+ is a function, f, on Φ^+ with values in $\mathbb{N}^*$ satisfying the condition,*

$$f(\beta) = \inf_{\alpha \in \text{supp}(\beta)} f(\alpha)$$

where $\text{supp}(\beta) = \{\gamma \in \Phi^+ | \beta = \gamma + \delta,$ *for some* $\delta \in \Phi^+\}$.

Clearly the W-functions are in bijective correspondence with the set of $\mathbb{N}^*$ valued functions on Δ.

We may associate to any W-function, f, a certain parabolic subgroup scheme of G. For each $n \geq 0$, let $\Delta_n = \{\alpha | \alpha \in \Delta, f(\alpha) \geq n\}$. Let Φ_n^+ denote the positive roots in the root system generated by Δ_n. Let P_n be the reduced parabolic subgroup scheme whose Levi factor has as its root system the root subsystem generated by Δ_n. Thus, $\Phi_0^+ = \Phi^+$ and $P_0 = G$.

The Frobenius kernel, $P_n^{(n)}$, has as its hyperalgebra the algebra generated by $D_B^{(n)}$ and the divided powers, $X_\beta^{[r]}$, for $\beta \in \Phi_n^+$ and $r < p^n$. Then the product of group schemes, $P_0^{(0)} \cdot P_1^{(1)} \cdot \ldots \cdot P_r^{(r)} \cdot P_{r+1}$ is a parabolic group scheme. For all large r,P_r is equal to P_∞.This is just B if f never attains the value , ∞, but in general it is a reduced parabolic subgroup scheme. Thus, the product stabilizes when $P_r = P_\infty$. Let P_f denote this product for r such that $P_r = P_\infty$.

THEOREM 6. *The set of parabolic subgroup schemes of G containing B is in bijective correspondence with the set of all W-functions on Φ^+. The correspondence is the one which assigns to the W-function, f, the group scheme, P_f.*

PROOF. Write D_f for the hyperalgebra of P_f. It is clear that D_f is the algebra spanned by monomials in the $X_\alpha^{[r]}$ for α negative, the binomial symbols, $\binom{H_\alpha}{r}$ for all values of r and the divided powers X_β^s for β positive and $r < p^{f(\beta)}$. Thus the algebras, D_f, are all distinct. Since two connected subgroup schemes of G are equal if and only if their hyperalgebras are equal, it follows that the groups P_f are all distinct and parabolic. What remains to be shown is that every parabolic subgroup scheme is of this type.

Let P be any parabolic subgroup scheme of G. Let $\Phi_\nu = \Phi^+({}^\nu\mathfrak{p})$ and define a function on Φ^+ by the equation

$$f(\beta) = \inf\{\nu | \beta \notin \Phi_\nu\} \tag{7}$$

Since $\Phi_\nu = \Phi^+({}^\nu\mathfrak{p})$, lemma 4 implies that $f(\beta) = \inf\{\nu|\beta \in \Phi^+, X_\beta^{[p^\nu]} \notin D_P\}$. Thus $f(\beta) \geq n$ if and only if $X_\beta^{[p^{n-1}]} \in D_P$

Now apply the construction above to this W-function. The set Δ_n is just the set of positive simple roots α such that $X_\alpha^{[p^{n-1}]} \in D_P$. The group scheme P_n is the reduced parabolic with Levi factor determined by the root subsystem spanned by the set of positive simple roots α, such that $X_\alpha^{[p^{n-1}]} \in D_P$. Thus the hyperalgebra of $P_n^{(n)}$ contains $D_B^{(n)}$ and all the divided powers $X_\beta^{[r]}$ where $r < p^n$ and β ranges over all those positive roots such that $X_\beta^{[s]} \in D_P$ for all $s < p^n$. It follows that the hyperalgebra of the group scheme P_f constructed from this f coincides with D_P. □

An observation is in order. Suppose that P and Q are closed subgroups of G defined by the ideals I_P and I_Q respectively. The scheme theoretic intersection of P and Q is defined by the ideal $I_P + I_Q$. In general if K is any closed subgroup of G with ideal I_K, its hyperalgebra is equal to the elements of D_G which vanish on I_K when viewed as elements of the dual of the coordinate ring of G. Thus $D_{P\cap Q} = D_P \cap D_Q$. It is hence a tautology to say that the ideal defining $P \cap Q$ is the set of regular functions on G on which the elements of $D_P \cap D_Q$ vanish.

Let P be the parabolic subgroup scheme associated to the W-function, f. Let U_P denote the unipotent radical of P_{red}. It is the product of the root subgroups corresponding to positive roots which are not in the root system of the Levi factor of P. Let U_P^0 be the product of the root subgroups U_β such that $U_{-\beta} \subseteq U_P$. Then U_P^0 is a T-stable reduced subgroup of G such that $U_P^0 \cap P_{red} = e$ and $U_P^0 \cdot P_{red}$ is dense open in G. By the remarks above $U_P^0 \cap P$ is a subgroup scheme of G with hyperalgebra, $D_P \cap D_{U_P^0}$.

PROPOSITION 7. *Let P be the parabolic subgroup scheme of G associated to the W-function, f. For each root, β, let $U_{\beta,n}$ denote the kernel of ϕ_n on U_β. That is, it is the unique subgroup scheme isomorphic, as a scheme, to* Spec $k[x]/(x^{p^n})$. *Let $\beta_1, \ldots, \beta_r$ be the set of negative roots such that $f(-\beta) < \infty$. Let U_f^0 denote the image in G under the multiplication map of the product of schemes, $Y = U_{\beta_1, f(\beta_1)} \times_k \cdots \times_k U_{\beta_r, f(\beta_r)}$. Then,*

(1) *U_f^0 is a subgroup scheme and it is independent of the order on the roots $\beta_1, \ldots, \beta_r$. Moreover, multiplication induces an isomorphism of schemes between Y and U_f^0.*
(2) *$U_f^0 = U_P^0 \cap P$. Its hyperalgebra is the subspace of D_G which has as a basis the monomials $X_{\beta_1}^{[m_1]} \cdots X_{\beta_r}^{[m_r]}$ where the m_i range over all possibilities such that $m_i < p^{f(-\beta_i)}$.*
(3) *The multiplication induces an isomorphism, $P \simeq U_f^0 \times_k P_{red}$.*

PROOF. Let $\bar{P} = P_{red}$. Consider the "parabolic" big cell, $U_P^0 \cdot \bar{P}$. This is an open subset of G containing the closed subscheme, $\bar{P}$. Hence it contains any closed subscheme with support, $\bar{P}$. Thus $P \subseteq U_P^0 \cdot \bar{P}$. Furthermore P, being a subgroup scheme of G containing $\bar{P}$, is $\bar{P}$ stable under right multiplication.

Let $\pi : G \to G/\bar{P}$ be the natural projection. Over $\bar{U} = \pi(U_P^0)$ there is a section $\sigma = \pi^{-1} : \bar{U} \to U_P^0$. Let $X = U_P^0 \cdot \bar{P}$ and let $g = \sigma \circ \pi$. If Z is any $\bar{P}$-stable subscheme of X, viewing g as a map of functors of points, we may always write $x = g(x) \cdot g(x)^{-1}x$. Since $g(x)^{-1}x \in \bar{P}$, $\bar{P}$-stability implies that $g(x) \in Z$ whenever $x \in Z$. Hence, $Z \simeq \sigma \circ \pi(Z) \times \bar{P}$, the isomorphism being the one induced by multiplication. Moreover, $\sigma \circ \pi$ induces an isomorphism between $Z \cap U_P^0$ and $\sigma \circ \pi(Z)$. Thus $P \simeq (P \cap U_P^0) \times_k \bar{P}$, the isomorphism just being the one induced by multiplication.

Now consider $P \cap U_P^0$. This is a closed subgroup scheme of G with hyperalgebra $D_P \cap D_{U_P^0}$. Let m be the number of positive roots and order them so that the first r of them are the ones on which the value of f is finite. Following the standard method, D_G admits a basis of monomials of the form:

$$X_{-\beta_1}^{[q_1]} \cdots X_{-\beta_m}^{[q_m]} \binom{H_{\alpha_1}}{r_1} \cdots \binom{H_{\alpha_\ell}}{r_\ell} X_{\beta_m}^{[s_m]} \cdots X_{\beta_1}^{[s_1]}$$

Examining the expressions of this type in D_P and in $D_{U_P^0}$ and comparing shows that $D_P \cap D_{U_P^0}$ has as a basis monomials in divided powers of $X_{\beta_1} \ldots, X_{\beta_r}$ only, with exponent strictly bounded strictly by $p^{f(-\beta_j)}$. Thus $D_P \cap D_{U_P^0}$, which is the hyperalgebra of $P \cap U_P^0$, is the linear dual of the coordinate ring of U_f. That U_f is a group scheme independent of the ordering chosen and equal to $P \cap U_P^0$ follows. The proof of the proposition is complete. □

2. Characters, Line Bundles

Let f be a W-function and let P_f be the corresponding parabolic subgroup scheme.We shall adopt certain conventions for the remainder of this paper. Let $n_1 < n_2 < \cdots < n_r$ be the increasing sequence of actual values, $f(\beta)$, for $\beta \in \Phi$. For any j let:

(1) $$\Phi^j = \{\beta | f(\beta) \geq n_j\}$$

Since f is a W-function, each of the Φ^j is a sub root system of $\Phi = \Phi^0$ and the Φ^j are a decreasing sequence of root systems in Φ. Write $P(j)$ for the parabolic subgroup that has Levi factor with root system Φ^j. For finite n_j, $P(j)^{(n_j)}$ is the corresponding Frobenius kernel but when $n_j = \infty$ let $P(j)^{(n_j)} = P(j)$. When the maximum value of f is finite let q be the integer for which n_q is the maximum. In this case we let $P(q+1) = B$ and we let $n_{q+1} = \infty$. Then, with this convention, we may always write $P_f = P(1)^{(n_1)} \cdot P(2)^{(n_2)} \cdot \ldots \cdot P(q+1)^{(\infty)}$. These conventions will remain fixed for the remainder of this paper. We will require terminology for these notions.

DEFINITION 2. *Let P be the parabolic subgroup scheme of G corresponding to the W-function, f. Let the sequence, $n_1 < n_2 < \cdots < n_r$ be the increasing sequence of values of f. Then n_r is called the altitude of P, r is called its width and the sequence n_j is called its characteristic sequence. If n_r is infinite, $P(r)$ is called the support of P; otherwise, B is called its support. When n_r is finite, the sequence $n_1 \ldots n_r, n_{r+1} = \infty$ is called the augmented sequence of P.*

Notice that if the minimum value, n_1, is not 0, then $\Phi^1 = \Phi$ and so $P(1)^{(n_1)} = G^{(n_1)}$. In most instances this will lead to a situation that can be trivially reduced to the case $n_1 = 0$.

It will also be convenient to give a very precise sense to the notion of a Levi factor. For each j, let T_j be the subgroup of T defined by

(3) $$T_j = \cap_{\beta \in \Phi^j} \operatorname{Ker}(\beta)$$

Then let $M_j = Z_G(T_j)$, the centralizer of T_j. This is what we refer to as the Levi factor of $P(j)$. In addition, it is possible to define a Levi factor for a general P.

DEFINITION 4. *Let P be a parabolic subgroup scheme of G with augmented sequence, $n_1, \ldots, n_{r+1}$. The Levi factor of P is the product of group schemes:*

$$T \cdot M_1^{(n_1)} \cdot \ldots \cdot M_r^{(n_r)}$$

We shall write M_P for this group scheme.

We wish to describe the group of characters of P. This, of course, is the group of group scheme theoretic homomorphisms, $Hom(P, \mathbb{G}_m)$. As is the case with a reduced parabolic, the characters are a submodule of the full group of weights. We write ω_α for the fundamental dominant weight corresponding to the simple root, $\alpha \in \Delta$.

PROPOSITION 5. *Let P be the parabolic subgroup scheme corresponding to the W-function f. Then the group of characters of P is the group,*

$$\mathbb{X}(P) = \coprod_{\alpha \in \Delta} \mathbb{Z} p^{f(\alpha)} \omega_\alpha$$

The expression, p^∞ is understood to be 0. Furthermore, for each $\lambda \in \mathbb{X}(P)$ which is dominant, there is a function on G, $h \in k[G]$ which is P right semi-invariant for λ and left B^0 semi-invariant for $-\lambda$. Moreover h may be chosen so that its restriction to P is the character λ.

PROOF. Let N be any algebraic group and let D_N denote its hyperalgebra. We shall describe the quotient, $N/[N, N]$, by describing its hyperalgebra. Let J_N denote the two sided ideal in D_N generated by all of the commutators $[u, v] = uv - vu$ for u and v arbitrary elements of D_N. It is trivially contained in the augmentation ideal and it is certainly preserved by the antipode. We will show it to be a Hopf ideal.

Let $\check{m} : D_N \to D_N \otimes D_N$ be the co-multiplication on D_N. The composition of $\check{m}$ and the natural surjection, $\pi \circ \check{m}$, maps D_N to the commutative algebra, $D_N/J_N \otimes D_N/J_N$. By the commutativity of this latter, $J_N \subseteq \mathrm{Ker}(\pi \circ \check{m})$. Thus $\check{m}(J_N) \subseteq J_N \otimes D_N + D_N \otimes J_N$. Thus J_N is a two sided ideal and a co-ideal preserved by the antipode and augmentation. Thus, $J_N^\perp$, the set of regular functions, on which the elements of J_N vanish when viewed as elements of the dual of the coordinate ring of N, is a Hopf sub algebra of the ring of regular functions on N. It is precisely the coordinate ring of $N/[N, N]$.

Consider D_P. By the proof of §1 ,theorem 6, D_P is the subring of D_G generated by D_B and the divided powers $X_\beta^{[r]}$ with $r < p^{f(\beta)}$. Thus J_P contains commutators of the form $[\binom{H_\alpha}{q}, X_\beta^{[r]}]$ where q can be arbitrary and $r < p^{\tilde{f}(\beta)}$. Now $\binom{H_\alpha}{q} X_\beta^{[s]} = X_\beta^{[s]} \binom{H_\alpha + s\check{\beta}(\alpha)}{q}$. On the other hand, $\binom{H_\alpha + s\check{\beta}(\alpha)}{q} = \sum_{j=0}^{q} \binom{H_\alpha}{j} \binom{s\check{\beta}(\alpha)}{q-j}$ In particular,

$$[\binom{H_\alpha}{q}, X_\beta^{[s]}] = \sum_{j=0}^{q-1} X_\beta^{[s]} \binom{H_\alpha}{j} \binom{s\check{\beta}(\alpha)}{q-j}$$

In particular, when $r = s = p^\nu$ for some ν, since $\binom{p^\nu m}{n} = 0$ in a field of characteristic p whenever $n < p^\nu$, $[\binom{H_\alpha}{p^\nu}, X_\beta^{[p^\nu]}] = \check{\beta}(\alpha) X_\beta^{p^\nu}$. Since every divided power, $X_\beta^{[r]}$, is a monomial in the powers, $X_\beta^{[p^s]}$ for $p^s < r$, this means that every divided power, $X_\beta^{[r]}$, such that $X_\beta^{[r]} \in D_P$ is also in J_P. (Here we use the hypothesis, $p \geq 5$.)

Moreover,J_P contains the commutators, $[X_\alpha^{[s]}, X_{-\alpha}^{[q]}]$ for $s < p^{f(\alpha)}, q < p^{f(\alpha)}$ and α simple. But when $s = q = p^n$, $[X_\alpha^{[p^n]}, X_{-\alpha}^{[p^n]}]$ is congruent to $\binom{H_\alpha}{p^n}$ modulo the kernel of the map induced by ϕ_n. This implies that $\binom{H_\alpha}{r} \in J_P$ for $r < p^{f(\alpha)}$. That J_P contains these elements together with the results of the last paragraph imply that $D_P/J_P \simeq D_T/J_P \cap D_T$. But then this means that the coordinate ring of $P/[P,P]$ can be identified with $(J_P \cap D_T)^{\perp} \cap k[T]$. In particular the characters of P, which are equal to the characters of $P/[P,P]$, are just the characters of T orthogonal to $D_T \cap J_P$. Since the elements, $\binom{H_\alpha}{r}$ for $r < p^{f(\alpha)}$ are in this intersection, and since the group of characters orthogonal to these elements is the group, $\sum_{\alpha \in \Delta} \mathbb{Z} p^{f(\alpha)} \omega_\alpha$, it follows that $X(P)$ is contained in this latter group.

Let $X_1 = \coprod_{\alpha \in \Delta} \mathbb{Z} p^{f(\alpha)} \omega_\alpha$. This is a free $\mathbb{Z}$-module with basis $\{p^{f(\alpha)} \omega_\alpha | \alpha \in \Delta\}$. In the previous paragraph we proved that $X(P) \subseteq X_1$. For each $\alpha \in \Delta$ let u_α be the highest weight matrix coefficient of the representation with highest weight ω_α regarded as a function on G. Recall that B^0 is the Borel subgroup opposite to B and let $P(\alpha)$ denote the maximal parabolic whose unique character restricts to ω_α on B. Then, if $\omega = \omega_\alpha$, we may always choose u_α so that it satisfies:

$$
\begin{aligned}
&u_\alpha(xy) = y^\omega u_\alpha(x) \quad && \text{whenever} \quad && y \in P(\alpha) \\
(6) \qquad &u_\alpha(yx) = y^\omega u_\alpha(x) && \text{whenever} && y \in B^0 \\
&u_\alpha(e) = 1
\end{aligned}
$$

These conditions insure that the restriction of u_α to $P(\alpha)$ is a character. Suppose that $\alpha \in \Phi^j$ but $\alpha \notin \Phi^{j+1}$. Then $f(\alpha) = n_j$. Let $h_\alpha = u_\alpha^{p^{n_j}}$. Since $\alpha \notin \Phi^{j+1}$, h_α is right translation invariant for M_{j+1}. Since it is a p^{n_j}-th power it is invariant for right (and left) translation under the action of the Frobenius kernels $M_r^{(n_r)}, r \leq j$ and even for $G^{(n_j)}$. By this last observation the restriction of h_α to $G^{(n_j)} \cdot P(\alpha)$ is a character and since P is contained in $G^{(n_j)} \cdot P(\alpha)$, the restriction of h_α to P is a character corresponding to $p^{n_j}\omega_\alpha$. Let $\bar{h}_\alpha = h_\alpha|P$. Since it is a character on P, $\bar{h}_\alpha$ is invertible. If $\sum_{\alpha \in \Delta} m_\alpha p^{n_\alpha} \omega_\alpha = \xi$ then $\prod_{\alpha \in \Delta} \bar{h}_\alpha^{m_\alpha}$ is a character on P which restricts to ξ on T. Thus $X_1 \subseteq X(P)$ and so $X_1 = X(P)$.

If $\xi \in X_1$ is dominant, then $\xi = \sum_{\alpha \in \Delta} m_\alpha p^{n_\alpha} \omega_\alpha$ and each of the m_α is positive. Let $h = \prod_{\alpha \in \Delta} h_\alpha^{m_\alpha}$. This is a right P semi-invariant of weight, ξ. Since its value at e is 1, it is non-zero. and it is satisfies the requirements of the proposition. □

COROLLARY 7. *Let P be the parabolic subgroup scheme of G associated to the W-function, f. Let $X = G/P$ be the homogeneous space with stabilizers conjugate to P. Then*

(1) $\mathrm{Pic}(X) = X(P) = \coprod_{\alpha \in \Delta} \mathbb{Z} p^{f(\alpha)} \omega_\alpha$. *The isomorphism associates to the character, λ, the induced line bundle, $\mathcal{L}_P(\lambda)$, on X.*
(2) *Let $\lambda = \sum_{\alpha \in \Delta} m_\alpha p^{f(\alpha)} \omega_\alpha$. Then $H^0(X, \mathcal{L}_P(\lambda)) \neq (0)$ if and only if $m_\alpha \geq 0$ for each $\alpha \in \Delta$.*
(3) *Let λ be as in 2). Then $\mathcal{L}_P(\lambda)$ is ample if and only if $m_\alpha > 0$ for each $\alpha \in \Delta$.*

PROOF. Let $\bar{P} = P_{red}$ and let $\pi : G/\bar{P} \to G/P$ be the natural map. Then π is a faithfully flat finite purely inseparable morphism. In particular, purely as a sheaf of abelian groups, $\mathcal{L}_P(\lambda) \subseteq \pi^*\mathcal{L}_P(\lambda) = \mathcal{L}_{\bar{P}}(\lambda)$. The first statement follows from the assumption that G is simply connected [**6**], 2.9. For the second, since $\mathcal{L}_P(\lambda) \subseteq \mathcal{L}_{\bar{P}}(\lambda)$, non-vanishing of $H^0(G/P, \mathcal{L}_P(\lambda))$ implies the same for $H^0(G/\bar{P}, \mathcal{L}_{\bar{P}}(\lambda))$ and hence non-negativity of λ. The function h constructed in the proposition is a section of $\mathcal{L}_P(\lambda)$. Hence non-negativity implies non-vanishing. As for ampleness, $\mathcal{L}_P(\lambda)$ is ample if and only if $\pi^*(\mathcal{L}_P(\lambda))$ is ([**5**], I.4.4 and [**7**], Proposition II.4,4). □

3. Local geometric structure, the canonical bundle

Let P be the parabolic subgroup scheme associated to the W-function, f, and let X be the homogeneous space G/P. In this section we will show that, if x_0 is the unique point with stabilizer P, then x_0 is contained in a T stable neighborhood isomorphic to affine space parametrized by coordinate functions, y_α, which are T-eigenfunctions of weight, $p^{f(\alpha)}\alpha$. A direct and very efficient proof of this is given in [**15**]. We include a geometric proof consistent with our methods. We then use this result to describe the canonical bundle on X. The terms, $\bar{P}$, U_P^0 and U_f denote what they did in §1, 7. Let $\pi : G/\bar{P} \to X$ be the natural map. Then, π is a homeomorphism and so $\pi(U_P^0 \cdot \tilde{x}_0) = U_P^0 \cdot x_0$ is a dense open set in X where $\tilde{x}_0$ is the identity coset in $G/\bar{P}$.

THEOREM 1. *Let P be the parabolic subgroup scheme with W-function, f. Let $\beta_1, \ldots, \beta_q$ be the negative roots such that $f(-\beta) < \infty$. Let $n_i = f(-\beta_i)$. Then there are mappings, $y_i : U_P^0 \cdot x_0 \to \mathbb{A}_k^1$ so that*

(1) $y_i(tu) = t^{p^{n_i}\beta_i} y_i(u)$
(2) *The product morphism, $y_1 \times \cdots \times y_q$, is a T-equivariant isomorphism from $U_P^0 \cdot x_0$ to $\mathbb{A}_k^q$. Here T acts on the i'th coordinate by the character $p^{n_i}\beta_i$*

PROOF. The proof will be by induction on the width of P. Let $n_1, \ldots n_r$ be the characteristic sequence of P. Assume that the statement is true for all parabolic subgroup schemes of width $r - 1$ or less in all possible groups.

Let P be a parabolic subgroup scheme of G of width r with characteristic sequence $n_1, \ldots, n_r$. First assume that $n_1 = 0$. Consider P_2, the parabolic with Levi factor having the root subsystem, Φ_2. Since $n_1 = 0$, $P \subseteq P_2$ and so there is a natural G-equivariant map, $\xi : X \to G/P_2$.Let U_2^0 be the maximal T-stable unipotent such that $U_2^0 \cap P_2 = e$ (*i.e.*the opposite unipotent radical) and let x_2 denote the identity coset. Then $U_2^0 \simeq U_2^0 \cdot x_2$ T-equivariantly and so there is a T-equivariant isomorphism induced by the action of U_2^0 on X, $\psi : U_2^0 \times \xi^{-1}(x_2) \to \xi^{-1}(U_2^0 \cdot x_2)$. This is a T-equivariant morphism. Notice that $k[U_2^0 \cdot x_2] = k[y_1, \ldots, y_s]$ where y_i is a T-eigenfunction of weight β_i and where $\{\beta_1, \ldots, \beta_s\}$ are the elements of $\Phi \setminus \Phi_2$.

Let T_2 be the joint kernel of Φ_2 and let $M_2 = Z_G(T_2)$ be the Levi factor of P_2. Then, $\xi^{-1}(x_2) = M_2/P \cap M_2$ and this isomorphism is T-equivariant. But then $P \cap M_2$ is a parabolic subgroup scheme of M_2 with characteristic sequence, $n_2, \dots, n_r$, and so of width $r-1$. Thus, the inductive hypothesis allows us to conclude the desired result provided $n_1 = 0$.

Suppose $n_1 > 0$. Let $n = n_1$ and let $f_1 = f - n$. Let $\phi_n : G \to G_n$ denote the Frobenius cover. Let Q denote the parabolic subgroup cheme of G_n with W-function, f_1. The previous paragraph allows us to conclude that G_n/Q contains a dense open set T_n isomorphic to $\mathrm{Spec}(k[\bar{y}_1, \dots, \bar{y}_m])$ where $\bar{y}_i$ is a T_n eigenfunction of weight $p^{f_1(-\alpha_i)}\alpha_i$ where the α_i are the negative roots.

But then, composing the G_n-action on G_n/Q with ϕ_n makes G_n/Q into a G homogeneous space so that X becomes isomorphic to the Frobenius cover of G_n/Q. Said otherwise, there is an equivariant commutative diagram:

$$\begin{array}{ccc} G & \longrightarrow & X \\ \downarrow{\scriptstyle \phi_n} & & \| \\ G_n & \longrightarrow & G_n/Q \end{array} \tag{2}$$

But then, an eigenfunction of weight β for T_n becomes, under composition with ϕ_n, an eigenfunction of weight $p^n\beta$ for T. This completes the proof that the inductive hypothesis implies the result for r.

The result for $r = 1$ remains. This is nothing more than the description of the maximal Bruhat cell up to a Frobenius twist. □

COROLLARY 3. *Let P be parabolic in G with W-function, f. Let λ be a dominant character in $X(P)$. Let γ be a character of T occurring as a T-weight in the G-representation, $H^0(G/P, \mathcal{L}_P(\lambda))$. Then there are positive integers, $r_\beta, \beta \in \Phi^+$ so that γ may be written in the form:*

$$\gamma = -\lambda + \sum_{\beta \in \Phi^+} r_\beta p^{f(\beta)}\beta \tag{4}$$

PROOF. Let $x_0 \in G/P$ be the identity coset. By §2, 5 there is a function, h, on G such that $h(xy) = y^\lambda h(x)$ in the sense of functors of points for $y \in P$ and $h(yx) = y^\lambda h(x)$ for $x \in B^0$. Now h is a unit on the big cell and it is of weight $-\lambda$ as a T-weight under left translation. Then h can be viewed as a nowhere vanishing section of $\mathcal{L}_P(\lambda)$ on the affine open set, $U_P^0 \cdot x_0$. Hence it may be taken as a free generator of $H^O(U_P^0 \cdot x_0, \mathcal{L}_P(\lambda))$. Now G/P is reduced and irreducible and $\mathcal{L}_P(\lambda)$ is a line bundle and so $H^0(G/P, \mathcal{L}_P(\lambda))$ is a T submodule of $H^0(U_P^0 \cdot x_0, \mathcal{L}_P(\lambda))$. By the theorem, $H^0(U_P^0 \cdot x_0, \mathcal{L}_P(\lambda)) = hk[y_1, \dots, y_q]$, and so its T-weights are sums of terms of the form $hy_1^{s_1} \cdot \dots \cdot y_q^{s_q}$. Since left translation requires insertion of an inverse, these weights are of the requisite form. □

REMARK 5. *When P is reduced we may apply the long element in the Weyl group to obtain weights of the usual type. In the present instance, the situation* *is complicated by the fact that the lattice,* $\coprod_{\alpha\in\Delta}\mathbb{Z}p^{n_\alpha}\alpha$ *is not stable under the Weyl group.*

DEFINITION 6. *Denote by* ρ_f *the weight* $\frac{1}{2}\sum_{\beta\in\Phi^+}p^{f(\beta)}\beta$*, where* f *is a W-function.*

It should be established that ρ_f is in the weight lattice. This is always the case for an odd prime. To see this notice that for p odd $2\rho_f$ is congruent to 2ρ modulo two times the weight lattice. Though we are not considering characteristic, 2, it is interesting to note that ρ_f need not be in the weight lattice in this case.

PROPOSITION 7. *Let* P *be a parabolic subgroup scheme of* G *with W-function,* f*. Then the canonical bundle on* G/P *is the induced bundle,* $\mathcal{L}_P(-2\rho_f)$*.*

PROOF. The proof has three main ingredients. Let $\mathcal{L}_P(V)$ denote the homogeneous bundle on $G/P = X$ induced by the B representation, V. First, for any r, $\Lambda^r_{\mathcal{O}_X}\mathcal{L}_P(V) = \mathcal{L}_P(\Lambda^r_k(V))$. Secondly, notice that the T weight of the fibre of $\mathcal{L}_P(\gamma)$ at the identity coset under the restriction of the natural representation of G to T is $-\gamma$. Lastly notice that if R is a finitely generated k-algebra and if $\mathfrak{m}$ is a maximal ideal corresponding to a simple point, then $(R/\mathfrak{m})\otimes_R\Omega_{R/k}\simeq\mathfrak{m}/\mathfrak{m}^2$. If the algebraic group, M is operating on R so that $\mathfrak{m}$ is M-stable, then this isomorphism, being the one induced by differentiation, may be taken to be M-equivariant.

The sheaf of Kähler differentials, $\Omega_{X/k}$,is certainly homogeneous. Let x_0 be the identity coset. By theorem 1, x_0 admits a neighborhood T-isomorphic to $\operatorname{Spec} k[y_1,\ldots,y_q]$ as in the theorem with x_0 identified with the origin. Thus, $\mathfrak{m}$, the functions vanishing at x_0, is identified with the ideal generated by the polynomial indeterminates, $y_1,\ldots,y_q$. The fiber of $\Omega_{X/k}$ at x_0, hence, is a vector space with basis, $\bar{y}_1,\ldots,\bar{y}_q$ where these latter are the residue classes of the y_j in $\mathfrak{m}/\mathfrak{m}^2$. Under left translation, y_i is a T-eigenvector of weight, $-p^{n_i}\beta_i$ and so the same is true of $\bar{y}_i$. Now, $\mathcal{K}_X = \Lambda^q_{\mathcal{O}_X}(\Omega_{X/k})$ has x_0 fiber $\Lambda^q_k\mathfrak{m}/\mathfrak{m}^2$. But then the restriction of $\Lambda^q_k\mathfrak{m}/\mathfrak{m}^2$ to T is one dimensional with weight $2\rho_f = \sum_{\alpha\in\Phi^+}p^{f(\alpha)}\alpha$. But this means that $\mathcal{K}_X = \mathcal{L}_P(-2\rho_f)$. □

We conclude this section with a character formula. This formula, first observed by one of the authors (see [**10**]) is an application of the algebraic version of the equivariant Lefschetz fixed point formula due to H. A. Nielsen [**12**]. This fixed point formula is used in [**6**] to give a proof of Weyl's character formula. Now let T be a split torus. Then the Grothendieck ring of the category of rational representations of T is just the group ring of the group of characters, $X(T)$. For the character, $\lambda\in X(T)$, e^λ denotes the class of the one dimensional module on which T acts with character, λ, in the Grothendieck ring of T. If V is a finite dimensional rational representation of T, write $\operatorname{ch}[V]$ for its class in the Grothendieck ring of T. Write K_T for this Grothendieck ring which we call the character ring of T.

If X is smooth and projective over k with a rational action, one may consider the category of locally free T-sheaves on X. Let $K_T(X)$ denote the Grothendieck group of this category. If $\mathcal{F}$ is a locally free T sheaf on X, then for each i, $H^i(X,\mathcal{F})$ is a finite dimensional rational representation of T. Write $\chi_T(X,\mathcal{F})$ for the Euler characteristic (in K_T):

$$\chi_T(X,\mathcal{F}) = \sum_i (-1)^i \operatorname{ch}[H^i(X,\mathcal{F})] \tag{8}$$

This depends, of course, only on the class of $\mathcal{F}$ in $K_T(X)$. For any closed k point $x \in X$, let $T_x(X)$ denote the Zariski tangent space at X. Finally, for a coherent sheaf, $\mathcal{F}$, let $\mathcal{F}(x)$ denote the fiber of $\mathcal{F}$ at X, that is, $\dfrac{\mathcal{F}_x}{\mathfrak{m}_x\mathcal{F}_x}$.

Suppose that T has only a finite fixed point set. Denote it X^T. In this case, the Nielsen formula, which is essentially a localization formula, becomes ([**12**], 4.7 p. 95):

$$\chi_T(X,\mathcal{F}) = \sum_{x\in X^T} \frac{\operatorname{ch}[\mathcal{F}(x)]}{\sum_i \operatorname{ch}[\Lambda^i T_x(X)^\vee]} \tag{9}$$

Notice that $T_x(X)^\vee = \Omega_{X/k}(x)$ (the check denotes the dual). Moreover, when $\operatorname{ch}[V] = \sum_i e^{\lambda_i}$, then:

$$\sum_j (-1)^j \operatorname{ch}[\Lambda^j V] = \prod_i (1 - e^{\lambda_i}) \tag{10}$$

Putting this all together and applying it to the case, $X = G/P$, $\mathcal{F} = \mathcal{L}_P(\gamma)$ produces the result:

THEOREM 11. *Let P be the parabolic subgroup scheme of G associated to the W-function, f, and let γ be a character of P. If $\Phi_\infty = \{\alpha \in \Phi | f(\alpha) = \infty\}$ and W_∞ denotes the Weyl group of Φ_∞, then*

$$\chi_T(G/P, \mathcal{L}_P(\gamma)) = \sum_{w\in W/W_\infty} \frac{e^{-w\lambda}}{\prod_{\alpha\in\Phi^+\setminus\Phi^+_\infty}(1 - e^{-p^{f(\alpha)}w\alpha})} \tag{12}$$

PROOF. We first observe that we may always assume that $P_{red} = B$. Suppose not. Define a new W-function f_1 by letting $f_1(\alpha) = f(\alpha)$ whenever $f(\alpha) < \infty$ and $f_1(\alpha) = 0$ otherwise. Let P_1 be the parabolic scheme associated to f_1. Then $P_1 \subseteq P$ and so there is a projection, $\pi : G/P_1 \to G/P$. If γ is a character on P, then $\mathcal{L}_{P_1}(\gamma)$ is trivial on the fibers of π. Thus the Leray spectral sequence for π degenerates ($R^i\pi_*\mathcal{L}_{P_1}(\gamma) = (0)$ for all $i > 0$). We may hence compute $\chi_T(G/P_1, \mathcal{L}_{P_1}(\gamma))$ rather than the same expression on G/P. Thus we assume that $P_{red} = B$.

But now G/P, as a topological space with T-action, is identical to G/B. Thus the T-fixed points are just the W translates of the identity coset. Let x_0 denote the identity coset. Suppose that $\mathcal{F}$ is a G-sheaf on G/P and that $s \in \mathcal{F}(x_0)$ is a T-eigenvector of weight, λ. Then w induces an isomorphism from $\mathcal{F}(x_0)$ to $\mathcal{F}(wx_0)$ and it certainly carries s to an element of weight $w\lambda$.

But now $\operatorname{ch}[\mathcal{L}_P(\gamma)(x_0)] = e^{-\gamma}$ and $\operatorname{ch}[T_{x_0}(X)^\vee] = \sum_{\alpha\in\Phi^+} e^{-p^{f(\alpha)}\alpha}$ by the calculation in proposition 7. Thus the denominator, corresponding to the fixed point, x_0, in formula (9), is $\prod_{\alpha\in\Phi^+}(1-e^{-p^{f(\alpha)}\alpha})$. The numerator is just $e^{-\gamma}$. As for the remaining summands in (12), these correspond to the fixed points, wx_0, by the observation above. □

4. Frobenius Splitting

Suppose P is a parabolic subgroup scheme and let $\bar{P}$ be the reduced part of P. Let π_P denote the canonical finite purely inseparable morphism

$$\pi_P : G/\bar{P} \to G/P$$

Recall that a k-variety X is called Frobenius split if the morphism $\mathcal{O}_X \to F_*\mathcal{O}_X$ of $\mathcal{O}_X$-modules admits a section, where $F : X \to X$ is the Frobenius map. Mehta and Ramanathan proved in [**11**] that the generalized flag varieties and their Schubert varieties are Frobenius split. A Frobenius split variety X has vanishing for ample line bundles and satisfies Kodaira's vanishing theorem (if X is smooth and projective) [**11**].

Because of the character formula Theorem 11, §3, a vanishing theorem for varieties of unseparated flags becomes interesting. It is therefore natural to ask if G/P is Frobenius split or if the morphism

$$\mathcal{O}_{G/P} \to (\pi_P)_*\mathcal{O}_{G/\bar{P}}$$

of $O_{G/P}$-modules is split. In the latter case we call G/P π_P-split. Since an ample line bundle pulls back to an ample line bundle under a finite morphism, G/P has vanishing for ample line bundles if it is π_P-split.

Our main theorem in this section is that G/P is Frobenius split and π_P-split only in the case where P is a uniform infinitesimal thickening of $\bar{P}$. This means that G/P is a Frobenius cover of $G/\bar{P}$ and thereby isomorphic to $G/\bar{P}$ as $G/\bar{P}$ is defined over $\mathbb{F}_p$. We will restrict ourselves to the case where $\bar{P} = B$. The theorem still holds in the general case under the added assumption, that $p > h$ where h denotes the Coxeter number of G - see [**10**], proposition 4.3.

Suppose $f : X \to Y$ is a finite morphism between two k-varieties X and Y. Notice that f is an affine morphism and that we have an equivalence of the category of quasi-coherent sheaves on X and the category of quasi-coherent sheaves on Y, which are sheaves of $f_*\mathcal{O}_X$-modules. Now $\mathcal{H}om_{\mathcal{O}_Y}(f_*\mathcal{O}_X, \mathcal{G})$ is a $f_*\mathcal{O}_X$-module through multiplication on $f_*\mathcal{O}_X$. Let $f^!\mathcal{G}$ denote the corresponding $\mathcal{O}_X$-module. Duality for f is the isomorphism induced by the natural map

$$f_*\mathcal{H}om_{\mathcal{O}_X}(\mathcal{F}, f^!\mathcal{G}) \to \mathcal{H}om_{\mathcal{O}_Y}(f_*\mathcal{F}, \mathcal{G})$$

where $\mathcal{F}$ is a coherent sheaf on X and $\mathcal{G}$ a coherent sheaf on Y.

We are primarily interested in the case, where f is a finite morphism between smooth projective varieties of the same dimension. In this setup we have the following derived from [**13**], 1.16.

PROPOSITION 1. *Let $f : X \to Y$ be a finite morphism between smooth projective varieties of the same dimension and let K_X and K_Y denote the canonical line bundles on X and Y respectively. Then there is a canonical isomorphism*

$$\mathcal{H}om_{\mathcal{O}_Y}(f_*\mathcal{O}_X, \mathcal{O}_Y) \to f_*(K_X \otimes f^* K_Y^{-1})$$

PROOF. Applying the duality for f to a vector bundle V and $f^! K_Y$ and then using Serre duality, it follows that $f^! K_Y$ is a dualizing sheaf for X and thereby isomorphic to K_X. The duality for f applied to $\mathcal{F} = f^* K_Y$ and $\mathcal{G} = K_Y$ gives

$$f_* \mathcal{H}om_{\mathcal{O}_X}(f^* K_Y, f^! K_Y) \cong \mathcal{H}om_{\mathcal{O}_Y}(f_* f^* K_Y, K_Y)$$

The right hand side is isomorphic to $\mathcal{H}om_{\mathcal{O}_Y}(f_*\mathcal{O}_X, \mathcal{O}_Y)$, since K_Y is locally free of rank 1. Combining these two isomorphisms gives the result. $\square$

The following proposition is the key to the main theorem in this section

PROPOSITION 2. *Let f be a finite valued W-function on Φ^+, where Φ is irreducible. If $p > 3$ then the weight*

$$\omega = \sum_{\gamma \in \Phi^+} p^{f(\gamma)} \gamma$$

is dominant if and only if f is constant.

PROOF. We prove that if f is non-constant then ω is not dominant. To this end we introduce the notion of the δ-matrix [**10,** §4] corresponding to a given ordering $\alpha_1, \dots, \alpha_l$ of the basis Δ of Φ. Define

$$\begin{aligned} \Gamma_1 &= [\alpha_1, \dots, \alpha_l]^+ \setminus [\alpha_2, \dots, \alpha_l]^+, & \theta_1 &= \sum_{\gamma \in \Gamma_1} \gamma \\ \Gamma_2 &= [\alpha_2, \dots, \alpha_l]^+ \setminus [\alpha_3, \dots, \alpha_l]^+, & \theta_2 &= \sum_{\gamma \in \Gamma_2} \gamma \\ &\vdots \\ \Gamma_l &= [\alpha_l]^+, & \theta_l &= \sum_{\gamma \in \Gamma_l} \gamma = \alpha_l \end{aligned}$$

where $[\alpha_{i_1}, \dots, \alpha_{i_r}]$ is the root system generated by the simple roots $\alpha_{i_1}, \dots, \alpha_{i_r}$. Now define the δ-matrix of $\alpha_1, \dots, \alpha_l$ to be

$$M = (M_{ij}) = \begin{pmatrix} \alpha_1^\vee(\theta_1) & \dots & \alpha_l^\vee(\theta_1) \\ \vdots & \alpha_j^\vee(\theta_i) & \vdots \\ \alpha_1^\vee(\theta_l) & \dots & \alpha_l^\vee(\theta_l) \end{pmatrix}$$

Since Φ^+ is the disjoint union of $\Gamma_1, \Gamma_2, \dots, \Gamma_l$ the sum of the rows in M is the row vector $(2, 2, \dots, 2)$ signifying the coordinates of δ in the basis of the fundamental dominant weights $\omega_{\alpha_1}, \dots, \omega_{\alpha_l}$. From $\alpha_i^\vee(\alpha_j) \leq 0$ if $i \neq j$ and $\alpha_i^\vee(\theta_i + \theta_{i+1} + \dots + \theta_l) = 2$ it follows that

1. $M_{ij} \leq 0$ if $i > j$
2. M is lower triangular

One crucial property following from the irreducibility of Φ is

3. For $i > 1$ there exists j, such that $1 \leq j < i$ and $M_{ij} < 0$

Now let $\alpha_1, \ldots, \alpha_l$ be ordered such that $f(\alpha_1) \leq \cdots \leq f(\alpha_l)$. As f is a W-function we get

$$\omega = p^{f(\alpha_1)}\theta_1 + p^{f(\alpha_2)}\theta_2 + \cdots + p^{f(\alpha_l)}\theta_l$$

The assumption that f is not constant amounts to the existence of an index i, $1 \leq i < l$ so that $f(\alpha_i) < f(\alpha_{i+1})$. According to (3) we can find j, $1 \leq j < i+1$ such that $M_{i+1,j} = \alpha_j^\vee(\theta_{i+1}) < 0$. Now we have the following string of inequalities

$$\begin{aligned}\alpha_j^\vee(\omega) &= p^{f(\alpha_j)}\alpha_j^\vee(\theta_j) + \cdots + p^{f(\alpha_i)}\alpha_j^\vee(\theta_i) + p^{f(\alpha_{i+1})}\alpha_j^\vee(\theta_{i+1}) + \ldots \\ &\leq p^{f(\alpha_j)}\alpha_j^\vee(\theta_j + \cdots + \theta_i) + p^{f(\alpha_{i+1})}\alpha_j^\vee(\theta_{i+1} + \cdots + \theta_l) \\ &\leq p^{f(\alpha_j)}\alpha_j^\vee(\theta_j + \ldots \theta_i) + p^{f(\alpha_j)+1}(2 - \alpha_j^\vee(\theta_j + \cdots + \theta_i)) \\ &\leq 3p^{f(\alpha_j)} - p^{f(\alpha_j)+1} \\ &= p^{f(\alpha_j)}(3-p) < 0\end{aligned}$$

Since the coordinate of ω with respect to the fundamental dominant weight ω_{α_j} is negative, ω is not dominant. □

We can now prove

THEOREM 3. *Let G be of simple type and let P be a parabolic subgroup scheme with reduced part B. If f denotes the W-function associated with P, then the following are equivalent*

(1) *G/P is Frobenius split*
(2) *G/P is π_P-split*
(3) *G/P is isomorphic to G/B as a variety*
(4) *f is constant*

PROOF. If f is constant say $= n$, then G/P is the Frobenius cover of G/B of order n and (1) and (2) are equivalent. As G/B is defined over $\mathbb{F}_p$, G/P and G/B are isomorphic. This proves that (4) implies (1), (2) and (3), since G/B is Frobenius split by **[11]**.

Now let ω and $\delta = 2\rho$ be the weights defining the inverse canonical line bundles on G/P and G/B respectively. To finish the proof of the theorem, it suffices to prove that if f is non-constant, then G/P is neither Frobenius split nor π_P-split. By Proposition 1 it suffices to prove that in this case the $(p-1)$-th power of the inverse canonical line bundle $(K_{G/P}^{-1})^{p-1}$ and $K_{G/B} \otimes \pi_P^* K_{G/P}^{-1}$ do not have any global sections. In terms of the weights defining these line bundles, we have to prove that ω and $\omega - \delta$ are not dominant (Corollary 7 (2), §2). It suffices to prove that ω is not dominant. This is exactly the content of Proposition 2, since the root system of G is irreducible. □

5. A vanishing theorem

A natural question already asked by Kempf in 1978 [**8**] is to what extent the vanishing theorem for effective line bundles on $G/\bar{P}$ generalizes to G/P. In the next section we shall give examples that the vanishing theorem for ample line bundles breaks down for varieties of unseparated flags. In this section we use that G/B is Frobenius split to give conditions for a homogeneous line bundle to have vanishing on a variety of unseparated flags. The main theorem is

THEOREM 1. *Let P be a parabolic subgroup scheme with reduced part B and let f be the W-function associated with P and $n = \max_{\alpha\in\Phi^+} f(\alpha)$. Suppose $\lambda \in X(P)$ is a dominant weight. If*

$$e^{\lambda+\rho} \prod_{\alpha\in\Phi^+} \frac{1-e^{-p^n\alpha}}{1-e^{-p^{f(\alpha)}\alpha}} \in \mathbb{Z}[X(T)_+]$$

where $X(T)_+$ denotes the dominant weights of $X(T)$, then

$$\mathrm{H}^i(G/P, \mathcal{L}(\lambda)) = 0$$

for $i > 0$.

PROOF. There is a triangle diagram with natural finite morphisms:

$$\begin{array}{ccc} G/B & \xrightarrow{f} & G/P \\ & \searrow^{F_n} & \downarrow g \\ & & G/G^{(n)}B \end{array}$$

Let $\mathcal{O}$ denote the structure sheaf of G/B and $\mathcal{O}_n$ that of $G/G^{(n)}B$. The injection $\mathcal{O}_n \hookrightarrow \mathcal{O}$ corresponds to the morphism $F_n : G/B \to G/G^{(n)}B$ and is a split injection of $\mathcal{O}_n$-modules, since G/B is Frobenius split. Using the projection formula, we get an injection:

$$0 \to H^i(G/G^{(n)}B, V) \to H^i(G/B, F_n^*V)$$

where V is a vector bundle on $G/G^{(n)}B$. Let $\mathcal{L}$ be a line bundle on G/P. It follows that $H^i(G/P, \mathcal{L}) = H^i(G/G^{(n)}B, g_*\mathcal{L})$, since g is a finite morphism. The vanishing of the pull back of $g_*\mathcal{L}_\lambda$ implies the vanishing of $g_*\mathcal{L}$.

Let E be a B-representation and $E = \bigoplus E_\chi$ the weight space decomposition with respect to $T \subseteq B$. It follows from Kempf's vanishing theorem that if all the weights χ in the weight space decomposition satisfy that $\chi + \rho$ is dominant, then the homogeneous bundle $\mathcal{L}(E)$ induced by E on G/B has vanishing.

We wish to compute the characters occuring in the B-representation inducing the homogeneous bundle $F_n^* g_*\mathcal{L}_\lambda$ It suffices to compute the characters of the $G^{(n)}B$-representation inducing the push-down $g_*\mathcal{L}_\lambda$. We take advantage of our explicit knowledge of the structure of the T-stable big cells U_P in G/P and U_n in $G/G^{(n)}B$. Let $\{\alpha_1, \ldots, \alpha_N\}$ be the positive roots of Φ, where $\alpha_1, \ldots, \alpha_s$ are the simple roots ordered so that $f(\alpha_1) \leq \cdots \leq f(\alpha_s)$. If $m_i = f(\alpha_i)$ then $U_P = \operatorname{Spec} k[X_{\alpha_1}^{p^{m_1}}, \ldots, X_{\alpha_N}^{p^{m_N}}]$ and $U_n = \operatorname{Spec} k[X_{\alpha_1}^{p^n}, \ldots, X_{\alpha_N}^{p^n}]$

Now $g_*\mathcal{L}_\lambda(U_n) = \mathcal{L}(g^{-1}(U_n)) = \mathcal{L}(U_P) = k[U_P]\, h$, where h is a highest weight function of weight λ, is a finite free module over $k[U_n]$ with basis

$$X_{\alpha_1}^{r_1 p^{m_1}} \ldots X_{\alpha_N}^{r_N p^{m_N}}\ h$$

where $0 < r_i < p^{n-m_i}$. The basis vectors are T-eigenvectors with weights:

$$\lambda - \sum_{\substack{0 \leq r_i < p^{n-m_i} \\ \alpha_i \in \Phi^+}} r_i p^{m_i} \alpha_i$$

The assumption in the theorem is that for the λ, all these weights are in $X(T)_+ - \rho$. Vanishing for the bundle $F_n^* g_* \mathcal{L}_\lambda$ on G/B follows immediately. $\square$

PROPOSITION 2. *With the notation from Theorem 1, there exist polynomials* $V_1, \ldots, V_s$ *in* p *given by* f *such that whenever*

$$\begin{aligned} \alpha_1^\vee(\lambda) + 1 &\geq V_1(p) \\ \alpha_2^\vee(\lambda) + 1 &\geq V_2(p) \\ &\vdots \\ \alpha_s^\vee(\lambda) + 1 &\geq V_s(p) \end{aligned}$$

the condition in Theorem 1 is satisfied.

PROOF. Define

$$\Gamma_i = [\alpha_i, \ldots, \alpha_s]^+ \setminus [\alpha_{i+1}, \ldots, \alpha_s]^+$$

for $i = 1, \ldots, s$. Notice that $\Phi^+ = \cup_i \Gamma_i$ and if $\alpha \in \Gamma_i$ then $m_\alpha = m_{\alpha_i}$. With this notation we have

$$\sum_{\substack{0 \leq r_\alpha < p^{n-m_\alpha} \\ \alpha \in \Phi^+}} r_\alpha p^{m_\alpha} \alpha = \sum_i \sum_{\substack{0 \leq r_\alpha < p^{n-m_\alpha} \\ \alpha \in \Gamma_i}} r_\alpha p^{m_\alpha} \alpha = \sum_i p^{m_i} \sum_{\substack{0 \leq r_\alpha < p^{n-m_i} \\ \alpha \in \Gamma_i}} r_\alpha\, \alpha$$

We also have the following bound

$$\begin{aligned} \alpha_j^\vee\Big(\sum_{\substack{0 \leq r_\alpha < p^{n-m_i} \\ \alpha \in \Gamma_i}} r_\alpha\, \alpha \Big) &= \sum_{\substack{0 \leq r_\alpha < p^{n-m_i} \\ \alpha \in \Gamma_i}} r_\alpha\, \alpha_j^\vee(\alpha) \\ &\leq (p^{n-m_i} - 1) \sum_{\alpha \in \Gamma_i} \max(\alpha_j^\vee(\alpha), 0) \end{aligned}$$

By choosing the r_α appropriately equality can be obtained. We define the matrix H with entries

$$H_{ij} = \sum_{\alpha \in \Gamma_i} \max(\alpha_j^\vee(\alpha), 0)$$

Observe that by multiplying the rows of H with $p^n - p^{m_i}$ respectively and summing up the columns, we get the maximal coordinates, with respect to the fundamental dominant weights, occuring in the weights

$$\sum_{\substack{0 \leq r_\alpha < p^{n-m_\alpha} \\ \alpha \in \Phi^+}} r_\alpha p^{m_\alpha} \alpha$$

The expressions for the maximal coordinates are polynomials in p given by f. If a weight λ satisfies the given inequalities then it obviously satisfies the condition in Theorem 1. $\square$

EXAMPLE 3. *Suppose G is of type D_5 with simple roots $\alpha_1, \ldots, \alpha_5$ such that $\alpha_1^\vee(\alpha_2) = \alpha_2^\vee(\alpha_3) = \alpha_3^\vee(\alpha_4) = \alpha_3^\vee(\alpha_5) = -1$ and $m_{\alpha_1} \leq m_{\alpha_2} \leq m_{\alpha_4} \leq m_{\alpha_5} \leq m_{\alpha_3} = n$. Then*

$$\begin{aligned}
\Gamma_1 = \{&\alpha_1, \, \alpha_1 + \alpha_2, \, \alpha_1 + \alpha_2 + \alpha_3, \\
&\alpha_1 + \alpha_2 + \alpha_3 + \alpha_4, \, \alpha_1 + \alpha_2 + \alpha_3 + \alpha_5, \, \alpha_1 + \alpha_2 + \alpha_3 + \alpha_4 + \alpha_5, \\
&\alpha_1 + \alpha_2 + 2\alpha_3 + \alpha_4 + \alpha_5, \, \alpha_1 + 2\alpha_2 + 2\alpha_3 + \alpha_4 + \alpha_5\} \\
\Gamma_2 = \{&\alpha_2, \, \alpha_2 + \alpha_3, \, \alpha_2 + \alpha_3 + \alpha_4, \, \alpha_2 + \alpha_3 + \alpha_5, \, \alpha_2 + \alpha_3 + \alpha_4 + \alpha_5, \\
&\alpha_2 + 2\alpha_3 + \alpha_4 + \alpha_5\} \\
\Gamma_3 = \{&\alpha_4, \, \alpha_3 + \alpha_4, \, \alpha_3 + \alpha_4 + \alpha_5\} \\
\Gamma_4 = \{&\alpha_5, \, \alpha_3 + \alpha_5\} \\
\Gamma_5 = \{&\alpha_3\}
\end{aligned}$$

and the matrix H is

$$\begin{pmatrix} 8 & 2 & 2 & 2 & 2 \\ 0 & 6 & 2 & 2 & 2 \\ 0 & 0 & 4 & 1 & 1 \\ 0 & 0 & 0 & 3 & 1 \\ 0 & 0 & 0 & 0 & 2 \end{pmatrix}$$

This gives the vanishing polynomials

$$\begin{aligned}
&V_1 = 8p^n - 8p^{m_1}, && V_2 = 8p^n - 6p^{m_2} - 2p^{m_1}, \\
&V_3 = 6p^n - p^{m_5} - p^{m_4} - 2p^{m_2} - 2p^{m_1}, && V_4 = 8p^n - 4p^{m_4} - 2p^{m_2} - 2p^{m_1}, \\
&V_5 = 8p^n - 3p^{m_5} - p^{m_4} - 2p^{m_2} - 2p^{m_1}
\end{aligned}$$

6. Twisted $\mathbb{P}^1$-fibrations

Let P_α denote the minimal parabolic subgroup generated by B and the root subgroup corresponding to the simple root α. In this section we focus on the special parabolic subgroup schemes $P_{\alpha,n} = P_\alpha^{(n)} B$, where $P_\alpha^{(n)}$ denotes the n-th Frobenius kernel of P_α. Notice that $P_{\alpha,n}$ has W-function with only non-zero value n occuring for the simple root α. From §2, 5 we know that $X(P_{\alpha,n}) =$

$\mathbb{Z}\omega_{\alpha_1}+\cdots+\mathbb{Z}p^n\omega_\alpha+\cdots+\mathbb{Z}\omega_{\alpha_l}$. For $\lambda \in X(P_{\alpha,n})$ we will denote the cohomology group $H^i(G/P_{\alpha,n}, \mathcal{L}(\lambda))$ by $H^i(G/P_{\alpha,n}, \lambda)$.

Since the canonical morphism $\pi : G/B \to G/P_\alpha$ is trivial [**13**], 2.9, 2.10, it follows that $H^i(G/P_\alpha, V) = H^i(G/B, \pi^* V)$ for a vector bundle V on G/P_α. Now suppose $\lambda \in X(P_{\alpha,n})$ and $\alpha^\vee(\lambda) = r\,p^n \geq 0$. The push-down $(\pi_{\alpha,n})_* \mathcal{L}(\lambda)$ under the canonical morphism $\pi_{\alpha,n}$

$$\begin{array}{ccc} G/B & \longrightarrow & G/P_{\alpha,n} \\ & \overset{\pi}{\searrow} & \downarrow \pi_{\alpha,n} \\ & & G/P_\alpha \end{array}$$

of the line bundle $\mathcal{L}(\lambda)$ on $G/P_{\alpha,n}$ is induced by the representation $V^r_{\alpha,n} = H^0(P_\alpha/P_{\alpha,n}, \lambda)$ of P_α. Twisting the situation of the usual $\mathbb{P}^1$-fibration $\pi : G/B \to G/P_\alpha$ [**1**], 1.2 with Frobenius we get

LEMMA 1.

(1) $V^r_{\alpha,n}$ *has a basis of* T*-eigenvectors* $e_0^{\alpha,n}, \ldots, e_r^{\alpha,n}$, *where* $t.e_i^{\alpha,n} = (\lambda - i\,p^n\,\alpha)(t)e_i^{\alpha,n}$

(2) $x_\alpha(z).e_i^{\alpha,n} = \sum_{j=0}^{i} \binom{i}{j} z^{p^n(i-j)} e_j^{\alpha,n}$

This leads to the following

PROPOSITION 2. *Let* $\lambda \in X(P_{\alpha,n})$ *be a dominant weight with* $\alpha^\vee(\lambda) = r\,p^n$.

(1) *The formal Euler character of* $\mathcal{L}(\lambda)$ *on* $G/P_{\alpha,n}$ *is given by*

$$\chi(\lambda) + \chi(\lambda - p^n\alpha) + \cdots + \chi(\lambda - r\,p^n\,\alpha)$$

where χ *denotes the Weyl character.*

(2) *If furthermore* $n = r = 1$ *then* $H^i(G/P_{\alpha,1}, \lambda) = 0$ *for* $i \geq 2$ *and there is an exact sequence*

$$0 \to H^0(G/P_{\alpha,1}, \lambda) \to H^0(G/B, \lambda) \to H^0(G/B, \lambda - \alpha) \to H^1(G/P_{\alpha,1}, \lambda) \to 0$$

PROOF. Since $R^i(\pi_{\alpha,n})_*(\mathcal{L}(\lambda)) = 0$ for $i > 0$ it follows from the Leray spectral sequence that for $i \geq 0$

$$H^i(G/P_{\alpha,n}, \lambda) = H^i(G/P_\alpha, (\pi_{\alpha,n})_*\mathcal{L}(\lambda)) =$$
$$H^i(G/P_\alpha, \mathcal{L}(V^r_{\alpha,n})) = H^i(G/B, \pi^*\mathcal{L}(V^r_{\alpha,n}))$$

The vector bundle $\mathcal{L}(V^r_{\alpha,n})$ on G/B has a filtration with quotient line bundles induced by the characters in Lemma 1 (1). This gives the statement in (1).

If $n = r = 1$, $V^1_{\alpha,1}$ is the middle term in the short exact sequence of B-modules

$$0 \to k_{\lambda - p\alpha} \to V^1_{\alpha,1} \to k_\lambda \to 0$$

This gives the long exact sequence on cohomology

$$\begin{aligned}&0 \to H^0(G/B, \mathcal{L}(V^1_{\alpha,1})) \to H^0(G/B, \lambda) \to \\ &H^1(G/B, \lambda - p\alpha) \to H^1(G/B, \mathcal{L}(V^1_{\alpha,1})) \to \cdots \to \\ &0 \to H^i(G/B, \lambda - p\alpha) \to H^i(G/B, \mathcal{L}(V^1_{\alpha,1})) \to 0 \to \dots\end{aligned}$$

by Kempf's vanishing theorem. Now since $\lambda - p\,\alpha = s_\alpha \cdot (\lambda - \alpha)$ and $0 \leq \alpha^\vee(\lambda - \alpha) < p$ the statement in (2) follows from the isomorphism [**1**], 2.3.(ii).

$$H^i(G/B, \lambda - \alpha) = H^{i+1}(G/B, s_\alpha \cdot (\lambda - \alpha))$$

□

The following example shows that vanishing for ample line bundles breaks down for varieties of unseparated flags.

EXAMPLE 3. *Let $G = SL_3(k)$, where k is of characteristic $p > 3$. Fix a Borel subgroup B associated with roots $\{-\alpha, -\beta, -\alpha - \beta\}$, where $\alpha = 2\omega_\alpha - \omega_\beta$ and $\beta = -\omega_\alpha + 2\omega_\beta$. For the character $\lambda = \omega_\alpha + p\,\omega_\beta$ inducing the ample line bundle $\mathcal{L}(\lambda)$ on $G/P_{\beta,1}$ we have*

$$\dim H^1(G/P_{\beta,1}, \lambda) = \frac{(p-2)(p-3)}{2}$$

PROOF. By the strong linkage principle [**2**] the highest weights of the possible composition factors of $H^0(G/B, \lambda)$ are

$$\{\omega_\alpha + p\,\omega_\beta, 2\omega_\alpha + (p-2)\omega_\beta, (p-4)\omega_\beta\}$$

On the other hand §3, 3 gives that the dominant weights in $H^0(G/P_{\beta,1}, \lambda)$ are

$$\{\omega_\alpha + p\,\omega_\beta, (p-1)\omega_\beta\}$$

It follows that $H^0(G/P_{\beta,1}, \lambda)$ is the irreducible G-module L_λ of highest weight λ. By Steinberg's tensor product theorem $\dim H^0(G/P_{\beta,1}, \lambda) = \dim L_\lambda = \dim L_{\omega_\alpha} \dim L_{\omega_\beta} = 9$.

From proposition 2 (2) we get

$$\begin{aligned}\dim H^1(G/P_{\beta,n}, \mathcal{L}(\lambda)) &= 9 - \dim H^0(G/B, \mathcal{L}(\lambda)) + \dim H^0(G/B, \mathcal{L}(\lambda - \beta)) \\ &= 9 - (p+1)(p+3) + \frac{3(p-1)(p+2)}{2} \\ &= \frac{(p-2)(p-3)}{2}\end{aligned}$$

□

It is interesting to notice that Kodaira's vanishing theorem is true for the variety of unseparated flags X in the above example. This follows from the fact that X embeds as a divisor in $\mathbb{P}^2 \times \mathbb{P}^2$ and is the zeros of a section in $\mathcal{O}(1) \times \mathcal{O}(p)$. Thereby X admits a flat lifting to $\mathbb{Z}$ and since $p > \dim X = 3$ it follows from [**3**], 2.8 that Kodaira's vanishing theorem holds for X.

Due to the intricate structure of canonical line bundle, it was suggested to us by Ramanathan, that the right generalization of Kempf's vanishing theorem to varieties of unseparated flags could be the condition on dominant weights imposed by Kodaira's vanishing theorem. Unfortunately counterexamples exist already for $G = SL_4(k)$.

EXAMPLE 4. *Let G be of type A_3 with simple roots $\alpha_1, \alpha = \alpha_2, \alpha_3$ such that $\alpha^\vee(\alpha_1) = \alpha^\vee(\alpha_3) = -1$.*

The canonical line bundle ω on $G/P_{\alpha,1}$ is given by $-2\omega_1 - 2\omega_2 - 2\omega_3 - (p-1)(-\omega_1 + 2\omega_2 - \omega_3) = (p-3)\omega_1 - 2p\omega_2 + (p-3)\omega_3$. Let $\mathcal{L}$ be the ample line bundle induced by the character $\lambda = \omega_1 + 3p\omega_2 + \omega_3$. Now $\mathcal{L} \otimes \omega$ is induced by $\mu = (p-2)\omega_1 + p\omega_2 + (p-2)\omega_3$ and we wish to prove that a G-homomorphism

$$\mathrm{H}^0(G/B, \mu) \to \mathrm{H}^0(G/B, \mu - \alpha)$$

is not surjective. In view of proposition 2 (2) this gives that $\mathrm{H}^1(G/P_{\alpha,1}, \mu) \neq 0$.

The affine Weyl group W_p of G is the group generated by the reflections $s_{\beta,mp}$, where $s_{\beta,mp}(\nu) = s_\beta(\nu) + mp\beta$ and s_β denotes the reflection for the root β. We let $s_{\beta,mp} \cdot \nu = s_{\beta,mp}(\nu + \rho) - \rho = \nu - (\beta^\vee(\nu + \rho) - mp)\beta$ denote the translated action of W_p on $X(T)$. Let s_1, s_2, s_3 denote the simple reflections and $\tilde{\alpha}$ the long root of A_3. By the strong linkage principle the highest weights of the possible composition factors of $\mathrm{H}^0(G/B, \mu)$ are μ, $s_2 \cdot \mu = (p-1)\omega_1 + (p-2)\omega_2 + (p-1)\omega_3$ and $(s_{\tilde{\alpha},2p}\, s_2) \cdot \mu = (p-2)\omega_2$. Since $\mu - \alpha = s_2 \cdot \mu$, the highest weights of the possible composition factors of $\mathrm{H}^0(G/B, \mu - \alpha)$ are $\mu - \alpha$ and $s_{\tilde{\alpha},2p} \cdot (\mu - \alpha) = (p-2)\omega_2$. We will prove that the multiplicity m of $L_{(p-2)\omega_2}$ in $H^0(G/B, \mu)$ is 0 and that the multiplicity n of $L_{(p-2)\omega_2}$ in $H^0(G/B, \mu - \alpha)$ is 1 (for $p > 2$).

Recall [**7**], *Proposition II.8.19, that $V(\mu) = H^0(G/B, -w_0\mu)^*$ has a filtration*

$$V(\mu) = V(\mu)^0 \supset V(\mu)^1 \supset V(\mu)^2 \supset \dots$$

with $V(\mu)/V(\mu)^1 \cong L_\mu$, such that

$$\sum_{i>0} \operatorname{ch} V(\mu)^i = \sum_{\gamma \in \Phi^+} \sum_{0 < mp < \gamma^\vee(\mu+\rho)} \nu_p(mp) \chi(s_{\gamma,mp} \cdot \mu)$$

Using this formula it follows for $p > 2$ that $\operatorname{ch} L_{(p-1)\omega_1 + (p-2)\omega_2 + (p-1)\omega_3} = \chi((p-1)\omega_1 + (p-2)\omega_2 + (p-1)\omega_3) - \chi((p-2)\omega_2)$ and that $n = 1$. Evaluating the formula with μ yields $\chi((p-1)\omega_1 + (p-2)\omega_2 + (p-1)\omega_3) - \chi((p-2)\omega_2)$. Now a simple computation gives the final result $m = 0$. For $p = 2$ the same computations give $m = 1$ and $n = 2$.

A counterexample to Kodaira's vanishing theorem with negative Euler characteristic can be found in [**9**]. Notice that unlike the ordinary flag manifolds, varieties of unseparated flags do not necessarily admit a flat lifting to $\mathbb{Z}$. This observation follows from [**3**], 2.8.

REFERENCES

1. Andersen, H. H., *Cohomology of line bundles on G/B*, Ann. Sci. Éc. Norm. Sup. (4) **12** (1979), 85–100.
2. Andersen, H. H., *The strong linkage principle*, J. reine angew. Math. **315** (1980), 53–59.
3. Deligne, P., Illusie, L., *Relèvements modulo p^2 et décomposition du complexe de de Rham*, Invent. Math. **89** (1987), 247–270.
4. Haboush, W. J., *Central differential operators on split semi-simple groups over fields of positive characteristic*, Lecture Notes in Mathematics **795** (1980), Springer Verlag, 35–85.
5. Hartshorne, R., *Ample Subvarieties of Algebraic Varieties*, Lecture Notes in Mathematics **156** (1970), Springer Verlag.
6. Iversen, B., *The Geometry of Algebraic Groups*, Adv. in Math. **20** (1976), 57–85.
7. Jantzen, J. C., *Representations of Algebraic Groups*, Academic Press, Orlando, 1987.
8. Kempf, G., *Algebraic Representations of Reductive Groups*, Proceedings of the International Congress of Mathematicians, Helsinki 1978 (1980), 575–577.
9. Lauritzen, N., *The Euler characteristic of a homogeneous line bundle*, C. R. Acad. Sc. Paris **315** (1992), 715–718.
10. Lauritzen, N., *Splitting Properties of Complete Homogeneous Spaces*, J. Algebra (to appear).
11. Mehta, V.B. and Ramanathan, A., *Frobenius splitting and cohomology vanishing for Schubert varieties*, Ann. of Math. **122** (1985), 27–40.
12. Nielsen, H. A., *Diagonalizably linearized coherent sheaves*, Bull. Soc. Math. France **102** (1974), 85–97.
13. Ramanathan, A., *Equations defining Schubert varieties and Frobenius splitting of diagonals*, Publ. Math. IHES **65** (1987), 61–90.
14. Wenzel, C., *Classification of all Parabolic Subgroup Schemes of a Reductive Linear Algebraic Group over an Algebraically Closed Field*, Trans. Amer. Math. Soc. (to appear).
15. Wenzel, C., *Rationality of G/P for a Non Reduced Parabolic Subgroup Scheme P*, Proc. Amer. Math. Soc. (to appear).

DEPARTMENT OF MATHEMATICS, UNIVERSITY OF ILLINOIS AT URBANA-CHAMPAIGN, 1409 W. GREEN ST. URBANA, IL 61801

E-mail address: haboush@symcom.math.uiuc.edu

MATEMATISK INSTITUT, AARHUS UNIVERSITET, NY MUNKEGADE, DK-8000 ÅRHUS C, DENMARK

Current address: Department of Mathematics, University of Illinois at Urbana-Champaign, 1409 W. Green St. Urbana, IL 61801

E-mail address: nielsl@odin.math.uiuc.edu

Contemporary Mathematics
Volume **153**, 1993

Bases for Demazure modules for symmetrizable Kac-Moody algebras

V. LAKSHMIBAI

Dedicated to Prof. R. Steinberg on his 70th birthday

ABSTRACT. Let $\underline{g}$ be a symmetrizable Kac-Moody algebra. Let λ be a dominant, integral weight, and $V(\lambda)$ the corresponding simple $\underline{g}$-module. Let $V_{\mathbf{Z}}(\lambda)$ be the canonical $\mathbf{Z}$-form of $V(\lambda)$. Let W be the Weyl group of $\underline{g}$. For $w \in W$, let $V_{\mathbf{Z},w}$ be the Demazure submodule of $V_{\mathbf{Z}}(\lambda)$ associated to w. We construct a $\mathbf{Z}$-basis for $V_{\mathbf{Z}}(\lambda)$ compatible with $\{V_{\mathbf{Z},w},\ w \in W\}$.

1. Introduction

Let $\underline{g}$ be a symmetrizable Kac-Moody algebra over $\mathbf{C}$, and U its universal enveloping algebra. Let U^+ (resp. U^-) be the subalgebra of U generated by E_α (resp. F_α), α simple. (here E_α, F_α are the usual elements in U). Let $U_{\mathbf{Z}}^+$ (resp. $U_{\mathbf{Z}}^-$) be the $\mathbf{Z}$-subalgebra of U generated by $\frac{E_\alpha^n}{n!}$ (resp. $\frac{F_\alpha^n}{n!}$), $n \in \mathbf{Z}$. (In the sequel, we shall denote $\frac{E_\alpha^n}{n!}$, $\frac{F_\alpha^n}{n!}$ by $E_\alpha^{(n)}$, $F_\alpha^{(n)}$ respectively.)

1980 Mathematics Subject Classification (1991 Revision), Primary 20G05, 20G10 , Secondary 14F05, 14M15

Partially supported by NSF Grant DMS 9103129

This paper is in final form and no version of it will be submitted for publication elsewhere.

Let λ be a dominant, integral weight, and $V(\lambda)$ the irreducible $\underline{g}$-module with highest weight λ. Let us fix a highest weight vector e in $V(\lambda)$. Let $V_{\mathbf{Z}}(\lambda) = U_{\mathbf{Z}}^{-}e$. Let W be the Weyl group of $\underline{g}$. For $\tau \in W$, let $e_\tau = \tau e$, $V_\tau = U^+ e_\tau$, $V_{\mathbf{Z},\tau} = U_{\mathbf{Z}}^+ e_\tau$. ($V_\tau$'s (resp. $V_{\mathbf{Z},\tau}$'s) are called the *Demazure submodules* of $V(\lambda)$ (resp. $V_{\mathbf{Z}}(\lambda)$).
In [L-S]$_2$, we conjectured a $\mathbf{Z}$-basis $B(\lambda)$ for $V_{\mathbf{Z}}(\lambda)$ compatible with $\{ V_{\mathbf{Z},\tau}, \tau \in W\}$.

This conjecture has two parts. The first part gives a character formula for V_τ, in terms of a nice indexing set I_τ consisting of certain "weighted chains" of elements of W (cf §3). The second part describes a basis B_τ for V_τ. The set B_τ is given as $\{e\} \cup \{F_{\beta_r}^{(n_r)} \ldots F_{\beta_1}^{(n_1)} e$, β_i simple, $n_i > 0$ (for some suitable n_i's), and $s_{\beta_r} \cdots s_{\beta_1}$ is a reduced expression $\leq \tau\}$. Recently, Littelmann (cf [Li]) has proved the first part of this conjecture. In this paper, we prove the second part of this conjecture (cf Theorem 4.1).

Let $\tau \in W$. To prove the results for V_τ, we fix a w such that $w < \tau, w = s_\alpha \tau$, with α simple. Let $sl(2,\alpha)$ be the copy of $sl(2)$ in $\underline{g}$ corresponding to α. Then V_τ is just the $sl(2,\alpha)$- span of V_w. This fact (together with induction on $l(\tau)$) enables us to write down a $\mathbf{Z}$-basis B_τ for $V_{\mathbf{Z},\tau}$, where B_τ looks like $B_\tau = B_w \cup \{F_\alpha^{(i)} u$, for some u 's $\in B_w$, and suitable i 's $\geq 1\}$ (here, B_w is the inductive basis for $V_{\mathbf{Z},w}$). We then use Demazure character formula, and the character formula as given by [L-S]$_2$ (and proved in [Li]) (see §3 for this formula) to show that B_τ is a basis for $V_{\mathbf{Z},\tau}$ compatible with $\{V_{Z,\theta},\ \theta \leq \tau\}$. Our results extend to the Quantum situation also (cf[L]).

The authour would like to thank P.Littelmann and C.S.Seshadri for some useful comments on a preliminary version of this paper.

2. Preliminaries

We preserve the notations of §1.

Definition 2.1: Let $X(w)$ be a Schubert divisor in $X(\tau)$. We say that $X(w)$ is a *moving divisor in* $X(\tau)$ *moved by the simple root* α if $w = s_\alpha \tau$.

Lemma 2.2 (cf $[\text{L-S}]_1$). *Let* $X(w)$ *be a moving divisor in* $X(\tau)$ *moved by* α. *Let* $X(\theta)$ *be a Schubert sub variety in* $X(\tau)$. *Then either*

(i) $X(\theta) \subseteq X(w)$ *or*

(ii) $X(\theta) = X(s_\alpha \theta')$ *for some* $X(\theta') \subseteq X(w)$

Definition 2.3: Let λ be a dominant integral weight of $\underline{g}$. Let $X(w)$ be a divisor in $X(\tau)$. Let $w = s_\beta \tau$ for some $\beta \in R^+$. We define $m_\lambda(w,\tau)$ as the non-negative integer $m_\lambda(w,\tau) = (w(\lambda), \beta^*)$ $(= -(\tau(\lambda), \beta^*))$, and call it the λ-*multiplicity of* $X(w)$ *in* $X(\tau)$. (here (,) is a W-invariant form on $\underline{h}^\vee$, $\underline{h}^\vee$ being the dual of the Cartan subalgebra $\underline{h}$ of $\underline{g}$).

2.4: For a simple root α, let U_α(resp.$U_{-\alpha}$) be the sub algebra of U generated by E_α(resp.F_α). Let $U_{\alpha,\mathbf{Z}}$ (resp. $U_{-\alpha,\mathbf{Z}}$) denote the $\mathbf{Z}$-submodule of U spanned by $E_\alpha^{(n)}$ (resp. $F_\alpha^{(n)}$), $n \in \mathbf{Z}^+$. Let $sl_{\mathbf{Z}}(2,\alpha)$ (resp.$sl(2,\alpha)$) denote the copy of $sl_2(\mathbf{Z})$(resp.sl_2) in $U_{\mathbf{Z}}$(resp.U) associated to α.

Lemma 2.5. *Let* $X(w)$ *be a moving divisor in* $X(\tau)$ *moved by a simple root* α. *Then* $V_{\mathbf{Z},\tau} = U_{-\alpha,\mathbf{Z}} V_{\mathbf{Z},w}$.

Proof. The result follows from the fact that $V_{\mathbf{Z},\tau}$ is the smallest $U_{-\alpha,\mathbf{Z}}$-stable submodule of $V_{\mathbf{Z}}(\lambda)$ containing $V_{\mathbf{Z},w}$.

2.6: In the sequel, we will be frequently using the following two formulae in $U_{\mathbf{Z}}$(cf[S]).

$$(1)\qquad E_\alpha^{(s)}F_\alpha^{(i)}=\sum_{j=0}^{\min(s,i)}F_\alpha^{(i-j)}\binom{H_\alpha-s-i+2j}{j}E_\alpha^{(s-j)}$$

$$(2)\qquad F_\alpha^{(s)}E_\alpha^{(i)}=\sum_{j=0}^{\min(s,i)}E_\alpha^{(i-j)}\binom{H_{-\alpha}-s-i+2j}{j}F_\alpha^{(s-j)}$$

2.7: Demazure Character Formula.

Let L_α be the Demazure operator on $\mathbf{Z}[X]$ given by

$$L_\alpha(e^\mu)=\frac{e^{\mu+\rho}-e^{s_\alpha(\mu+\rho)}}{1-e^{-\alpha}}e^{-\rho}$$

(here X is the weight lattice). Then we have (cf [D],[K],[M]),

$$\text{char } V_\tau=L_\alpha(\text{char } V_w)$$

where w,τ are as in Lemma 2.5 The following is easily checked:

$$L_\alpha(e^\mu)\quad=\quad\begin{cases}\sum_{i=0}^{n}e^{\mu-i\alpha}, & \text{if } (\mu,\alpha^*)=n\geq 0\\ 0, & \text{if } (\mu,\alpha^*)=-1\\ -\sum_{i=1}^{m}e^{\mu+i\alpha}, & \text{if } (\mu,\alpha^*)=-(m+1)\leq -2\end{cases}$$

2.8. Let E, F, H denote the usual elements in sl_2. Let $\underline{h}$ be the Cartan subalgebra generated by H, and $\underline{b}$ the Borel subalgebra generated by $\{E,H\}$. Let N be a finite dimensional $\underline{h}$-module. For a weight vector u in N, we shall denote its weight by r_u (note that r_u is given by $Hu=r_u u$).

Lemma 2.9. *Let M be a finite dimensional sl_2-module, and N a $\underline{b}$-submodule of M such that $M=sl_2$-span of N. Let B be a basis for N consisting of weight vectors. Let $B_1=\{u\in B\mid r_u\geq 0\}$, and $B_2=\{u\in B\mid r_u<0\}$. Let V_1 =* span of *$\{F^{(i)}u,\ u\in B_1,\ 0\leq i\leq r_u\}$,* and *$V_2$* = span of *$\{u\in B_2\}$. Then $M=V_1+V_2$.*

Proof. Let us denote V_1+V_2 by V'. Then $V'\subset M$ (obviously). In view of the hypothesis, it suffices to show that $F^{(i)}u\in V'$, $i\geq 1$, $u\in B$. We first

Claim. *V' is stable under the action of E.*

To prove the claim, we first observe that $V' = N + V_1$. Now N being stable for the action of E, it suffices to prove that $E^{(s)}F^{(i)}u \in V'$, for $u \in B_1$, and $1 \leq i \leq r_u$. Now we have (cf §2.6, (1))

$$E^{(s)}F^{(i)}u = \sum_{j=0}^{\min(s,i)} F^{(i-j)}\binom{H-s-i+2j}{j}E^{(s-j)}u$$

Now $E^{(s-j)}u \in$ the span of B_1 (since, weight of $E^{(s-j)}u \geq 0$), and $i - j \leq r_u + 2(s-j)$ ($=$ weight of $E^{(s-j)}u$). Hence each term in the summation belongs to V_1. The claim follows from this.

Let $m = \max\{|r_u|,\ u$ is a weight vector in $M\}$.

I. Let $u \in B_1$. We are required to show that $F^{(i)}u \in V'$, for $i > r_u$. We distinguish the following two cases:

Case 1. $r_u = m - 2r$, for some $r \geq 0$.

We have

$$F^{(i)}u = 0,\ i > m - r \tag{1}$$

(by weight considerations and the definition of m). Thus it remains to check that $F^{(i)}u \in V'$, $m - 2r + 1 \leq i \leq m - r$. For $s, t \in \mathbf{Z}^+$, We have (cf §2.6, (2)),

$$(*) \qquad F^{(s)}E^{(t)}u = \sum_{j=0}^{\min(s,t)} E^{(t-j)}\binom{-H-s-t+2j}{j}F^{(s-j)}u = \sum u_j, \text{ say}$$

Fix k, $0 \leq k \leq r-1$. Take $s = m - 2k, t = r - k$. If L.H.S.$\neq 0$, then weight of $E^{(t)}u = m - 2k > 0$, and $s = m - 2k$. Hence, L.H.S. of (*) belongs to V'. On R.H.S. $u_t = \binom{-H-s+t}{t}F^{(m-r-k)}u = \binom{t}{t}F^{(m-r-k)}u = F^{(m-r-k)}u$. For $j < t$ (note that $\min(s,t) =$

$t = r - k$), if $u_j \neq 0$, then F occurs with a power $> m - r - k$ (in u_j). Hence by decreasing induction on $m - r - k$, we obtain $F^{(m-2k-j)}u \in V'$ and hence $u_j \in V'$ (in view of claim). Hence $u_t \in V'$, and the result follows from this. (For $k = 0$, we have

$$F^{(m)}E^{(r)}u = \binom{-H - m + r}{r} F^{(m-r)}u = F^{(m-r)}u$$

(in view of (*) and (1)). As above, the element on L.H.S. belongs to V').

Case 2. $r_u = m - (2r - 1)$, for some $r > 0$.

As in Case 1 we have

$$F^{(i)}u = 0, \ i \geq m - r + 1. \tag{2}$$

Thus, it is required to show that $F^{(i)}u \in V'$, $m-2r+2 \leq i \leq m-r$. We fix k, $1 \leq k \leq r - 1$, and take $s = m - (2k - 1), t = r - k$ (in (*)). Then $\min(s,t) = r - k = t$. As in Case 1, we find that $u_t = F^{(m-r-k+1)}u$. Rest of the argument is as in Case 1.

II. Let $u \in B_2$. We are required to show that $F^{(i)}u \in V'$, for all $i \geq 1$. Here again, we distinguish the following two cases.

Case 1. $r_u = -m + 2r$ (< 0)

We have (by weight considerations and the definition of m), $F^{(i)}u = 0$, $i > r$. Thus we are required to show that $F^{(i)}u \in V'$, $1 \leq i \leq r$. Fix k, $1 \leq k \leq r$. Take $s = m - 2(k-1)$, $t = m - r - (k-1)$, in (*) (cf I, Case 1). If L.H.S.$\neq 0$ then weight of $E^{(m-r-k+1)}u = m - 2k + 2 > 0$, and $s = m - 2k + 2$. Hence we obtain that L.H.S. belongs to V'. Now $\min(s,t) = t$, and $u_t = F^{(r-k+1)}u$. For $j < t$, if $u_j \neq 0$, then F occurs with a power $> r + 1 - k$. Hence by decreasing induction on $r + 1 - k$, we obtain $F^{(s-j)}u \in V'$, and hence $u_j \in V'$ (in view of Claim). Hence $u_t \in V'$, and the result follows from this. (For $k = 1, F^{(m)}E^{(m-r)}u = F^{(r)}u$. As above we conclude that $F^{(r)}u \in V'$).

Case 2. $r_u = -m + (2r - 1)$ (< 0).

We have, $F^{(i)}u = 0$, $i \geq r$. Thus, it is required to show that $F^{(i)}u \in V'$, $1 \leq i \leq r-1$. Fix k, $1 \leq k \leq r-1$. Take $s = m - 2k + 1, t = m - r + 1 - k$, in (*). Then $\min(s,t) = t$. As in Case 1, we find that $u_t = F^{(r-k)}u$. The rest of the argument is as above.

This completes the proof of Lemma 2.9.

Remark 2.10. Let M, N be as in Lemma 2.9. Let $N_{\mathbf{Z}}$ be a $\underline{b}_{\mathbf{Z}}$-stable $\mathbf{Z}$- submodule of N such that $N = N_{\mathbf{Z}} \otimes_{\mathbf{Z}} \mathbf{Q}$. Let $M_{\mathbf{Z}}$ be the $sl_2(\mathbf{Z})$-span of $N_{\mathbf{Z}}$. Let B be a $\mathbf{Z}$-basis for $N_{\mathbf{Z}}$. Let B_1, B_2 be as in Lemma 2.9. Let $V_1 = \mathbf{Z}$-span of $\{F^{(i)}u,\ u \in B_1,\ 0 \leq i \leq r_u\}$, and $V_2 = \mathbf{Z}$-span of $\{u \in B_2\}$. Then the above proof shows that $M_{\mathbf{Z}} = V_1 + V_2$.

Lemma 2.11. *Let w, τ be as in Lemma 2.5. There exists a decomposition $V_\tau = \oplus_{i \in I} M_i$, where each M_i is $sl(2,\alpha)$-irreducible, $i \in I$ (for some finite indexing set I) with the following property. Let u_i be a highest weight vector in M_i. Let*

$$A_1 = \{i \in I \mid M_i \subseteq V_w\}$$

$$A_2 = \{i \in I \mid M_i \not\subseteq V_w\}$$

Then

$$V_w = \oplus_{i \in A_1} M_i \oplus_{j \in A_2} (u_j),$$

(u_j) being the line through u_j.

The assertion follows from the results in [K] (by specialising q to 1).

Corollary 2.12. *Let w, τ be as in Lemma 2.5. With notations as in Lemma 2.11, let*

$$M_w = \oplus_{i \in A_1} M_i,$$

$$L_w = \oplus_{j \in A_2} (u_j),$$

$$V'_w = \{v \in V_w \mid sl(2,\alpha)v \subseteq V_w\}.$$

Then $V'_w = M_w$.

Proof. The inclusion $M_w \subseteq V'_w$ is clear. Let $v \in V_w$, and $v \notin M_w$. Then

$$v = \sum_{i \in A_1} v_i + \sum_{j \in A_2} c_j u_j,$$

where $v_i \in M_i, c_j \neq 0$, for at least one j.
Let $m = \max_{\{j,\ c_j \text{ non-zero}\}} \{r_j\}$, where r_j = weight of u_j (for the H_α-action). Note that $m > 0$, since $r_j > 0,\ j \in A_2$. Then

$$F_\alpha^{(i)} v \notin V_w,\ 1 \leq i \leq m.$$

This implies that $sl(2,\alpha)v \not\subseteq V_w$, and hence $v \notin V'_w$. This completes the proof of Corollary 2.12.

Corollary 2.13. *Let notations be as in Corollary 2.12. Let* u *be a weight vector (for the* H_α*-action) in* L_w *of weight* r_u. *Then* $E_\alpha u = 0,\ r_u > 0,\ sl(2,\alpha)u \not\subseteq V_w$, *and* $sl(2,\alpha)u \cap V_w = (u)$.

Proof. By hypothesis, we can write

$$u = \sum_{j \in A_2} c_j u_j,$$

where at least one $c_j, j \in A_2$ is non-zero. Now $r_j = r_u, j \in A_2$, with $c_j \neq 0$. Hence $r_u > 0$ (since $r_j > 0, j \in A_2$), $E_\alpha u = 0$, and $F_\alpha^{(m)} u \notin V_w,\ 1 \leq m \leq r_u$ (by Lemma 2.11). Hence we obtain, $sl(2,\alpha)u \not\subseteq V_w$, $sl(2,\alpha)u \cap V_w = (u)$. This completes the proof of Corollary 2.13.

Corollary 2.14. *Let* w, τ *be as in Lemma2.5. Let* u *be a weight vector (for the* H_α*-action) in* V_w *such that* $E_\alpha u = 0$. *Let* r_u

= weight of u. Then either $sl(2,\alpha)u \subseteq V_w$ or $sl(2,\alpha)u \not\subseteq V_w$, $sl(2,\alpha)u \cap V_w = (u)$, and $r_u > 0$.

Proof. Let $sl(2,\alpha)u \not\subseteq V_w$. This implies that $r_u > 0$; further we can write

$$u = \sum_{i \in A_1} c_i u_i + \sum_{j \in A_2} c_j u_j,$$

where at least one $c_j, j \in A_2$ is non-zero (since $sl(2,\alpha)u \not\subseteq V_w$). Then for $1 \leq m \leq r_u$,

$$F_\alpha^{(m)} u = \sum_{i \in A_1} c_i F_\alpha^{(m)} u_i + \sum_{j \in A_2} c_j F_\alpha^{(m)} u_j.$$

Hence in view of of Lemma 2.11, we obtain , $F_\alpha^{(m)} u \notin V_w$, $m \geq 1$. Thus we obtain that if $sl(2,\alpha)u \not\subseteq V_w$, then $F_\alpha^{(m)} u \notin V_w$, $1 \leq m \leq r_u$. All the assertions follow from this.

2.15 Let n^+ be the subalgebra of $\underline{g}$ generated by E_β, β simple. Let b^+ be the Borel subalgebra given by

$$b^+ = n^+ \oplus \underline{h},$$

where $\underline{h}$ is the Cartan subalgebra. For a simple root α, let

$$\underline{p}_\alpha = n^+ \oplus \underline{h} \oplus (F_\alpha)$$

Lemma 2.16. *Let V be a $\underline{p}_\alpha$-module, and V_1, a b^+-submodule of V. Let*

$$V_1' = \{v \in V_1 \mid sl(2,\alpha)v \subseteq V_1\},$$

$sl(2,\alpha)$ being as before. Then V_1' is $\underline{p}_\alpha$ -stable and is the largest $\underline{p}_\alpha$ -stable submodule of V_1.

Proof. Let M be any $\underline{p}_\alpha$ -stable submodule of V_1. Then for any $v \in M$, $sl(2,\alpha)v \subseteq V_1$ (by $\underline{p}_\alpha$-stability of M). Hence $M \subseteq V_1'$. Now taking M to be the b^+-submodule of V_1 generated by V_1', we obtain that V_1' is $\underline{p}_\alpha$- stable (note that V_1' is $sl(2,\alpha)$-stable).

Corollary 2.17. *Let w, τ be as in Lemma 2.5. Let*

$$V^w = \sum_{\{\theta \ | s_\alpha \theta < \theta < w\}} V_\theta$$

Then $M_w = V^w$ (here M_w is as in Corollary 2.12).

Proof. It is easiliy seen that V^w is the largest $\underline{p}_\alpha$-stable submodule of the b^+-module V_w. This together with Corollary 2.12 and Lemma 2.16 implies the required result.

Definition 2.18: For a weight vector u in $V(\lambda)$, we define the *α-weight of u* as the integer (weight of u, α^*), and denote it by $r_u(\alpha)$ or just r_u.

Lemma 2.19. *Let w, τ be as in Lemma 2.5. Let A_w be a* **Z**-*basis for $V_{\mathbf{Z},w}$ consisting of weight vectors. Let $B_1 = \{u \in A_w \mid r_u \geq 0\}$, and $B_2 = \{u \in A_w \mid r_u < 0\}$. Let $V_1 =$* **Z**-*span of $\{F_\alpha^{(i)} u,\ u \in B_1,\ 0 \leq i \leq r_u\}$, and $V_2 =$* **Z**-*span of $\{u \in B_2\}$. Then $V_{\mathbf{Z},\tau} = V_1 + V_2$.*

The result follows from Lemma 2.9 (see also Remark 2.10).

3. The set $I(\lambda)$ and the L-S character formula.

Let $\underline{c} = \{\mu_0, \mu_1, \ldots, \mu_r\}$ be a λ-chain in W, i.e. $\mu_i \geq \mu_{i-1}$, $l(\mu_i) = l(\mu_{i-1}) + 1$ (if $\lambda = \sum d_i \omega_i$, ($\omega_i$ being the fundamental weights), and $d_t = 0$ for $t = j_1, \ldots, j_k$, then we shall work with W^Q, the set of minimal representatives of W_Q in W, W_Q being the subgroup of W generated by the set of simple reflections $\{s_t, t = j_1, \ldots, j_k\}$). Let $\mu_{i-1} = s_{\beta_i} \mu_i, \beta_i \in R^+$. Let $m_\lambda(\mu_{i-1}, \mu_i) = m_i$.

Definition 3.1: A λ-chain $\underline{c}$ is called *simple* (resp. *non-simple*) if all (resp. some) β_i's are simple (resp. non-simple).

Definition 3.2: By a *weighted λ-chain* , we shall mean $(\underline{c}, \underline{n})$ where $\underline{c} = \{\mu_0, \ldots, \mu_r\}$ is a λ-chain and $\underline{n} = \{n_1, \ldots, n_r\}$, $n_i \in \mathbf{Z}^+$.

Definition 3.3: A weighted λ-chain $(\underline{c}, \underline{n})$ is said to be *admissible* if $1 \geq \frac{n_1}{m_1} \geq \frac{n_2}{m_2} \cdots \geq \frac{n_r}{m_r} \geq 0$.

Let $(\underline{c}, \underline{n})$ be admissible. Let us denote the unequal values in $\{\frac{n_1}{m_1}, \ldots, \frac{n_r}{m_r}\}$ by $a_1, \ldots, a_s$ so that $1 \geq a_1 > a_2 > \ldots > a_s \geq 0$. Let $i_0, \ldots, i_s$ be defined by

$$i_0 = 0, \ i_s = r, \ \frac{n_j}{m_j} = a_t, \ i_{t-1} + 1 \leq j \leq i_t.$$

We set

$$D_{\underline{c},\underline{n}} = \{(a_1, \ldots, a_s); (\mu_{i_0}, \ldots, \mu_{i_s})\}$$

Definition 3.4: Let $(\underline{c}, \underline{n})$, $(\underline{c}', \underline{n}')$ be two admissible weighted λ-chains. We say $(\underline{c}, \underline{n}) \sim (\underline{c}', \underline{n}')$, if $D_{\underline{c},\underline{n}} = D_{\underline{c}',\underline{n}'}$.

3.5. Let C_λ = {all admissible weighted λ-chains}, and $I(\lambda) = C_\lambda / \sim$. In [Li], the elements of $I(\lambda)$ are called the Lakshmibai-Seshadri λ-paths. Hence keeping up with the terminology in [Li], in the sequel we shall refer to the elements of $I(\lambda)$ as L-S λ-paths.

3.6. With notation as above, let $\pi \in I(\lambda)$, and let $(\underline{c}, \underline{n})$ be a representative of π. Keeping up with the notation in [Li], we denote

$$\phi(\pi) = \mu_{i_s}, \nu(\pi) = \sum_{t=0}^{s} (a_t - a_{t+1}) \mu_{i_t}(\lambda),$$

where $a_0 = 1, a_{s+1} = 0$ (note that $\phi(\pi)$ and $\nu(\pi)$ depend only on π and not on the representative chosen).

3.7. L-S character formula. (cf [Li]). For $\tau \in W$, let $I_\tau(\lambda)$ (or just I_τ) be defined as

$$I_\tau = \{\pi \in I(\lambda) \mid \tau \geq \phi(\pi)\}$$

Then

$$\text{char } V_\tau = \sum_{\pi \in I_\tau} e^{\nu(\pi)}$$

$$\text{char } V(\lambda) = \sum_{\pi \in I(\lambda)} e^{\nu(\pi)}$$

4. A Z-basis for $V_{\mathbf{Z}}(\lambda)$ compatible with $\{V_{\mathbf{Z},\tau},\ \tau \in W\}$

We preserve the notations of the previous sections.

Theorem 4.1. *Let $\tau \in W$. The $\mathbf{Z}$-module $V_{\mathbf{Z},\tau}$ has a basis $B_\tau = \{Q_\pi,\ \pi \in I_\tau\}$ with the following properties:*

(i) *Q_π is a weight vector of weight $\nu(\pi)$.*

(ii) *$B_\tau = \{e\} \cup \{F_{\gamma_r}^{(n_r)} \dots F_{\gamma_1}^{(n_1)} e,\ \gamma_i$ simple, $n_i > 0$ (for some suitable n_i's), and $s_{\gamma_r} \cdots s_{\gamma_1}$ is a reduced expression $\leq \tau\}$.*

(iii) *For $\tau' \leq \tau, \{Q_\pi \mid \tau' \geq \phi(\pi)\}$ is a $\mathbf{Z}$-basis for $V_{\mathbf{Z},\tau'}$.*

(iv) *If $Q_\pi = F_{\gamma_r}^{(n_r)} \dots F_{\gamma_1}^{(n_1)} e$, and $\phi(\pi) = x$, then $x = s_{\gamma_r} \cdots s_{\gamma_1}$ (reduced).*

Proof. (by induction on $l(\tau)$). If $l(\tau) = 0$, then $\{Q_\pi = e\}$ is a basis for $V_{\mathbf{Z},\tau}$ ($= \mathbf{Z}e$). (here, recall that e is the highest weight vector in $V(\lambda)$ that we have fixed). Let then $l(\tau) > 0$. Let us fix a moving divisor $X(w)$ in $X(\tau)$ moved by a simple root α. By induction, $V_{\mathbf{Z},w}$ has a $\mathbf{Z}$-basis $B_w = \{Q_\pi, \pi \in I_w\}$, such that

(A) Q_π is a weight vector of weight $\nu(\pi)$

(B) $B_w = \{e\} \cup \{F_{\beta_s}^{(m_s)} \dots F_{\beta_1}^{(m_1)} e,\ \beta_i$ simple, $m_i > 0$ (for some suitable m_i's), and $s_{\beta_s} \cdots s_{\beta_1}$ is a reduced expression $\leq w\}$.

(C) For $w' \leq w, \{Q_\pi,\ \pi \in I_w,\ w' \geq \phi(\pi)\}$ is a $\mathbf{Z}$-basis for $V_{\mathbf{Z},w'}$.

(D) If $Q_\pi = F_{\beta_s}^{(m_s)} \dots F_{\beta_1}^{(m_1)} e$, and $\phi(\pi) = y$, then $y = s_{\beta_s} \cdots s_{\beta_1}$ (reduced).

For $\pi \in I_w$, let $r_\pi = (\nu(\pi), \alpha^*)$. Let

$$B_1 = \{\pi \in I_w \mid r_\pi \geq 0\}$$

$$B_2 = \{\pi \in I_w \mid r_\pi < 0\}$$

Then by Lemma 2.19, $\{Q_\pi,\ \pi \in B_2\} \cup \{F_\alpha^{(i)} Q_\pi,\ 0 \leq i \leq r_\pi,\ \pi \in B_1\}$ generates the $\mathbf{Z}$-module $V_{\mathbf{Z},\tau}$. Let

(I) $$C_1 = \{\pi \in I_w \mid w \not\geq s_\alpha \phi(\pi)\},$$

(I₁) $$C_2 = \{\pi \in I_w \mid w \geq s_\alpha \phi(\pi)\}.$$

Then we have (by (C) above)

(1) $V^w =$ span of $\{Q_\pi, \pi \in C_2\}$
(here, V^w is as in Corollary 2.17). Hence,

$$(2) \qquad C_1 = \{\pi \in I_w \mid sl(2,\alpha)Q_\pi \not\subseteq V_w\},$$

$$(3) \qquad C_2 = \{\pi \in I_w \mid sl(2,\alpha)Q_\pi \subseteq V_w\}.$$

(cf.Corollaries 2.12 and 2.17). Now (2) implies that

$$(4) \qquad Q_\pi = u + \sum c_j u_j, \ \pi \in C_1$$

where $u \in V^w$, $j \in A_2$, and $c_j \neq 0$ for at least one j (notations being as in Lemma 2.11). Hence, we obtain by Lemma 2.11

$$(5) \qquad r_\pi > 0, \text{ and } F_\alpha^{(i)} Q_\pi \notin V_w, \ 1 \leq i \leq r_\pi, \ \pi \in C_1$$

(since $r_j > 0$, for $j \in A_2$). In particular we obtain

$$(6) \qquad C_1 \subset B_1.$$

Now (2), (3), and (5) together with Lemma 2.19 imply that

$$(7) \quad \{Q_\pi, \ \pi \in I_w\} \dot\cup \{F_\alpha^{(i)} Q_\pi, \ 1 \leq i \leq r_\pi, \ \pi \in C_1\} = A_\tau, \text{ say}$$

generates the $\mathbf{Z}$-module $V_{\mathbf{Z},\tau}$.

Linear Independence of A_τ.

Let L_α be the Demazure operator on $\mathbf{Z}[X]$ (cf §2.7), namely,

$$L_\alpha(e^\mu) = \frac{e^{\mu+\rho} - e^{s_\alpha(\mu+\rho)}}{1 - e^{-\alpha}} e^{-\rho}$$

Then we have (cf §2.7),

$$\text{char } V_\tau = L_\alpha(\text{char } V_w).$$

The following is easily checked:

$$L_\alpha(e^\mu) = \begin{cases} \sum_{i=0}^{n} e^{\mu-i\alpha}, & \text{if } (\mu,\alpha^*) = n \geq 0 \\ 0, & \text{if } (\mu,\alpha^*) = -1 \\ -\sum_{i=1}^{m} e^{\mu+i\alpha}, & \text{if } (\mu,\alpha^*) = -(m+1) \leq -2 \end{cases}$$

Now

$$L_\alpha(\text{char} \quad V^w) = \text{char} \quad V^w$$

(since V^w is $sl(2,\alpha)$-stable). Hence we obtain (cf.(5) above)

$$\text{(II)} \qquad \text{char } V_\tau - \text{char } V_w = \sum_{\theta\in C_1} \sum_{i=1}^{r_\theta} e^{\nu(\theta)-i\alpha}$$

This implies in particular that

$$\dim V_\tau = \dim V_w + \sum_{\theta \in C_1} r_\theta$$

Hence we obtain $\dim V_\tau = \# A_\tau$ (cf (7) for definition of A_τ). This implies that A_τ is linearly independent over $\mathbf{Q}$ and hence over $\mathbf{Z}$. Thus
(III): A_τ is a $\mathbf{Z}$-basis for $V_{\mathbf{Z},\tau}$.
Let

$$\text{(IV)} \qquad I = \{\pi \in I_w \mid \phi(\pi) = w\}$$

Then

$$I \subset C_1$$

(since $w \not\geq s_\alpha w$). Let

$$C_1' = \{\pi \in C_1 \mid \phi(\pi) < w\}$$

Then

$$\text{(V)} \qquad C_1 = C_1' \dot{\cup} I$$

For $\theta \in I(\lambda)$ such that $e_\alpha \theta = 0$, let $S_\theta = \{f_\alpha^i \theta,\ 0 \le i \le r_\theta\}$ (here, the operators e_α, f_α on $I(\lambda)$ are as in [Li]). Let

$$E_1 = \{\theta \in I_w \mid r_\theta > 0,\ e_\alpha \theta = 0,\ S_\theta \cap I_w = \{\theta\}\}$$

$$E_1' = \{\theta \in E_1 \mid \phi(\theta) < w\},\ J = \{\theta \in E_1 \mid \phi(\theta) = w\}.$$

Then

(VI) $$E_1 = E_1' \dot{\cup} J, \text{ and } J \subseteq I$$

Claim 1: $E_1 = C_1$.

Prrof of Claim 1 Let $\pi \in E_1'$, and $\phi(\pi) = w'$. Let μ= the bigger of $\{w',\ s_\alpha w'\}$. Then the facts that $S_\pi \subseteq I_\mu,\ S_\pi \not\subseteq I_w$ imply that $\mu = s_\alpha w'$, and $w \not\geq \mu$; hence $\pi \in C_1'$. Thus we obtain $E_1' \subseteq C_1'$, and hence

(*) $$E_1 \subseteq C_1$$

(cf.(V) and (VI)). Now, in view of L-S character formula, we have (cf [Li])

(VII) $$\text{char } V_\tau - \text{char } V_w = \sum_{\theta \in E_1} \sum_{i=1}^{r_\theta} e^{\nu(\theta) - i\alpha}$$

Comparing (II) and (VII), we obtain $A' = B'$, where

$$A' = \sum_{\theta \in C_1} \sum_{1 \le i \le r_\theta} e^{\nu(\theta) - i\alpha},$$

$$B' = \sum_{\theta \in E_1} \sum_{1 \le i \le r_\theta} e^{\nu(\theta) - i\alpha}$$

(cf (V) and (VI) above). This implies in particular that

$$\sum_{\theta \in C_1} r_\theta = \sum_{\eta \in E_1} r_\eta$$

Now each $r_\theta, r_\eta > 0$, and $E_1 \subseteq C_1$ (cf (*) above). Hence we obtain

$$(**) \qquad C_1 = E_1.$$

This completes the proof of Claim 1. Now set

$$Q_{f_\alpha^i \theta} = F_\alpha^{(i)} Q_\theta, \; \theta \in C_1, \; 1 \leq i \leq r_\theta$$

Then $\{Q_\theta, \; \theta \in I_w\} \dot{\cup} \{Q_\pi, \; \pi = f_\alpha^i \theta, \; \theta \in C_1, \; 1 \leq i \leq r_\theta\}$ is a **Z**-basis for $V_{\mathbf{Z},\tau}$ (cf (7) and (III) above). Note that this set is simply A_τ (cf. (7) above). Hence we obtain

$$A_\tau = \{Q_\pi, \; \pi \in I_\tau\}$$

(note that $I_\tau = I_w \dot{\cup} \{f_\alpha^i \theta, \; \theta \in E_1, \; 1 \leq i \leq r_\theta\}$ (cf [Li])).
We shall now show that A_τ has the properties (i) through (iv). Property (i) follows in view of the fact that weight of $Q_\pi = \nu(\pi)$ (this is clear for $\phi(\pi) \leq w$ (by induction hypothesis). If $\phi(\pi) \not\leq w$, then $\phi(\pi) = s_\alpha w'$, for some $w' \leq w$. This implies that $Q_\pi = F_\alpha^{(i)} Q_\theta$, for some $\theta \in C_1$. Hence weight of $Q_\pi = \nu(\theta) - i\alpha = \nu(\pi)$, since $\pi = f_\alpha^i \theta$, and $\nu(\pi) = \nu(\theta) - i\alpha$ (cf [Li])).

Properties (ii) and (iv) follow by induction hypothesis, and the fact that a Q_π such that $\pi \in I_\tau - I_w$ is given by $Q_\pi = F_\alpha^{(i)} Q_\theta$, for some $\theta \in C_1$ (note that if $\theta \in C_1$, and $\phi(\theta) = x$, then $s_\alpha x > x$ and $\phi(f_\alpha^i \theta) = s_\alpha x, 1 \leq i \leq r_\theta$).

Finally, it remains to prove (iii). Let $\tau' \leq \tau$, and let $A_{\tau'} = \{Q_\pi \in A_\tau \mid \pi \in I_{\tau'}\}$. We are required to show that $A_{\tau'}$ is a **Z**-basis for $V_{\mathbf{Z},\tau'}$. If $\tau' \leq w$, then $A_{\tau'} = \{Q_\pi \in B_w \,|\tau' \geq \phi(\pi)\}$ and the result follows by induction hypothesis. Let then $\tau' \not\leq w$. We have (cf Lemma 2.2), $s_\alpha\tau' \leq w$. Let us denote $s_\alpha\tau'$ by w'. Then by Lemma 2.5, we have

$$V_{\mathbf{Z},\tau'} = U_{-\alpha,\mathbf{Z}} V_{\mathbf{Z},w'}$$

Let us denote $V = \mathbf{Z}$-span of $\{Q_\pi \mid \phi(\pi) \leq w'\}$. Then since $w' \leq w$,

$$V_{\mathbf{Z},w'} = V$$

(by induction hypothesis). Let W_1 be the $\mathbf{Z}$-submodule of V spanned by $\{Q_\pi \mid \phi(\pi) \leq w',\ \pi \in C_1(w')\}$, where $C_1(w') = \{\pi \in I_{w'} \mid w' \not\geq s_\alpha\phi(\pi)\}$. Then as above we get

$$V_{\mathbf{Z},\tau'} = V + U_{-\alpha,\mathbf{Z}} W_1$$

and the union of

$$\{Q_\pi \mid \phi(\pi) \leq w',\ \pi \notin C_1(w')\}$$

and

$$\{F_\alpha^{(i)} Q_\pi,\ 0 \leq i \leq r_\pi \mid \phi(\pi) \leq w',\ \pi \in C_1(w')\}$$

is a basis for $V_{\mathbf{Z},\tau'}$.

Now, $\tau' > s_\alpha\tau'$ $(= w')$ implies that for any $\mu \leq \tau'$, $\tau' \geq$ the bigger of $\{\mu, s_\alpha\mu\}$.

Claim 2. For $\theta \in I_\tau$, $Q_\theta \in V_{\mathbf{Z},\phi(\theta)}$.

Proof of Claim 2: By our construction, we have either

$$\phi(\theta) \leq w \tag{8}$$

(or)

$$Q_\theta = F_\alpha^{(i)} Q_{\theta'}, \quad \text{for some } \theta' \in C_1 \tag{9}$$

Further, if (9) holds then $\phi(\theta') < s_\alpha\phi(\theta') = \phi(\theta)$, and $\phi(\theta) \not\leq w$. Hence we have

$$Q_\theta \in V_{\mathbf{Z},\phi(\theta)}, \quad \text{if (8) holds,}$$

(by induction hypothesis), and

$$Q_\theta \in V_{\mathbf{Z},s_\alpha\phi(\theta')}, \quad \text{if (9) holds.}$$

Thus $Q_\theta \in V_{\mathbf{Z},\phi(\theta)}$. This proves Claim 2.
Now Claim 2 implies that $A_{\tau'} \subseteq V_{\mathbf{Z},\tau'}$.

Generation of $V_{\mathbf{Z},\tau'}$ by $A_{\tau'}$: To prove that $V_{\mathbf{Z},\tau'}$ is generated by $A_{\tau'}$, it suffices to show that $F_\alpha^{(i)}Q_\pi$, $1 \leq i \leq r_\pi$, $\pi \in C_1(w')$ belongs to the span of $A_{\tau'}$. This is clear if $\pi \in C_1$ (for, $\pi \in C_1$ implies that $F_\alpha^{(i)}Q_\pi = Q_{f_\alpha^i\pi}$, $\phi(f_\alpha^i\pi) = s_\alpha\phi(\pi)$; and hence $F_\alpha^{(i)}Q_\pi \in A_{\tau'}$, since $\tau' \geq s_\alpha\phi(\pi)$). Let then $\pi \in C_2$. This implies that $s_\alpha\phi(\pi) \leq w$. Let μ = the bigger of $\{\phi(\pi),\ s_\alpha\phi(\pi)\}$. We have, $\tau' \geq \mu$. Now $F_\alpha^{(i)}Q_\pi \in V_{\mathbf{Z},\mu}$ (since $Q_\pi \in V_{\mathbf{Z},\phi(\pi)}$ (cf. Claim 2 above). Hence $F_\alpha^{(i)}Q_\pi$, $1 \leq i \leq r_\pi$, belongs to the span of A_μ (by induction hypothesis (note that $w \geq \mu$)). This implies $F_\alpha^{(i)}Q_\pi, 1 \leq i \leq r_\pi$, belongs to the span of $A_{\tau'}$ (note that $A_\mu \subseteq A_{\tau'}$, since $\tau' \geq \mu$). Thus we obtain $A_{\tau'}$is a $\mathbf{Z}$-basis for $V_{\mathbf{Z},\tau'}$.
This completes the proof of Theorem 4.1.

Corollary 4.2. *Let $\tau \in W$, and α a simple root such that $\tau > s_\alpha\tau = w$, say. Let A_τ be the basis $\{Q_\pi,\ \pi \in I_\tau\}$ for $V_{\mathbf{z},\tau}$ as constructed in Theorem 4.1. Let $\pi \in I_\tau$, $\phi(\pi) = w'$, and $w \not\geq s_\alpha w'$. Then we have*

(1). $sl(2,\alpha)Q_\pi \not\subseteq V_w$

(2). $r_\pi > 0$

(3). $e_\alpha\pi = 0$

(4). $S_\pi \cap I_w = \{\pi\}$

Proof. The hypothesis that $w \not\geq s_\alpha w'$ implies that $\pi \in C_1$. Hence the assertions (1) and (2) follow from (2) and (5) in the proof of Theorem 4.1. Now $\pi \in C_1$ implies that $\pi \in E_1$ (since $C_1 = E_1$ (cf.(**)in the proof of Theorem 4.1)), and the assertions (3) and (4) follow from this.

Corollary 4.3. *Let τ, w, w', α be as in Corollary 4.2. Let v be a weight vector (for the H_α-action) in the span of $\{Q_\pi, \phi(\pi) = w'\}$. Let r_v = weight of v. Then*
(1). $sl(2,\alpha)v \not\subseteq V_w$
(2). $r_v > 0$
(3). $F_\alpha^{(i)} v \in$ span of $\{Q_\pi, \phi(\pi) = s_\alpha w'\}$, $1 \leq i \leq r_v$

Proof. By hypothesis we can write

$$(a) \qquad v = \sum_{i \in J} a_i Q_{\pi_i}$$

where $\phi(\pi_i) = w'$. Assertion (2) follows from the fact that $\pi_i \in C_1, i \in J$ (cf.(5) in the proof of Theorem 4.1). Also,

$$(b). \qquad F_\alpha^{(m)} v = \sum_{i \in J} a_i F_\alpha^{(m)} Q_{\pi_i},\ 1 \leq m \leq r_v,$$

Now, since $\pi_i \in C_1$, we have by our construction, $F_\alpha^{(m)} Q_{\pi_i} = Q_{f_\alpha^m \pi_i}$, and $\phi(f_\alpha^m \pi_i) = s_\alpha \phi(\pi_i)$ $(= s_\alpha w')$. Assertion (3) follows from this; we also obtain,

$$(c) \qquad F_\alpha^{(m)} v \notin V_w,\ 1 \leq m \leq r_v,$$

(since $w \not\geq s_\alpha w'$). Assertion (1) follows from (c) (in view of (2)).

References

[D] M.Demazure, *Desingularisation des variétiés de Schubert generaliseés*, Ann. Sci. École. Norm, Sup. **t.7** (1974), 53-88

[K] M. Kashiwara, *Crystal Base and Littelmann's Refined Demazure CharacterFormula*, (preprint, RIMS_880, 1992)

[K] S.Kumar, *Demazure character formula in arbitrary Kac-Moody setting*, Inv. Math., **89** (1987), 395-423

[L] V. Lakshmibai, *Bases for Quantum Demazure modules -II* (in preparation).

[Li] P.Littelmann, *Littlewood-Richardson rule for symmetrizable Kac-Moody algebras*, preprint, 1992

[L-S]$_1$ V.Lakshmibai-C.S.Seshadri, *Geometry of $G/P - V$*, J. Algebra **100** (1986) , 462-557

[L-S]$_2$ V.Lakshmibai-C.S.Seshadri, *Standard monomial theory*, Proceedings of the Hyderabad Conference on ALgebraic Groups, Manoj Prakashan, Madras, (1991), 279-323

[M] O.Mathieu, *Formules de caractères pour les algèbres de Kac-Moody générales*, Astérisque **159-160** (1988)

[M-R] V.B.Mehta-A.Ramanathan, *Frobenius splitting and cohomology vanishing for Schubert varieties*, Ann. Math **122** (1985), 27-40

[S] R.Steinberg, *Lectures on Chevalley groups*, Lecture Notes in Mathematics, Yale University, 1967

V. Lakshmibai
Mathematics Department
Northeastern University
Boston, MA 02115
e-mail address: lakshmibai@northeastern.edu

Contemporary Mathematics
Volume **153**, 1993

REPRESENTATIONS OF QUANTUM AFFINE LIE ALGEBRAS AND CRYSTAL BASES

KAILASH C. MISRA*

1. Introduction

The theory of Kac-Moody Lie algebras has been developing in a rapid pace during the last two decades with several interesting and important connections with other fields of mathematics and physics. Most of these connections are in fact with affine Lie algebras - an important family of infinite dimensional Kac-Moody Lie algebras - and their integrable highest weight representations. The integrable highest weight representations (also known as standard representations) of Kac-Moody Lie algebras are infinite dimensional analogues of the finite dimensional irreducible highest weight representations of finite dimensional semisimple Lie algebras (see [5]).

More recently, Drinfeld [1] and Jimbo [2] have discovered another fundamental algebraic object known as *quantized universal enveloping algebra* (or *quantum group*) $U_q(\mathfrak{g})$ associated with a symmetrizable Kac-Moody Lie algebra $\mathfrak{g}$ which can be thought of as a q-analogue or q-deformation of the universal enveloping algebra of $\mathfrak{g}$. The quantized universal enveloping algebra associated with an affine Lie algebra is known as a *quantum affine Lie algebra.* It is known [12,16] that for generic q (i.e. q not a root of unity) the integrable representations of a symmetrizable Kac-Moody Lie algebra can be deformed consistently to those of the corresponding quantized universal enveloping algebra. In particular, the internal structures of the integrable highest weight representations of an affine Lie algebra is essentially the same as that of the corresponding quantum affine Lie algebra. However, working in the larger context of a quantum affine Lie algebra it is easier to extract more informations about the internal structures of these representations.

The eminent role of the quantized universal enveloping algebras in two dimensional solvable lattice models is widely known. The R-matrices, which are the intertwiners of tensor product representations, give the Boltzmann weights of the lattice models with commuting transfer matrices (see [3]). The quantum parameter 'q' corresponds to temperature in the lattice model. In particular, 'q=0' corre-

1991 Mathematics Subject Classifications: Primary 17B37, 17B10, 17B67; Secondary 82B23
*Supported in part by NSA/MSP grant #MDA92-H-3076.
This paper is in final form and no version of it will be submitted for publication elsewhere.

sponds to the absolute temperature zero in the lattice model. So, one can expect that the quantized universal enveloping algebra has a simpler structure at $q = 0$. Motivated by this, Kashiwara [6,7] introduced a canonical base for the integrable representation of a quantized universal enveloping algebra called *crystal base* and proved the existence and uniqueness of this base. The crystal base can be roughly thought of as a base at $q = 0$ for the representation. Unlike the *global base* [7] or *canonical base* [13,14], the crystal base is not a base for the representation, instead it is a parameterization of a base for the corresponding representation. However, because of it's simpler structure the crystal base provides a powerful combinatorial tool to study the internal structures of the integrable representations.

In [4,15], using the Fock space representations of $U_q(\hat{sl}(n))$, we gave explicit descriptions of the crystal bases for the integrable highest weight representations of the quantum affine Lie algebra $U_q(\hat{sl}(n))$ in terms of certain infinite Young diagrams which are parametrized by certain paths that arise naturally in solvable lattice models. Recently, the study of crystal bases for the integrable highest weight representations of other quantum affine Lie algebras led us to the theory of affine and perfect crystals [8,9]. Using the perfect crystals we have obtained path realizations of the crystal bases for the integrable highest weight representations of the quantum affine Lie algebras $U_q(\mathfrak{g})$, where $\mathfrak{g} = A_n^{(1)}, B_n^{(1)}, C_n^{(1)}, D_n^{(1)}, A_{2n}^{(2)}, A_{2n-1}^{(2)}$, or $D_{n+1}^{(2)}$ [10]. More recently, we have obtained path realizations of the crystal bases of Verma modules for the above mentioned quantum affine Lie algebras [11]. In this paper we will summarize some of the results obtained in [9,10]. It is an expanded version of our talk at the conference on "Linear algebraic groups" held at UCLA in honor of Professor Steinberg. We thank the organizers for their invitation and support.

2. Quantum affine Lie algebras and crystal bases

Let $\mathfrak{g}$ be an indecomposable affine Kac-Moody Lie algebra generated by e_i, f_i ($i \in I = \{0, 1, \dots, n\}$) and the Cartan subalgebra $\mathfrak{h}$ over $\mathbb{Q}$. Let $\{\alpha_i \mid i \in I\} \subset \mathfrak{h}^*$ and $\{h_i \mid i \in I\} \subset \mathfrak{h}$ denote the simple roots and simple coroots respectively. Note that $\{\alpha_i \mid i \in I\}$ and $\{h_i \mid i \in I\}$ are linearly independent and $\dim \mathfrak{h} = n + 2$. Let $(\, , \,)$ denote the normalized nondegenerate symmetric invariant bilinear form on $\mathfrak{h}^*$ so that $(\alpha_i, \alpha_i) \in \mathbb{Z}_{>0}$. Let $Q = \sum_i \mathbb{Z}\alpha_i$, $Q_+ = \sum_i \mathbb{Z}_{\geq 0}\alpha_i$ and $Q_- = -Q_+$. Let $\delta \in Q_+$ be the generator of null roots and let $c \in \sum_i \mathbb{Z}_{\geq 0} h_i$ be the generator of the center. Set $\mathfrak{h}_{cl} = \bigoplus_i \mathbb{Q}h_i \subset \mathfrak{h}$ and $\mathfrak{h}_{cl}^* = (\bigoplus_i \mathbb{Q}h_i)^*$. Let $cl : \mathfrak{h}^* \to \mathfrak{h}_{cl}^*$ denote the canonical morphism. Then we have an exact sequence

$$0 \to \mathbb{Q}\delta \to \mathfrak{h}^* \xrightarrow{cl} \mathfrak{h}_{cl}^* \to 0.$$

Take an integer d such that $\delta - d\alpha_0 \in \sum_{i \neq 0} \mathbb{Z}\alpha_i$. Note that $d = 1$ unless $\mathfrak{g} = A_{2n}^{(2)}$ in which case $d = 2$. Let $af : \mathfrak{h}_{cl}^* \to \mathfrak{h}^*$ be a map satisfying $cl \circ af = id$ and $af \circ cl(\alpha_i) = \alpha_i$ for $i \neq 0$. Let Λ_i be the element of $\mathfrak{h}_{cl}^*$ such that $\langle h_j, \Lambda_i \rangle = \delta_{ij}$. Hence $\alpha_i = \sum_j \langle h_j, \alpha_i \rangle af(\Lambda_j) + \delta_{i,0} d^{-1}\delta$. We take $P = \sum_i \mathbb{Z}af(\Lambda_i) + \mathbb{Z}d^{-1}\delta \subset \mathfrak{h}^*$ and $P_{cl} = cl(P) \subset \mathfrak{h}_{cl}^*$. An element of P is called an *affine weight* and an element of P_{cl} is called a *classical weight.* Let $P^* = Hom(P, \mathbb{Z})$, $P^*_{cl} = Hom(P_{cl}, \mathbb{Z})$ denote the lattices dual to P and P_{cl} respectively. Note that $h_i \in P^*$ for all $i \in I$.

The quantum affine Lie algebra $U_q(\mathfrak{g})$ ('q' not a root of unity) is then the $\mathbb{Q}(q)$-algebra generated by the symbols $e_i, f_i (i \in I)$ and $q^h (h \in P^*)$ satisfying the following relations:

$$q^0 = 1,\ q^h q^{h'} = q^{h+h'} \text{ for all } h, h' \in P^*, \tag{2.1}$$

$$q^h e_i q^{-h} = q^{<h,\alpha_i>} e_i,\ q^h f_i q^{-h} = q^{-<h,\alpha_i>} f_i \text{ for all } i \in I,\ h \in P^*, \tag{2.2}$$

$$[e_i, f_j] = \delta_{ij} \frac{t_i - t_i^{-1}}{q_i - q_i^{-1}}, \text{ where } q_i = q^{(\alpha_i, \alpha_i)} \text{ and } t_i = q_i^{h_i}, \tag{2.3}$$

$$\sum_{k=0}^{b} (-1)^k e_i^{(k)} e_j e_i^{(b-k)} = 0 = \sum_{k=0}^{b} (-1)^k f_i^{(k)} f_j f_i^{(b-k)} \text{ for all } i \neq j \tag{2.4}$$

where $b = 1- < h_i, \alpha_j >, e_i^{(k)} = e_i^k/[k]_i!,\ f_i^{(k)} = f_i^k/[k]_i!,\ [k]_i! = \Pi_{m=1}^k [m]_i,\ [m]_i = \frac{q_i^m - q_i^{-m}}{q_i - q_i^{-1}}$.

We define the quantized universal enveloping algebra $U'_q(\mathfrak{g})$ to be the $\mathbb{Q}(q)$-algebra generated by $e_i,\ f_i\ (i \in I)$ and $q^h (h \in P^*_{cl})$ satisfying (2.1) - (2.4) with P^* replaced by P^*_{cl}.

It is well-known that the algebra $U_q(\mathfrak{g})$ has a Hopf algebra structure with comultiplication Δ defined by

$$\begin{cases} \Delta(e_i) & = e_i \otimes t_i^{-1} + 1 \otimes e_i, \\ \Delta(f_i) & = f_i \otimes 1 + t_i \otimes f_i, \\ \Delta(q^h) & = q^h \otimes q^h, \end{cases} \tag{2.5}$$

for all $i \in I,\ h \in P^*$. The tensor product of two $U_q(\mathfrak{g})$-modules has a structure of $U_q(\mathfrak{g})$-module via this comultiplication. For $i \in I$, let $U_q(\mathfrak{g}_i)$ denote the subalgebra of $U_q(\mathfrak{g})$ generated by e_i, f_i, t_i and t_i^{-1}.

A $U_q(\mathfrak{g})$-module M is said to be *integrable* if $M = \bigoplus_{\lambda \in P} M_\lambda$, $M_\lambda = \{u \in M \mid q^h u = q^{<h,\lambda>} u \text{ for all } h \in P^*\}$ and M is a union of finite-dimensional $U_q(\mathfrak{g}_i)$-modules for any $i \in I$. By $U_q(sl(2))$ representation theory any element $u \in M_\lambda$ can be uniquely written as

$$u = \sum_{k \geq 0} f_i^{(k)} u_k, \tag{2.6}$$

where $u_k \in ker(e_i) \cap M_{\lambda + k\alpha_i}$. Define the endomorphisms $\tilde{e}_i$ and $\tilde{f}_i$ on M by

$$\tilde{e}_i u = \sum_{k>0} f_i^{(k-1)} u_k, \tag{2.7}$$

$$\tilde{f}_i u = \sum_{k \geq 0} f_i^{(k+1)} u_k, \tag{2.8}$$

for $u \in M_\lambda,\ \lambda \in P$.

Let A be the subring of $\mathbb{Q}(q)$ consisting of the rational functions regular at $q = 0$. For a $\mathbb{Q}$-vector space V, a subset B of V is called a *pseudo-base* if there exists a base B' of V such that $B = B' \cup (-B')$.

A *crystal base* (resp. *crystal pseudo-base*) of an integrable $U_q(\mathfrak{g})$-module M is a pair (L, B) such that

$$L \text{ is a free } A\text{-submodule of } M \text{ such that } M \cong \mathbb{Q}(q) \otimes_A L, \tag{2.9}$$

$$B \text{ is a base (resp. pseudo-base) of the } \mathbb{Q}\text{-vector space } L/qL, \tag{2.10}$$

$$L = \bigoplus_{\lambda \in P} L_\lambda, \text{ where } L_\lambda = L \cap M_\lambda, \tag{2.11}$$

$$B = \bigsqcup_{\lambda \in P} B_\lambda \text{ where } B_\lambda = B \cap (L_\lambda / qL_\lambda), \tag{2.12}$$

$$\tilde{e}_i L \subset L,\ \tilde{f}_i L \subset L,\ \tilde{e}_i B \subset B \sqcup \{0\}, \text{ and } \tilde{f}_i B \subset B \sqcup \{0\}, \tag{2.13}$$

$$\text{for } b,\ b' \in B,\ i \in I,\ b' = \tilde{f}_i b \text{ if and only if } b = \tilde{e}_i b'. \tag{2.14}$$

Note that if (L, B) is a crystal base for M then $(L, B \cup (-B))$ is a crystal pseudo-base. Associated with a crystal base (L, B) is a oriented colored (by $i \in I$) graph known as *crystal graph* (or *crystal*) with B as the set of vertices and for $b,\ b' \in B$, $b \xrightarrow{i} b'$ if $b' = \tilde{f}_i b$. The crystal graph describes completely the action of $\tilde{e}_i$ and $\tilde{f}_i$ on $B \sqcup \{0\}$. For $b \in B$, set

$$\epsilon_i(b) = \max\{k \geq 0 \mid \tilde{e}_i b \neq 0\}, \tag{2.15}$$

$$\phi_i(b) = \max\{k \geq 0 \mid \tilde{f}_i^k b \neq 0\}. \tag{2.16}$$

Then for $b \in B_\lambda$, we have

$$< h_i, \lambda >= \phi_i(b) - \epsilon_i(b). \tag{2.17}$$

Furthermore, as a graph B is connected if and only if M is irreducible.

For $\lambda \in P^+ = \{\lambda \in P \mid < h_i, \lambda > \geq 0 \text{ for all } i \in I\}$ let $V(\lambda)$ denote the irreducible integrable $U_q(\mathfrak{g})$-module with highest weight λ. Let u_λ denote the highest weight vector of $V(\lambda)$, and let $L(\lambda)$ be the smallest A-submodule of $V(\lambda)$ containing u_λ stable under $\tilde{f}_i$'s. Set $B(\lambda) = \{b \in L(\lambda)/qL(\lambda) \mid b = \tilde{f}_{i_1} \dots \tilde{f}_{i_r} u_\lambda \bmod (qL(\lambda)) \neq 0\}$. Then $(L(\lambda), B(\lambda))$ is the crystal base of $V(\lambda)$ (see [6,7]). The goal of this note is to realize the crystals $B(\lambda)$ in terms of certain paths that arise in associated lattice models naturally [9,10].

3. Affine Crystals, Perfect Crystals and paths:

Let $\mathfrak{g}$ be an affine Lie algebra with simple roots $\{\alpha_i \mid i \in I\}$, weight lattice P and $U_q(\mathfrak{g})$ be the corresponding quantum affine Lie algebra. Recall that $P_{cl} = cl(P)$ is the weight lattice for $U'_q(\mathfrak{g})$. Abstracting the properties of crystal base we define *P-weighted* (resp. *P_{cl}-weighted*) *crystals* which are known as *affine* (resp. *classical*) *crystals* [8,9] as follows.

A P-(resp. P_{cl}-)*weighted crystal* B is a set B equipped with maps

$$\begin{aligned} &wt : B \longrightarrow P(\text{resp.}P_{cl}),\\ &\epsilon_i : B \longrightarrow \mathbb{Z} \sqcup \{-\infty\}, \phi_i : B \longrightarrow \mathbb{Z} \sqcup \{-\infty\},\\ &\tilde{e}_i : B \longrightarrow B \sqcup \{0\}, \tilde{f}_i : B \longrightarrow B \sqcup \{0\} \end{aligned}$$

for all $i \in I$ satisfying the following conditions:

$$\phi_i(b) = \epsilon_i(b) + < h_i, wt(b) >, \tag{3.1}$$

$$\begin{aligned} \text{if } b \in B \text{ and } \tilde{e}_i b \in B, \text{ then } wt(\tilde{e}_i b) = wt(b) + \alpha_i,\\ \epsilon_i(\tilde{e}_i b) = \epsilon_i(b) - 1 \text{ and } \phi_i(\tilde{e}_i b) = \phi_i(b) + 1, \end{aligned} \tag{3.2}$$

$$\begin{aligned} \text{if } b \in B \text{ and } \tilde{f}_i b \in B, \text{ then } wt(\tilde{f}_i b) = wt(b) - \alpha_i,\\ \epsilon_i(\tilde{f}_i b) = \epsilon_i(b) + 1 \text{ and } \phi_i(\tilde{f}_i b) = \phi_i(b) - 1, \end{aligned} \tag{3.3}$$

$$\text{for } b, b' \in B, i \in I, b' = \tilde{f}_i b \text{ if and only if } b = \tilde{e}_i b', \tag{3.4}$$

$$\text{for } b \in B, \text{ if } \phi_i(b) = -\infty, \text{ then } \tilde{e}_i b = 0 = \tilde{f}_i b. \tag{3.5}$$

Observe that if B is a P-weighted crystal then $B = \bigsqcup_{\lambda \in P} B_\lambda$. Furthermore, a crystal B can be viewed as a oriented colored (by $i \in I$) graph with the set B as the set of vertices and for $b, b' \in B$, we draw i-arrows $b \xrightarrow{i} b'$ if $b' = \tilde{f}_i b$. Thus if (L, B) is a crystal base for the integrable $U_q(\mathfrak{g})$-module M then the set B has the structure of a P-weighted crystal.

For two (P- or P_{cl}-weighted) crystals B_1 and B_2, a *morphism* of crystals from B_1 to B_2 is a map $\psi : B_1 \sqcup \{0\} \longrightarrow B_2 \sqcup \{0\}$ such that:

$$\psi(0) = 0, \tag{3.6}$$

$$\psi(\tilde{e}_i b) = \tilde{e}_i \psi(b) \text{ for } b, \tilde{e}_i b \in B_1, \text{ and } \psi(\tilde{f}_i b) = \tilde{f}_i \psi(b) \text{ for } b, \tilde{f}_i b \in B_1, \tag{3.7}$$

$$\text{for } b \in B_1,\ \epsilon_i(b) = \epsilon_i(\psi(b)), \phi_i(b) = \phi_i(\psi(b)) \text{ if } \psi(b) \in B_2, \tag{3.8}$$

$$\text{for } b \in B_1, \ wt(b) = wt(\psi(b)) \text{ if } \psi(b) \in B_2. \tag{3.9}$$

A morphism of crystals $\psi : B_1 \longrightarrow B_2$ is called an *embedding* (resp. *isomorphism*) if ψ is *injective* (resp. *bijective*). An affine crystal B is said to a *crystal with highest weight* $\lambda \in P^+$ if B is isomorphic to the crystal $B(\lambda)$ where $(L(\lambda), B(\lambda))$ is the crystal base for the irreducible integrable $U_q(\mathfrak{g})$-module $V(\lambda)$. The crystals and their morphisms form a tensor category. For two crystals B_1 and B_2 their tensor product $B_1 \otimes B_2$ is defined as follows.

The underlying set in $B_1 \otimes B_2$ is the set $B_1 \times B_2$. For $b_1 \in B_1, \ b_2 \in B_2$, write $b_1 \otimes b_2$ for (b_1, b_2). We understand $b_1 \otimes 0 = 0 \otimes b_2 = 0$. Then $B_1 \otimes B_2$ equipped with the maps $wt, \epsilon_i, \phi_i, \tilde{e}_i, \tilde{f}_i, \ (i \in I)$ defined below is a crystal. For $b_1 \in B_1, \ b_2 \in B_2$,

$$wt(b_1 \otimes b_2) = wt(b_1) + wt(b_2) \tag{3.10}$$

$$\epsilon_i(b_1 \otimes b_2) = \max(\epsilon_i(b_1), \epsilon_i(b_2) - < h_i, wt(b_1) >), \tag{3.11}$$

$$\phi_i(b_1 \otimes b_2) = \max(\phi_i(b_2), \phi_i(b_1) + < h_i, wt(b_2) >), \tag{3.12}$$

$$\tilde{e}_i(b_1 \otimes b_2) = \begin{cases} \tilde{e}_i b_1 \otimes b_2 \text{ if } \phi_i(b_1) \geq \epsilon_i(b_2), \\ b_1 \otimes \tilde{e}_i b_2 \text{ if } \phi_i(b_1) < \epsilon_i(b_2), \end{cases} \tag{3.13}$$

$$\tilde{f}_i(b_1 \otimes b_2) = \begin{cases} \tilde{f}_i b_1 \otimes b_2 \text{ if } \phi_i(b_1) > \epsilon_i(b_2), \\ b_1 \otimes \tilde{f}_i b_2 \text{ if } \phi_i(b_1) \leq \epsilon_i(b_2). \end{cases} \tag{3.14}$$

Now let B be a classical crystal. For $b \in B$, set $\epsilon(b) = \sum_{i \in I} \epsilon_i(b)\Lambda_i$ and $\phi(b) = \sum_{i \in I} \phi_i(b)\Lambda_i$, so that $wt(b) = \phi(b) - \epsilon(b)$. Let $P_{cl}^+ = \{\lambda \in P_{cl} \mid < h_i, \lambda > \geq 0$ for all $i \in I\}$ and for $\ell \in \mathbb{Z}_{\geq 0}$ let $(P_{cl}^+)_\ell = \{\lambda \in P_{cl}^+ \mid < c, \lambda >= \ell\}$. We say B is a *perfect crystal of level* ℓ if it satisfies the following conditions:

$$B \otimes B \text{ is connected,} \tag{3.15}$$

(3.16)
there exists $\lambda_0 \in P_{cl}$ such that $wt(B) \subset \lambda_0 + \sum_{i \neq 0} \mathbb{Z}_{\leq 0} cl(\alpha_i)$ and that $\sharp B_{\lambda_0} = 1$,

(3.17)
there is a finite dimensional $U_q'(\mathfrak{g})$-module with a crystal pseudo-base(L, B_{ps}) such that B is isomorphic to $B_{ps}/\pm 1$,

$$\text{for any } b \in B, \ < c, \epsilon(b) > \geq \ell, \text{ and} \tag{3.18}$$

(3.19) the maps $\epsilon, \phi : B^{\min} = \{b \in B \mid < c, \epsilon(b) >= \ell\} \to (P_{cl}^{+})_{\ell}$ are bijective.

The elements of $B^{\min} = \{b \in B \mid < c, \epsilon(b) >= \ell\}$ are called *minimal elements.*

A $\mathbb{Z}$-valued function H on $B \otimes B$ is called an *energy function* on B if for any $i \in I$ and $b \otimes b' \in B \otimes B$ such that $\tilde{e}_i(b \otimes b') \neq 0$, we have

$$H(\tilde{e}_i(b \otimes b')) = \begin{cases} H(b \otimes b') \text{ if } i \neq 0, \\ H(b \otimes b') + 1 \text{ if } i = 0 \text{ and } \phi_0(b) \geq \epsilon_0(b'), \\ H(b \otimes b') - 1 \text{ if } i = 0 \text{ and } \phi_0(b) < \epsilon_0(b'). \end{cases} \tag{3.20}$$

It has been proved in [9] that for any perfect crystal B of level ℓ, the energy function H exists and is unique. From now on, assume that $\mathfrak{g}$ has rank ≥ 3 (i.e. $\sharp I \geq 3$). For $\lambda \in P^{+}$, let $B(\lambda)$ be the affine crystal with highest weight λ. Let u_{λ} denote the highest weight element of $B(\lambda)$.

Theorem 1. [9]. Let B be a perfect crystal of level ℓ and let b be an element of B. Then we have an isomorphism of classical crystals,

$$B(af(\epsilon(b))) \otimes B \cong B(af(\phi(b)))$$

where $u_{af(\epsilon(b))} \otimes b$ maps to $u_{af(\phi(b))}$.

For $\mu \in (P_{cl}^{+})_{\ell}$, let b_{μ} be the unique element of B such that $\phi(b_{\mu}) = \mu$. Define the isomorphism σ of $(P_{cl}^{+})_{\ell}$ by $\epsilon(b_{\mu}) = \sigma(\mu)$. Extend σ to an isomorphism of $af(P_{cl}^{+})_{\ell}$. Then by Theorem 1, for $\lambda \in af(P_{cl}^{+})_{\ell}$, we have

$$B(\lambda) \cong B(\sigma\lambda) \otimes B$$

given by $u_{\lambda} \longmapsto u_{\sigma\lambda} \otimes b_{cl(\lambda)}$.

Now set $\lambda_k = \sigma^k \lambda$ and $b_k = b_{cl(\lambda_{k-1})}$ for $k \geq 1$. Applying Theorem 1 repeatedly we have an isomorphism of classical crystals

$$\psi_k : B(\lambda) \cong B(\lambda_k) \otimes B^{\otimes k}$$

given by $\psi_k(u_{\lambda}) = u_{\lambda_k} \otimes b_k \otimes \ldots \otimes b_1$.

The sequence $(b_1, b_2, b_3, \ldots\ldots)$ is called the *ground-state path of weight* λ. Note that for $\lambda \epsilon af(P_{cl}^{+})_{\ell}$, the ground-state path $(b_k)_{k\geq 1}$ of weight λ is periodic with period less than or equal to $\sharp(P_{cl}^{+})_{\ell}$. A λ-path in B is a sequence $p = (p(k))_{k\geq 1}$ in B such that $p(k) = b_k$ for $k >> 0$. Let $\mathcal{P}(\lambda, B)$ denote the set of all λ-paths in B. Then $\mathcal{P}(\lambda, B)$ has the structure of a crystal. In particualr, the action of $\tilde{e}_i$ and $\tilde{f}_i$ on a path $p = (p(k))_{k\geq 1}$ is given inductively by (3.13) and (3.14). The following theorem gives a path realization of the crystal $B(\lambda)$ associated with the irreducible integrable $U_q(\mathfrak{g})$-module $V(\lambda)$.

Theorem 2. [9]. The crystal $B(\lambda)$ is isomorphic to $P(\lambda, B)$ given by $B(\lambda) \ni b \longmapsto p = (p(k))_{k\geq 1} \in \mathcal{P}(\lambda, B)$ where $\psi_k(b) = u_{\lambda_k} \otimes p(k) \otimes p(k-1) \otimes \ldots \otimes p(1)$ for $k >> 0$. Furthermore,

$$\begin{aligned} wt(p) = wt(b) = \lambda + \sum_{k=1}^{\infty}(af(wt(p(k))) - af(wt(b_k))) \\ - (\sum_{k=1}^{\infty} k(H(p(k+1) \otimes p(k)) - H(b_{k+1} \otimes b_k)))d^{-1}\delta. \end{aligned} \tag{3.21}$$

Note that for each perfect crystal B of level ℓ we have a path realization of the affine crystal $B(\lambda)$ of the irreducible integrable $U_q(\mathfrak{g})$-modulé $V(\lambda)$. Furthermore,

$$\begin{aligned} chV(\lambda) &= \sum_{\mu}(\dim(V(\lambda)_{\mu})e^{\mu} \\ &= \sum_{p\in\mathcal{P}(\lambda,B)} e^{wt(p)}. \end{aligned}$$

As seen in [10] for $\lambda \in af(P_{cl}^{+})\ell$ the perfect crystal B of level ℓ is by no means unique. An interesting open problem is to classify these perfect crystals.

4. Examples:

In [10] at least one perfect crystal of level ℓ has been given for the quantum affine Lie algebras $U_q(\mathfrak{g})$ where $\mathfrak{g} = A_n^{(1)}, B_n^{(1)}, C_n^{(1)}, D_n^{(1)}, A_{2n}^{(2)}, A_{2n-1}^{(2)}, D_{n+1}^{(2)}$. In this exposition we will discuss two perfect crystals of level ℓ for the quantum affine Lie algebras $U_q(A_n^{(1)})$ and $U_q(A_{2n-1}^{(1)})$ respectively.

4.1. $U_q(A_n^{(1)}), n \geq 2$:

For any positive integer ℓ, take

$$B_\ell = \{(x_1, x_2, \ldots, x_{n+1}) \in \mathbb{Z}^{n+1} \mid x_i \geq 0, \sum_{i=1}^{n+1} x_i = \ell\}.$$

For $b = (x_1, x_2, \ldots, x_{n+1}) \in B_\ell$ define

$$wt(b) = (x_{n+1} - x_1)\Lambda_0 + \sum_{i=1}^{n}(x_i - x_{i+1})\Lambda_i \tag{4.1}$$

$$\phi_0(b) = x_{n+1}, \phi_i(b) = x_i, \; i = 1, 2, \ldots, n, \tag{4.2}$$

$$\epsilon_0(b) = x_1, \; \epsilon_i(b) = x_{i+1}, \; i = 1, 2, \ldots, n, \tag{4.3}$$

$$\begin{aligned} &\tilde{e}_0 b = (x_1 - 1, \; x_2, \ldots, x_{n+1} + 1), \\ &\tilde{e}_i b = (x_1, \ldots, x_i + 1, x_{i+1} - 1, \ldots, x_{n+1}), i = 1, 2, \ldots, n, \end{aligned} \tag{4.4}$$

$$\begin{aligned} &\tilde{f}_0 b = (x_1 + 1, x_2, \ldots, x_{n+1} - 1), \\ &\tilde{f}_i b = (x_1, \ldots, x_i - 1, x_{i+1} + 1, \ldots, x_{n+1}), \; i = 1, 2, \ldots, n. \end{aligned} \tag{4.5}$$

Then B_ℓ equipped with the maps $wt, \phi_i, \epsilon_i, \tilde{e}_i, \tilde{f}_i (i \in I = \{0, 1, \ldots, n\})$ given by (4.1) - (4.5) is a perfect crystal of level ℓ for the algebra $U_q(A_n^{(1)})$. In this case $B_\ell^{\min} = B_\ell$ and for $\lambda = \sum_{i=0}^{n} k_i\Lambda_i \in af(P_{cl}^{+})_\ell$, we have $\sigma\lambda = k_1\Lambda_0 + k_2\Lambda_1 + \ldots + k_n\Lambda_{n-1} + k_0\Lambda_n$. Also note that as crystals for $U_q(A_n)$, the crystal B_ℓ is isomorphic to $B(\ell\Lambda_1)$. Furthermore, for $b = (x_1, x_2, \ldots, x_{n+1}) \in B_\ell$ and $b' = (x'_1, x'_2, \ldots, x'_{n+1}) \in B_\ell$, the energy function $H : B_\ell \otimes B_\ell \longrightarrow \mathbb{Z}$ is given by:

$$H(b \otimes b') = \max\{\theta_j(b \otimes b') \mid 0 \leq j \leq n\}, \tag{4.6}$$

where $\theta_j(b \otimes b') = \sum_{k=1}^{j}(x'_k - x_k) + x'_{j+1}$.

In particular, when $n = 2$, the perfect crystal B_1 of level 1 for $U_q(A_2^{(1)})$ is :

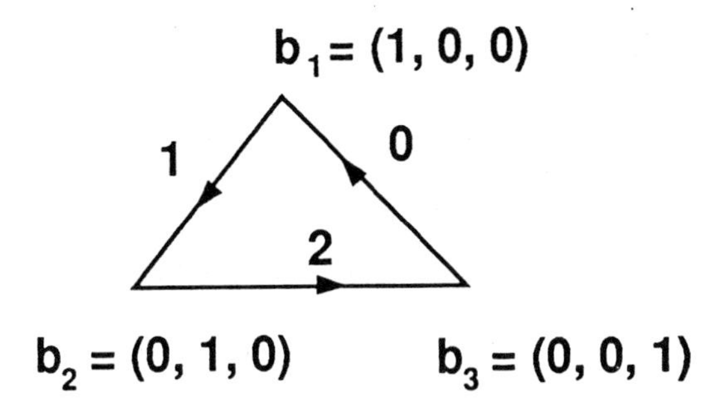

In this case, we have three dominant integral weights $\Lambda_0, \Lambda_1, \Lambda_2$ of level 1 and the respective ground-state paths are:

Λ_0-ground-state path $= (b_3, b_2, b_1, b_3, b_2, b_1, \ldots)$,
Λ_1-ground-state path $= (b_1, b_3, b_2, b_1, b_3, b_2, \ldots)$,
Λ_2-ground-state path $= (b_2, b_1, b_3, b_2, b_1, b_3, \ldots)$.

For instance, the path realization of the crystal $B(\Lambda_1)$ for the level one $A_2^{(1)}$-module $V(\Lambda_1)$ is :

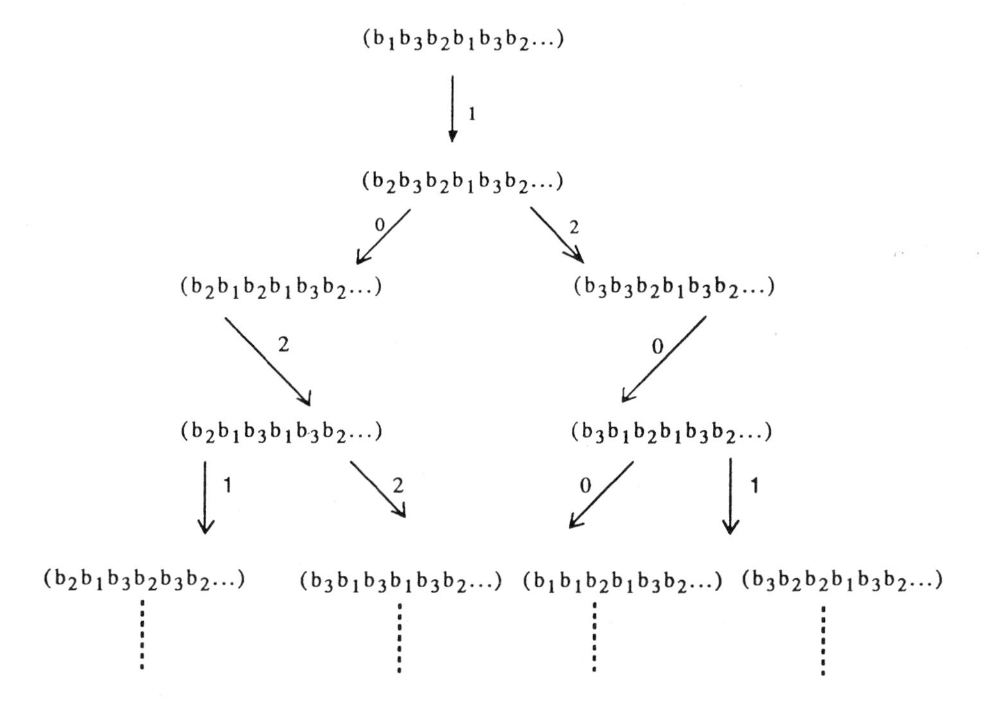

We can easily obtain the weight of a given path by using (3.21) and (4.6). For example, let us determine the weight of the path

$$p = (p(k))_{k\geq 1} = (b_1, b_1, b_1, b_1, b_3, b_2, b_1, b_3, b_2, \ldots)$$

in the crystal $B(\Lambda_1)$. Using (3.21) and neglecting the zero terms

$$\begin{aligned} wt(p) = \Lambda_1 &+ (af(wt\, b_1) - af(wt\, b_1)) \\ &+ (af(wt\, b_1) - af(wt\, b_3)) \\ &+ (af(wt\, b_1) - af(wt\, b_2)) \\ &- \delta[(H(b_1 \otimes b_1) - H(b_3 \otimes b_1)) \\ &+ 2(H(b_1 \otimes b_1) - H(b_2 \otimes b_3)) \\ &+ 3(H(b_1 \otimes b_1) - H(b_1 \otimes b_2)]. \end{aligned}$$

Since $af(wt\, b_1) = \Lambda_1 - \Lambda_0$, $af(wt\, b_2) = \Lambda_2 - \Lambda_1$ $af(wt\, b_3) = \Lambda_0 - \Lambda_2$ and by (4.6),

$$H(b_1 \otimes b_1) = 1,\ H(b_3 \otimes b_1) = 1,\ H(b_2 \otimes b_3) = 0,\ H(b_1 \otimes b_2) = 0,$$

we have

$$\begin{aligned} wt(p) &= \Lambda_1 + 0 + (\Lambda_1 - 2\Lambda_0 + \Lambda_2) + (2\Lambda_1 - \Lambda_0 - \Lambda_2) \\ &\quad - \delta[(1-1) + 2(1-0) + 3(1-0)] \\ &= \Lambda_1 + (-\alpha_0 + \delta) + \alpha_1 - 5\delta \\ &= \Lambda_1 - 5\alpha_0 - 3\alpha_1 - 4\alpha_2. \end{aligned}$$

4.2. $U_q(A^{(2)}_{2n-1}), n \geq 3:$
For any positive integer ℓ, take

$$B_\ell = \{(x_1, \dots, x_n, \overline{x}_n, \dots, \overline{x}_1) \in \mathbb{Z}^{2n} \mid x_i, \overline{x}_i \geq 0, \sum_{i=1}^{n} (x_i + \overline{x}_i) = \ell\}$$

For $b = (x_1, \dots, x_n, \overline{x}_n, \dots \overline{x}_1) \in B_\ell$ define

$$\begin{aligned} wt(b) = (\overline{x}_1 - x_1 + \overline{x}_2 - x_2)\Lambda_0 + \sum_{i=1}^{n-1} (x_i - \overline{x}_i + \overline{x}_{i+1} - x_{i+1})\Lambda_i \\ + (x_n - \overline{x}_n)\Lambda_n, \end{aligned} \tag{4.7}$$

$$\begin{aligned} \phi_0(b) = \overline{x}_1 + (\overline{x}_2 - x_2)_+, \phi_i(b) = x_i + (\overline{x}_{i+1} - x_{i+1})_+, \text{ for} \\ i = 1, 2, \dots, n-1,\ \phi_n(b) = x_n, \text{ where } x_+ = \max(x, 0), \end{aligned} \tag{4.8}$$

$$\begin{aligned} \epsilon_0(b) = x_1 + (x_2 - \overline{x}_2)_+,\ \epsilon_i(b) = \overline{x}_i + (x_{i+1} - \overline{x}_{i+1})_+, \text{ for} \\ i = 1, 2, \dots, n-1,\ \ \epsilon_n(b) = \overline{x}_n, \end{aligned} \tag{4.9}$$

$$\tilde{e}_0 b = \begin{cases} (x_1, x_2 - 1, \ldots, \overline{x}_2, \overline{x}_1 + 1) \text{ if } x_2 > \overline{x}_2, \\ x_1 - 1, x_2, \ldots, \overline{x}_2 + 1, \overline{x}_1) \text{ if } x_2 \leq \overline{x}_2, \end{cases}$$

$$\tilde{e}_i b = \begin{cases} (x_1, \ldots, x_i + 1, x_{i+1} - 1, \ldots, \overline{x}_1) \text{ if } x_{i+1} > \overline{x}_{i+1}, \\ (x_1, \ldots, \overline{x}_{i+1} + 1, \overline{x}_i - 1, \ldots, \overline{x}_1) \text{ if } x_{i+1} \leq \overline{x}_{i+1}, \end{cases} \tag{4.10}$$

for $i = 1, 2, \ldots, n-1$, and

$$\tilde{e}_n b = (x_1, \ldots x_n + 1, \overline{x}_n - 1, \ldots, \overline{x}_1),$$

$$\tilde{f}_0 b = \begin{cases} (x_1, x_2 + 1, \ldots, \overline{x}_2, \overline{x}_1 - 1) \text{ if } x_2 \geq \overline{x}_2, \\ (x_1 + 1, x_2, \ldots, \overline{x}_2 - 1, \overline{x}_1) \text{ if } x_2 < \overline{x}_2, \end{cases}$$

$$\tilde{f}_i b = \begin{cases} (x_1, \ldots x_i - 1, x_{i+1} + 1, \ldots, \overline{x}_1) \text{ if } x_{i+1} \geq \overline{x}_{i+1}, \\ (x_1, \ldots, \overline{x}_{i+1} - 1, \overline{x}_i + 1, \ldots, \overline{x}_1) \text{ if } x_{i+1} < \overline{x}_{i+1}, \end{cases} \tag{4.11}$$

for $i = 1, 2, \ldots, n-1$, and

$$\tilde{f}_n b = (x_1, \ldots, x_n - 1, \overline{x}_n + 1, \ldots, \overline{x}_1).$$

Then B_ℓ equipped with the maps $wt, \phi_i, \epsilon_i, \tilde{e}_i, \tilde{f}_i (i \in I)$ given by (4.7) - (4.11) is a perfect crystal of level ℓ for the algebra $U_q(A^{(2)}_{2n-1})$. In this case, we have

$$B_\ell^{\min} = \{(m_1, \ldots, m_n, m_n, \ldots, m_2, \overline{m}_1) \in \mathbb{Z}^{2n} \mid m_i, \overline{m}_1 \geq 0, m_1 + \overline{m}_1 + 2\sum_{i=2}^{n} m_i = \ell\},$$

and for $\lambda = \sum_{i=0}^{n} k_i \Lambda_i \in af(P^+_{cl})_\ell$, we have $\sigma\lambda = k_1\Lambda_0 + k_0\Lambda_1 + \sum_{i=2}^{n} k_i \Lambda_i$. Also note that as crystals for $U_q(C_n)$, the crystal B_ℓ is isomorphic to $B(\ell\Lambda_1)$. Furthermore, for $b = (x_1, \ldots, x_n, \overline{x}_n, \ldots, \overline{x}_1) \in B_\ell$ and $b' = (x'_1, \ldots, x'_n, \overline{x}'_n, \ldots, \overline{x}_1) \in B_\ell$, the energy function $H : B_\ell \otimes B_\ell \longrightarrow \mathbb{Z}$ is given by

$$H(b \otimes b') = \max\Big(\{\theta_j(b \otimes b'),\ \theta'_j(b \otimes b') \mid 1 \leq j \leq n-1\} \cup \{\eta_j(b \otimes b'),\ \eta'_j(b \otimes b') \mid 1 \leq j \leq n\} \Big), \tag{4.12}$$

where

$$\theta_j(b \otimes b') = \sum_{k=1}^{j} (\overline{x}_k - \overline{x}'_k),\ \theta'_j(b \otimes b') = \sum_{k=1}^{j} (x'_k - x_k),$$

$$\eta_j(b \otimes b') = \sum_{k=1}^{j} (\overline{x}_k - \overline{x}'_k) + (\overline{x}'_j - x_j), \text{ and}$$

$$\eta'_j(b \otimes b') = \sum_{k=1}^{j} (x'_k - x_k) + (x_j - \overline{x}'_j).$$

In particular, when $n = 3$, the perfect crystal B_1 of level 1 for $U_q(A_5^{(2)})$ is:

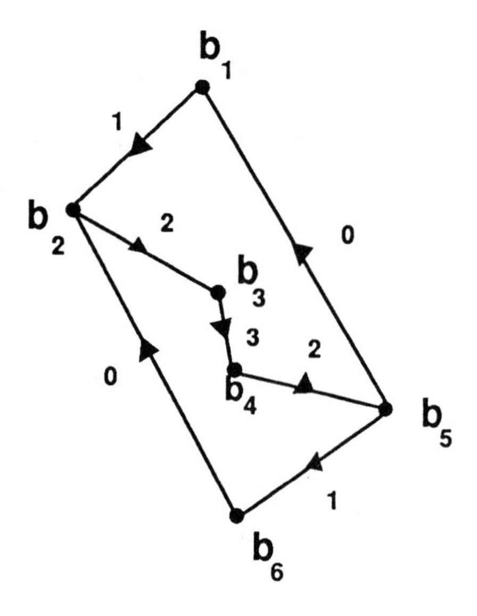

where

$$b_1 = (1,0,0,0,0,0),\ b_2 = (0,1,0,0,0,0),$$
$$b_3 = (0,0,1,0,0,0),\ b_4 = (0,0,0,1,0,0),$$
$$b_5 = (0,0,0,0,1,0),\ b_6 = (0,0,0,0,0,1).$$

In this case, we have only two dominant integral weights Λ_0 and Λ_1 of level 1. The respective ground-state paths are:

$$\Lambda_0\text{-ground-state path} = (b_6, b_1, b_6, b_1, \ldots), \text{ and}$$
$$\Lambda_1\text{-ground-state path} = (b_1, b_6, b_1, b_6, \ldots).$$

For instance, the path realization of the crystal $B(\Lambda_0)$ for the level one $A_5^{(2)}$-module $V(\Lambda_0)$ is:

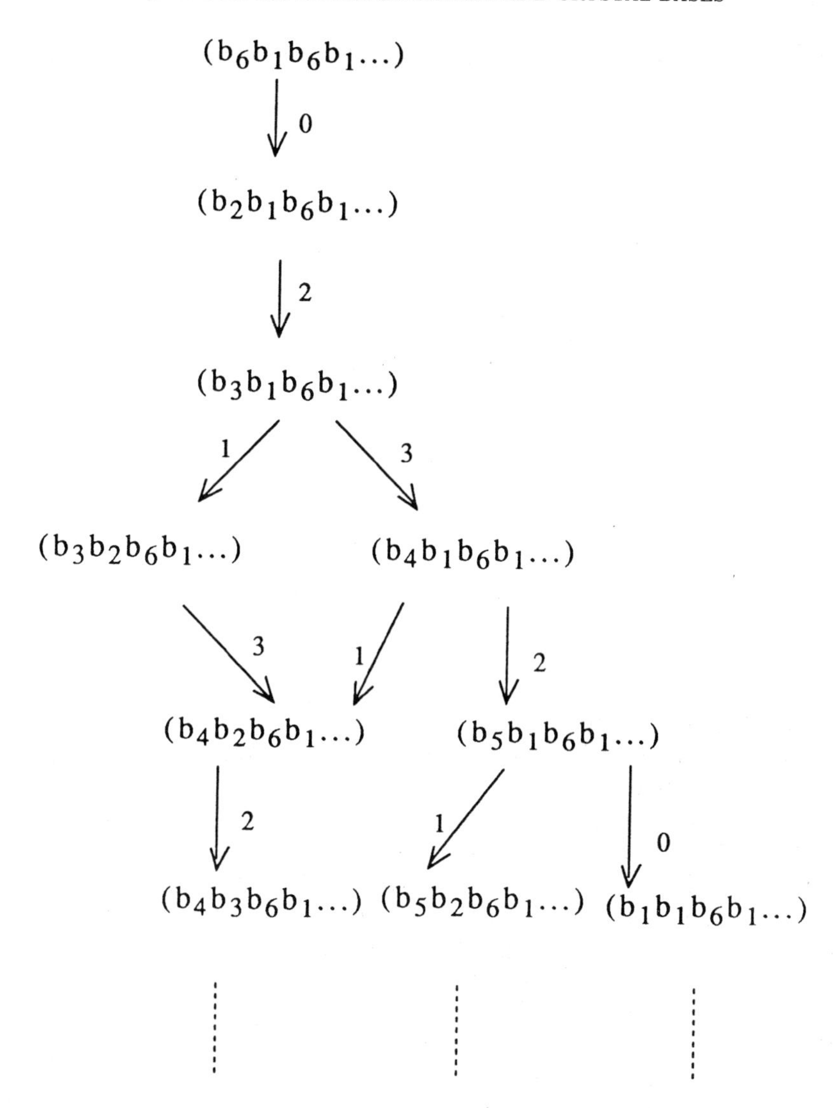

As before we can easily obtain the weight of any path in $B(\Lambda_0)$ by using (3.21) and (4.12). For example let us determine the weight of the path

$$p = (p(k))_{k\geq 1} = (b_1, b_1, b_6, b_1, b_6, b_1, \ldots)$$

in $B(\Lambda_0)$. By using (3.21) and neglecting the zero terms we have

$$\begin{aligned} wt(p) = \Lambda_0 &+ (af(wt\, b_1) - af(wt\, b_6)) \\ &- \delta(H(b_1 \otimes b_1) - H(b_1 \otimes b_6)). \end{aligned}$$

Since $af(wt\, b_1) = \Lambda_1 - \Lambda_0$, $af(wt\, b_6) = \Lambda_0 - \Lambda_1$, and by (4.12),

$$H(b_1 \otimes b_1) = 1, \quad H(b_1 \otimes b_6) = -1,$$

we have

$$\begin{aligned} wt(p) &= \Lambda_0 + (2\Lambda_1 - 2\Lambda_0) - 2\delta \\ &= \Lambda_0 + (2\Lambda_1 - \Lambda_2) - (2\Lambda_0 - \Lambda_2) - 2\delta \\ &= \Lambda_0 + \alpha_1 - (\alpha_0 - \delta) - 2\delta \\ &= \Lambda_0 + \alpha_1 - \alpha_0 - \delta \\ &= \Lambda_0 + \alpha_1 - \alpha_0 - \alpha_0 - \alpha_1 - 2\alpha_2 - \alpha_3 \\ &= \Lambda_0 - 2\alpha_0 - 2\alpha_2 - \alpha_3 \end{aligned}$$

which is also clear from the above weight diagram of the crystal $B(\Lambda_0)$.

References

[1] Drinfeld, V. G.: Hopf algebra and the Yang-Baxter equation. Soviet Math. Dokl. **32**, 254–258 (1985).

[2] Jimbo, M.: A q-difference analogue of $U(\mathfrak{g})$ and the Yang-Baxter equation. Lett. Math. Phys. **10**, 63–69 (1985).

[3] Jimbo, M.: Introduction to the Yang-Baxter equation. Inter. J. Mod. Phys. A, **4**, 3759–3777 (1989).

[4] Jimbo, M., Misra, K.C., Miwa, T., Okado, M.: Combinatorics of representations of $U_q(\hat{sl}(n))$ at $q = 0$. *Commun. Math. Phys.* **136**, 543–566 (1991).

[5] Kac, V. G.: Infinite dimensional Lie algebras. Cambridge Univ. Press. (1990).

[6] Kashiwara, M.: Crystalizing the q-analogue of universal eveloping algebras. *Commun. Math. Phys.* **133**, 249–260 (1990).

[7] Kashiwara, M.: On crystal bases of the q-analogue of universal enveloping algebras. *Duke Math. J.* **63**, 465–516 (1991).

[8] Kang, S.-J., Kashiwara, M., Misra, K.C., Miwa, T., Nakashima, T., Nakayashiki, A.: Vertex models and crystals. *C. R. Acad. Sci. Paris*, **315**, 375–380 (1992).

[9] Kang, S.-J., Kashiwara, M., Misra, K.C., Miwa, T., Nakashima, T., Nakayashiki, A.: Affine crystals and vertex models. *Int. J. Mod. Phys. A.*, **7(1A)**, 449–484 (1992).

[10] Kang, S.-J., Kashiwara, M., Misra, K.C., Miwa, T., Nakashima, T., Nakayashiki, A.: Perfect crystals of quantum affine Lie algebras. *Duke Math. J.*,**68(3)**, 499–607 (1992).

[11] Kang, S.-J., Kashiwara, M., Misra, K.C.: Crystal bases of Verma modules for quantum affine Lie algebras. *RIMS preprint* **887** (1992).

[12] Lusztig, G. I.: Quantum deformation of certain simple modules over enveloping algebras. Adv. in Math. **70**, 237–249 (1988).

[13] Lusztig, G.I.: Canonical bases arising from quantized enveloping algebra. *J. Amer. Math. Soc.* **3**, 447–498 (1990).

[14] Lusztig, G.I.: Canonical bases arising from quantized enveloping algebra II. *Progr. Theor. Phys. Suppl.* **102**, 175–201 (1990).

[15] Misra, K.C., Miwa, T.: Crystal base for the basic representation of $U_q(\hat{sl}(n))$. *Commun. Math. Phys.* **134**, 79–88 (1990).

[16] Rosso, M.: Finite dimensional representations of the quantum analogue of the enveloping algebra of a complex semisimple Lie algebra. *Commun. Math. Phys.* **117**, 581–593 (1988).

North Carolina State University
Mathematics Department
Raleigh, NC 27695-8205

Contemporary Mathematics
Volume **153**, 1993

THE FIRST BETTI NUMBER OF ARITHMETIC GROUPS AND THE CONGRUENCE SUBGROUP PROBLEM

M.S.RAGHUNATHAN AND T.N.VENKATARAMANA

INTRODUCTION

Kazhdan (in [Kazhdan]) showed that there exist arithmetic subgroups Γ of $SU(f)$, where f is an anisotropic Hermitian form over a totally imaginary quadratic extension of a totally real number field K, of Witt index 1 at one archimedean completion of K and zero at all others, such that $H^1(\Gamma; \mathbf{C}) \neq 0$. In [Millson], the corresponding result for orthogonal groups of quadratic forms was established. More recently Li (in [Li]) showed that if Γ is any cocompact congruence subgroup of $SO(n,1), n \geq 6 (n \neq 7)$, then $H^1(\Gamma'; \mathbf{C}) \neq 0$ for a suitable congruence subgroup Γ' of Γ. One of the principal aims of this paper is to establish the following result which, while dependent on Kazhdan's theorem, subsumes all the other non vanishing theorems for H^1 referred to above. Also, our theorem covers the case of $SO(5,1)$ which is not covered in the earlier works above (while this paper was in preparation, we received a preprint of Millson and Li dealing with this case as well). It may also be remarked that unlike the works of Li and Millson and Li the present paper does not use representation theoretic techniques (although it relies on Kazhdan's theorem which was proved by these techniques).

Theorem 1: *Let Δ_o be a cocompact arithmetic congruence subgroup of $SO(n,1)$ (where $n \geq 5$ and $n \neq 7$).Then $H^1(\Delta; \mathbf{C}) \neq 0$ for some congruence subgroup Δ of Δ_o.*

The case $n = 7$ is excluded from the above statement. However in the main body of the paper we cover also a whole class of arithmetic subgroups of $SO(7,1)$; the arithmetic groups that we are unable to handle are the ones arising from outer $K-$forms of a split D_4 which become inner only over a cubic or a sextic extension of the number field K. Our methods also yield additional information contained in the following two theorems.

1991 *Mathematics Subject Classification.* Primary 22E40; Secondary 20H05, 20H10.

Let Δ_o be a cocompact arithmetic congruence subgroup of $SO(2m-1,1)$ with $m \geq 3$ (if $m=4$, the exceptional arithmetic structures on $SO(7,1)$ are excluded). Then Δ_o may be extended in a natural way to a cocompact arithmetic lattice Γ_o in $SU(2m-1,1)$ coming from a hermitian form over a totally imaginary quadratic extension L of K (where K is the number field over which the arithmetic structure on $SO(2m-1,1)$ is defined).

Theorem 2: *Let $G_o = SU(2m-1,1), H' = SO(2m-1,1)$ be as above, with $m \geq 4$. Then the restriction map*

$$H^1(\Gamma_o; \mathbf{C}) \to \prod_{\sigma \in (AutG)(K)} H^1(\sigma(\Gamma_o) \cap H'; \mathbf{C})$$

is injective.

(Here Res : $H^1(\Gamma_o; \mathbf{C}) \to \prod_{\sigma \in (AutG)(K)} H^1(\sigma(\Gamma_o) \cap H'; \mathbf{C})$
refers to the map $\prod_\sigma Res_\sigma$, where

$$\text{Res}_\sigma \ : \ H^1(\Gamma_o; \mathbf{C}) \stackrel{\sigma}{\to} H^1(\sigma(\Gamma_o); \mathbf{C}) \stackrel{restriction}{\to} H^1(\sigma(\Gamma_o) \cap H'; \mathbf{C})$$

is the composite map).
The lattice $\Delta_o \subset H = SO(2m-1,1)$ for $m \geq 3$ may also be extended to a cocompact arithmetic congruence subgroup $\Gamma \subset G_1 = SO(2m+1,1)$. Then the following holds:

Theorem 3. *Let $G_1 = SO(2m+1,1), H' = SO(2m-1,1)$ be as above with $m \geq 3$. Then the restriction map, defined as above,*

$$\text{Res} \ : \ H^1(\Gamma; \mathbf{C}) \to \prod_{\sigma \in (AutG)(K)} H^1(\sigma(\Gamma) \cap H'; \mathbf{C})$$

is injective.

These two theorems are in turn deduced from a result, the main theorem below, which relates the "Congruence subgroup kernel" of H with that of G_o and G_1. We will now assume that the groups under consideration are $SO(n,1)$ with $(n+1)$ *even* and $n \geq 5$. Let Δ_o and H' be as before and G be the simply connected covering group of any one of the two groups G_o and G_1. We think of these groups as algebraic groups. We also take H to be the simply connected covering group of H'. Let $\widehat{G}(a)$ (resp. $\widehat{H}(a)$) denote the arithmetical completion of $G(K)$ (resp. $H(K)$) obtained by declaring arithmetic subgroups of $G(K)$ (resp. $H(K)$) as open. If $\widehat{G}(c)$ (resp. $\widehat{H}(c)$) denotes the congruence completion of $G(K)$ (resp. $H(K)$) we have an exact sequence $1 \to C_G \to \widehat{G}(a) \to \widehat{G}(c) \to 1$ where C_G is the 'congruence subgroup kernel'. Moreover the natural map $f : H \to G$ induces a natural homomorphism

$f_a : \widehat{H}(a) \to \widehat{G}(a)$ which takes C_H into C_G. Also, the group (Aut G)(K) acts on C_G in a natural fashion. We have then the

Main Theorem. *Let H and G be as above. Then, the closed subgroup $C_{G,H}$ generated (as a topological group) by the collection*

$$\{\sigma(f_a(C_H)); \sigma \in (AutG)(K)\}$$

of subgroups, has finite index in C_G.

The structure of the paper is as follows. In § 1, we fix the notation used in the paper. We consider the group $\hat{G}$ defined by the exact sequence

$$1 \to C_{G,H} \to \widehat{G}(a) \to \widehat{G} \to 1.$$

We will show in proposition (1.3) that the finiteness of $C(G,H) = C = C_G/C_{G,H}$ implies the injectivity of the map

$$\text{Res: } H^1(\Gamma; \mathbf{C}) \to \prod_{\sigma\in(AutG)(K)} H^1(\sigma(\Gamma)\cap H; \mathbf{C}).$$

In § 2 we prove the main theorem. We will show that the exact sequence $1 \to C \to \widehat{G} \to \widehat{G}(c) \to 1$ has a natural splitting over $f_c(\widehat{H}(c))$ enabling us to treat $f_c(\widehat{H}(c))$ as a subgroup of $\hat{G}$. This allows us to construct, for each finite prime p of K, certain subgroups $\widehat{G}_p$ of $\widehat{G}$ which contain $f_c(H(K_p))$ where $H(K_p)$ is considered as a subgroup of $\widehat{H}(c)$ and K_p is the completion of K at p. Using the (obvious) fact that $H(K_p)$ and $H(K_q)$ commute in $\widehat{H}(c) \simeq H(A_f)$ for distinct primes p and q, we show, by exploiting the existence of a large number of reflections in (Aut G)(K), that $\widehat{G}_p$ and $\widehat{G}_q$ commute. This proves that the extension $1 \to C \to \widehat{G} \to \widehat{G}(c) \to 1$ is central and by [Raghunathan], finite. Now the main theorem follows.

1. Notation and Conventions

(1.1). Let K be a numberfield and G a connected absolutely almost simple algebraic group defined over K. We denote the automorphism group of G by Aut G. Let V_∞ (resp V_f) denote the set of equivalence classes of archimedean (resp. non-archimedean) valuations of K. If $p \in V_\infty \cup V_f, K_p$ denotes the completion of K with respect to p. Let A_f denote the ring of finite adeles of K. A subgroup $\Gamma \subset G(K)$ is called a *congruence subgroup* if there is an open compact subgroup Ω of $G(A_f)$ such that $\Gamma = G(K) \cap \Omega$ (for the diagonal inclusion of $G(K)$ in $G(A_f)$). A subgroup Γ of $G(K)$ is *arithmetic* if there is a congruence subgroup $\Gamma' \supset \Gamma$ such that Γ'/Γ is finite. The collection of all congruence (resp. arithmetic) subgroups is a fundamental system of neighbourhoods of the identity for a unique topological group structure on $G(K)$. For either of these topological group structures, it is easily checked that Cauchy sequences for the right and the left uniform structures coincide so that we can identify the completions with respect to these two uniform structures with each other. We denote the completion with respect to the topology defined by congruence (resp. arithmetic) subgroups $\widehat{G}(c)$ (resp.$\widehat{G}(a)$). Then $\widehat{G}(c)$ and $\widehat{G}(a)$ are locally compact groups and one has a natural continuous map

$$\pi_G : \widehat{G}(a) \to \widehat{G}(c)$$

whose kernel we denote by $C(G)$. It is well known and easily proved that $C(G)$ is compact and totally disconnected and that π_G is surjective. The group $\widehat{G}(c)$, as is easily seen, is isomorphic to the closure of $G(K)$ in the adéle group $G(A_f)$. If G is *simply connected* and $G(K_p)$ is *not compact* for some $p \in V_\infty$ then $G(K)$ is dense in $G(A_f)$.

(1.2). Evidently $G \mapsto C(G)$ is a functor on the category of $K-$algebraic groups into the category of compact totally disconnected groups. Thus if $f : H \to G$ is a $K-$ morphism. One has a continuous homomorphism

$$C(f) : C(H) \to C(G).$$

We denote by $C(f, G, H)$ or simply $C(G, H)$ if there is no ambiguity about f, the quotient of $C(G)$ by the closed subgroup generated by $\{\sigma \text{ (Image } C(f)) \mid \sigma \in \text{(Aut } G)(K)\}-$ note that Aut $G(K)$ preserves the two topologies on $G(K)$ and hence induces automorphisms of $\widehat{G}(a)$ and $\widehat{G}(c)$ compatible with π_G. We establish below the *finiteness* of $C(G, H)$ in the specific situations already described briefly in the introduction. But before we proceed to the precise formulation of the main result, we prove the following :

(1.3) Proposition. *If G and H are $K-$ groups, with G semisimple, simply connected and f a morphism of H in G such that $C(f, G, H)$ is finite, then for any congruence group Γ, the natural map*

$$H^1(\Gamma; \mathbf{C}) \to \prod_{\sigma \in (AutG)(K)} H^1(f^{-1}\sigma^{-1}(\Gamma) \cap H(K); \mathbf{C})$$

is injective. (here $H^1(\Gamma; \mathbf{C}) \to H^1(f^{-1}\sigma^{-1}(\Gamma) \cap H(K); \mathbf{C})$ is the homomorphism induced by the map $\gamma \mapsto \sigma(F(\gamma), \gamma \in f^{-1}\sigma^{-1}(\Gamma) \cap H(K))$.

Proof. Let $w : \Gamma \to \mathbf{C}$ be any homomorphism of Γ in $\mathbf{C}$ such that $w(\sigma f(\gamma)) = 0$ for all $\sigma \in$ (Aut $G)(K)$ and $\gamma \in f^{-1}\sigma^{-1}(\Gamma) \cap H(K)$. Since Γ is finitely generated, $w(\Gamma) \simeq \mathbf{Z}^\ell$ for some ℓ. We then obtain a continuous homomorphism $\widehat{w} : \widehat{\Gamma} \to \widehat{\mathbf{Z}^\ell}$ where $\widehat{\Gamma}$ (resp. $\widehat{\mathbf{Z}}$) is the profinite completion of Γ (resp. $\mathbf{Z}$). Now $\widehat{\Gamma}$ has a natural identification with the closure of Γ in $\widehat{G}(a)$ and, as is easily seen, $\widehat{\Gamma}$ under this identification contains $C(G)$. Now since w is assumed to be trivial on $f^{-1}\sigma^{-1}(\Gamma) \cap H(K)$, we see that $\widehat{w}(f^{-1}\sigma^{-1}(\Gamma) \cap H(K)) = \{0\}$, leading to the conclusion that

$\widehat{w}(\sigma($ Image $C(f))) = \{0\}$ for all $\sigma \in (\text{Aut } G)(K)$. Since $C(f,G,H)$ is finite by assumption, and $\widehat{\mathbf{Z}^\ell}$ is torsion free, we see that $\widehat{w}$ factors through $\widehat{\Gamma}/C(G)$ and this may be identified with the closure $\overline{\Gamma}$ of Γ in $\widehat{G}(c)$ or equivalently in $G(A_f)$. Now the commutator subgroup of $\overline{\Gamma}$ has finite index in $\overline{\Gamma}(\overline{\Gamma}$ is an open compact subgroup of $G(A_f))$, so that $\overline{\Gamma}^{ab}$ is finite. Since $\widehat{\mathbf{Z}}$ has no torsion, we conclude that $\widehat{w}$ is zero. Hence the proposition.

(1.4). From now K will be assumed to be *totally real*. Let V_∞ (resp. V_f) denote a complete set of inequivalent archimedean (resp. non archimedean) valuations of K. For $p \in V_\infty \cup V_f$, we denote by K_p the completion of K with respect to p; the topology on K_p or K determined by p will be referred to as the $p-$adic topology. Let D be a central simple algebra over K such that $D \otimes_K K_p = M_2(K_p)$ for all $p \in V_\infty$ (so that D has degree 2 over K). Let tr_{red} be the reduced trace on D and $\sigma : D \to D$ the standard involution defined by $\sigma(x) = tr_{red}(x) - x$, for all $x \in D$. We fix a free right D-module W of rank ≥ 3 and a $\sigma-$ skew Hermitian form $h : W \times W \to D$ on W. We assume that h is non-degenerate and that $h(x,x) \neq 0$ for $x \neq 0$ in W. We denote the unitary group (resp. the special unitary group) $U(h)$ (resp. $SU(h)$) by H^* (resp. H') in the sequel. The group H' is a connected semisimple K-algebraic group of index 2 in the reductive K-group H^*. If E is an extension field of K we continue to denote by h the natural extension of h to $W_E \times W_E$ where $W_E = W \otimes_K E$. When $E = K_p$ we denote W_E also by W_p (here $p \in V_f \cup V_\infty$). We will also assume in the sequel that there is a unique valuation denoted ∞ in V_∞ such that H' has K_∞-rank 1 while K_p-rank $H' = 0$ for $p \in V_\infty \setminus \{\infty\}$. Our assumptions on h guarantee that H' is anisotropic over K. We denote by H the simply connected covering group of H' and by $\pi : H \to H'$ the covering projection. As is well known, kernel (π) is a group of order 2. It is also a known fact that under our assumptions on D and h, H^* is naturally isomorphic to the orthogonal group of a quadratic form over K_p for all $p \in V_\infty$. Moreover, over K_p for $p \neq \infty, p \in V_\infty$, the quadratic form is positive or negative definite while over K_∞ it is of signature $(2m-1,1)$ or $(1,2m-1)$ (note that since K is totally real, $K_p \simeq \mathbf{R}$ for all $p \in V_\infty$). Let det: $GL_D(W) \to K^*$ denote the Dieudonné determinant. Then $H' = \{x \in H^*, \det x = 1\}$. Our assumption that $dim_D(W) \geq 3$ guarantees that there is an element $g \in H^*(K)$ with det $g = -1$ so that det yields an isomorphism of $H^*(K)/H'(K)$ with the group $\{\pm 1\}$.

(1.5). We are interested in two kinds of simply connected $K-$ algebraic groups that contain H' or H as a $K-$ subgroup. Groups of either kind will be denoted G in the sequel. In order to treat both kinds simultaneously we will adopt certain notational conventions which will be explained below. Thus we will introduce a K-vector space V which will stand for two different kinds of vector spaces depending on the kind of G we are dealing with. The descriptions of the two kinds of groups are separated into two cases referred to as Case 1 and Case 2 in the sequel.

Case 1. Let V be the free right D module $V = W \oplus D$.
Let $\lambda \in D^\sigma = \{\alpha \in D; \sigma(\alpha) = -\alpha\}$ and $\underline{\lambda} : D \times D \to D$ the skew-Hermitian form on D defined by

$$\underline{\lambda}(\alpha, \beta) = \overline{\alpha}\lambda\beta \text{ for } \alpha, \beta \in D.$$

We assume that the unitary group $U(\underline{\lambda})$ of $\underline{\lambda}$ is anisotropic over K_p for all $p \in V_\infty$ (hence also over K). Let $h_V : V \times V \to D$ denote the skew hermitian form on V defined by

$$h_V((x, \alpha), (y, \beta)) = h(x, y) + \overline{\alpha}\lambda\beta$$

where $(x, y) \in W \times W$ and $(\alpha, \beta) \in D \times D-$ in other words h_V is the orthogonal direct sum $h \perp \underline{\lambda}$ of h and $\underline{\lambda}$. We denote by G^* (resp. G') the unitary group of h_V (resp. the special unitary group of h_V). Then G^* is semisimple and G' is the connected component of the identity in G^*; also G' has index 2 in G^*. We denote by G the simply connected covering group of G'. We assume that $\underline{\lambda}$ is so chosen that G' has K_p-rank 1 if $p = \infty$ and K_p-rank 0 for $p \in V_\infty \setminus \{\infty\}$: such a choice is easily seen to exist. Once again, as with H^*, G^* is isomorphic over $K_p, p \in V_\infty$, to the orthogonal group of a quadratic form q_p; further q_p is definite if $p \neq \infty$ while q_∞ has signature $(2m+1, 1)$ or $(1, 2m+1)$. One has clearly the commutative diagram below:

$$\begin{array}{ccccc} H & \xrightarrow{\underline{\pi}} & H' & \overset{i}{\hookrightarrow} & H^* \\ \downarrow j & & \downarrow j' & & \downarrow j^* \\ G & \xrightarrow{\underline{\pi}} & G' & \overset{\underline{i}}{\hookrightarrow} & G^* \end{array}$$

where $i, \underline{i}$ are the natural inclusions and $\pi, \underline{\pi}$ are the covering projections while j' (resp. j^*) is the inclusion which associates to each $h \in H'$ (resp. H^*) the unique element $j'(h)$ (resp. $j^*(h)$) which fixes the direct summand D in V pointwise and whose restriction to (the D-submodule) W (of V) equals h; also j is the morphism induced by j'.
Case 2. In this case we consider only W of D-rank ≥ 4. We fix a quadratic extension L of K so chosen that L splits D and $L \otimes_K K_p \simeq \mathbf{C}$ for all $p \in V_\infty$. Such a choice of L is possible as is easily seen. Let τ denote Galois conjugation in L fixing K. Let $D_L = D \otimes_K L (\simeq M_2(L))$ and
let $\tilde{\sigma} : D_L \to D_L$ be the involution of the second kind defined by $\tilde{\sigma}(\alpha \otimes \ell) = \sigma(\alpha) \otimes \tau(\ell)$ for $\alpha \in D, \ell \in L$. We now set $V = W \otimes_D D_L$ and denote by h_V the $\tilde{\sigma}-$ skew Hermitian form defined by setting

$$h_V(x \otimes \ell, x' \otimes \ell') = \tilde{\sigma}(\ell) h(x, x') \ell'$$

for $x, x' \in W$ and $\ell, \ell' \in D_L$. We now set G^* = the unitary group of h_V and take $G = G'$ = special unitary group of h_V (unlike in case 1, the special unitary group of h_V is simply connected which explains why we have set $G = G'$). Over $K_p \simeq \mathbf{R}, p \in V_\infty, G^*$ (resp G) is isomorphic to the unitary (resp. special unitary) group of a hermitian form in $2m$ variables which is definite if $p \neq \infty$ while it has signature $(2m-1, 1)$ or $(1, 2m-1)$ over K_∞.

We see thus that V, G, G' and G^* stand for different things in the two cases. We will also adopt the convention that when we are in case 1, L is taken to be K and D_L is taken to be D. This convention will enable us to treat both cases simultaneously.

With these notational conventions, we see that on $C(j, G, H)$ there is a natural action of $G(K)$. We can now formulate our main result precisely.

(1.6) Theorem: *$C(j, G, H)$ is a finite abelian group on which $G(K)$ acts trivially.*

(1.7). We introduce additional definitions and notation. We emphasise again that the same notation will mean different things depending on the case one is dealing with. A D- submodule M of V is *regular* if it satisfies the following conditions: (i) M is $D-$ free (ii) $h_V(M \times M) \subset D$ and (iii) h_V restricted to M - denoted h_M in the sequel - is nondegenerate (note that (ii) holds for *any* D- submodule in case 1 but not so in case 2). A *regular* D-submodule M of V is *admissible* if there is an element $g \in G^*(K)$ such that $g^{-1}(M) \subset W$. An element x is *regular* (resp. *admissible*) if the module xD is regular (resp. admissible). Note that if x is regular, $h(x, x)$ is invertible in D. Moreover, since dim $W \geq 3$, it follows from the Hasse principle and Witt's theorem that every regular element $x \in V$ is admissible. For $y \in V$, let $y^\perp = \{z \in V, h_v(y, z) = 0\}$ and $r(y) = \dim_K(y^\perp \cap W)$. Let $r = \inf\{r(y); y \in V\}(r = 4(m-1)$ if we are in case 1 while in case 2, $r = 4(m-2))$. An element $x \in V$ is *W-non-singular* if $r(x) = r$. Note that any regular element in $V \setminus \Delta$ is W-nonsingular in case 1. In case 2 however, a necessary and sufficient condition for $x \in V$ to be W-nonsingular is that $W \cap (xD_L + (1 \otimes \tau)(x)D_L) = M$ be D-free of rank 2 and h_M be nondegenerate. For a regular D-submodule M of V, we denote by $H^*(M)$ (resp. $H'(M)$) the unitary group (resp. special unitary group) of $h_M = h_{V|M}$. Also $H(M)$ will denote the 2-sheeted "spin" covering group of $H'(M)$. The inclusion of M in V induces an inclusion of $H^*(M)$ (resp. $H'(M)$)) in G^* (resp. G') denoted j_M^* (resp. j'_M) in the sequel. For $g \in H^*(M), j_M^*(g)$ is the extension of g to V obtained by setting it equal to identity on $M^\perp = \{x \in V; h_V(x, m) = 0$ for all $m \in M\}$. We denote by π_M the covering projection of $H(M)$ on $H'(M)$ and by j_M the K-morphism of $H(M)$ in G induced by j'_M.

(1.8) For each admissible M we have a homomorphism $\widehat{j_M}(c) : \widehat{H}(M)(c) \to \widehat{G}(c)$ inducing a homomorphism $C(j_M) : C(H(M)) \to C(G)$. If M, M' are two admissible D-submodules of V and $g \in G^*(K)$ is such that $g(M) = M'$, the inner conjugation Int (g) on G^* induces an automorphism of G' and hence also G; this automorphism carries the subgroup $H'(M)$ isomorphically onto $H'(M')$ and hence also induces a K-isomorphism of $H(M)$ on $H(M')$. We denote all these automorpisms and isomorphisms by $\mathbf{g}$: evidently the diagrams below are commutative:

$$\begin{array}{ccccccc} H(M) & \to & H'(M) & \to & G' & \to & G^* \\ \downarrow \mathbf{g} & & \downarrow \mathbf{g} & & \downarrow \mathbf{g} & & \downarrow \mathrm{Int}(g) \\ H(M') & \to & H'(M') & \to & G' & \to & G^* \end{array} \qquad \begin{array}{ccc} H(M) & \to & G \\ \downarrow \mathbf{g} & & \downarrow \mathbf{g} \\ H(M') & \to & G \end{array}$$

It follows that $\widehat{j_{M'}}(C(H(M')) = \widehat{\mathbf{g}} j_M(C(H(M)))$. This means that for *any admissible* $M \subset V$ the composite map $\widehat{H}(M)(a) \to \widehat{G}(a) \to \widehat{G}$, where $\widehat{G}$ is the quotient of $\widehat{G}(a)$ by the closed normal subgroup generated by the $\widehat{u}(\widehat{j}(C(H))), u \in$ (Aut $G)(K)$, contains in its kernel the group $C(H(M))$ and hence factors through $\widehat{H}(M)(c)$ to yield a continuous homomorphism $\varphi_M : \ \widehat{H}(M)(c) \to \widehat{G}$. The image $\varphi_M(\widehat{H}(M)(c))$ is easily seen to be a closed subgroup of $\widehat{G}$ and will be denoted $\widetilde{H}(M)$, in the sequel (Note that we have defined $\widetilde{H}(M)$ only for admissible M). If $\mathrm{rank}_D(M) \geq 2$, the group $H(M)$ is *semisimple* and *simply connected.* It follows that strong approximation property holds for $H(M)$ ([Platonov]). Consequently, if M is, in addition, noncompact at $\infty, \widehat{H}(M)(c)$ is naturally isomorphic to the adéle group $H(M)(A_f)$ of $H(M)$. For each $p \in V_f, H(M)(K_p)$ has a natural identification as a subgroup of $H(M)(A_f)$ and we denote its image under φ_M by $\widetilde{H}(M)(K_p)$: $\widetilde{H}(M)(K_p)$ is defined only for M admissible, noncompact at ∞ and of D-rank ≥ 2. We will now extend the definition of $\widetilde{H}(M)$ and of $\widetilde{H}(M)(K_p)(p \in V_f)$ to admissible ∞-noncompact D-rank 1 modules as well.

(1.9) Proposition-Definition. *Let E be an admissible D-rank 1 submodule of V which is noncompact at ∞. Let M be any admissible submodule of V of D-rank ≥ 2 containing E and $j_{E,M}$ the inclusion of E in M. Then we have a natural map $j_{E,M,\mathbf{A}_f} : H(E)(\mathbf{A}_f) \to H(M)(\mathbf{A}_f)$. Then the images $\widetilde{H}(E) = \varphi_M \circ j_{E,M,\mathbf{A}_f}(H(E)(\mathbf{A}_f))$ and $\widetilde{H}(E)(K_p) = \varphi_M \circ j_{E,M,\mathbf{A}_f}(H(E)(K_p))$ depend only on the module E and not on the choice of M.*

Proof. Note that since $M \supset E, M$ is noncompact so that φ_M is defined. Let $g \in G^*(K)$ be an element such that $g^{-1}(M) \subset W$ and let $g(W) = W'$. Evidently, W' is admissible and one has $E \subset M \subset W'$. It is also clear that $\varphi_{W'}$ restricted to $H(M)(A_f)$ coincides with φ_M. Thus we may assume, without loss of generality, that $M = g(W)$ for some $g \in G^*(K)$. Since the diagram

$$\begin{array}{ccc} H(W)(A_f) & \overset{\varphi_M}{\to} & \widehat{G} \\ \downarrow \mathbf{g} & & \downarrow \widehat{\mathbf{g}} \\ H(M)(A_f) & \overset{\varphi_M}{\to} & \widehat{G} \end{array}$$

is commutative, one has

$$\varphi_M \circ j_{E,M,A_f}(H(E)(A_f)) = \widehat{\mathbf{g}} \circ \varphi_W \circ j_{E,W,A_f}(H(E)(A_f))$$

where $F = g^{-1}(E)$. It follows now that we need only prove the following: let $g \in G^*(K)$ be such that $g(x) = x$ for all $x \in F$; then $\widehat{\mathbf{g}} \circ \varphi_W \circ j_{F,W,A_f}(u) =$

$\varphi_W \circ j_{F,W,A_f}(u)$ for all $u \in H(F)(A_f)$. This last assertion is proved as follows. Let g' denote the restriction of g to $F^\perp = \{y \in V; h_V(y,x) = 0 \text{ for all } x \in F\}$. Then as is well known, g' can be expressed as a product $r'_1 \cdots r'_t$ where for $1 \leq i \leq t, r'_t$ is an element of the unitary group of $h_{V|F'}$ which *fixes pointwise a free* D_L*- submodule* Z_i of $F^\perp$ of *co-rank 1.* Let r_i be the element of $G^*(K)$ defined by setting $r_i(x) = x$ if $x \in F$ while $r_i(y) = r'_i(y)$ for $y \in F^\perp$. Then the D-submodule $Y_i \stackrel{\text{def}}{=} \{y \in V; r_i(y) = y\} \cap W$ is *admissible* of *D-rank* ≥ 2 and *is non-compact at* ∞ since $Y_i \supset F$. Clearly r_i acts as the identity on $H(Y_i)(K)$ and hence the induced automorphism $\widehat{r_i}$ acts as identity on the closure of the image of $H(Y_i)(K)$ in $\widehat{G}$. Since Y_i is noncompact at ∞, the closure of the image of $H(Y_i)(K)$ contains the image under φ_{Y_i} of $H(Y_i)(\mathbf{A}_f)$ (strong approximation). Also, for $u \in H(F)(\mathbf{A}_f)$, $\varphi_W \circ j_{F,W,A_f}(u) = \varphi_{Y_i} \circ j_{F,Y_i,A_f}(u)$ for all $i, 1 \leq i \leq t$. Thus $\widehat{r_i}$ is identity on the image of $H(F)(\mathbf{A}_f)$ under $\varphi_W \circ j_{F,W,A_f}$, and since this holds for all $i, 1 \leq i \leq t$ and $\widehat{g} = \widehat{r_1} \circ \cdots \circ \widehat{r_t}$, the proposition is proved.

2. Proof of the Main Theorem

The following proposition holds the key to the proof of the main theorem.

(2.1) Proposition: *In the notation of (1.9), let* $\widehat{G}(p)$ *denote the closed subgroup of* $\widehat{G}$ *generated by* $\{g\widetilde{H}(W)(K_p)g^{-1} : g \in \widehat{G}\}$. *Let* T *be the finite set of valuations in* V_f *at which* H *remains anisotropic (Note that* $T = \phi$ *if dim* $(W) \geq 4$ *- in particular if we are in case 2). Then for any* $p \in V_f$ *and any* $q \in V_f \backslash T$ *with* $p \neq q, \widehat{G}(p)$ *and* $\widehat{G}(q)$ *commute.*

(2.2). We will now deduce theorem (1.6) from Proposition (2.1). (The proof of the proposition will be dealt with at the end). To prove Theorem (1.6) we first show that the natural map $\widehat{G} \to \widehat{G}(c) = G(A_f)$ is a central extension. For this, fix any finite subset $S \subset V_f$ containing T and consider the closed subgroup $\widehat{G}^S$ generated by the $\widehat{G}(p), p \notin S$; then since the $\widehat{G}(p)$ map into $G(K_p)$ in $G(\mathbf{A}_f)$ for all $p \in V_f$, we see that the composite map $\widehat{G} \to \widehat{G}(c) \to \prod_{p \in S} G(K_p)$ factors through $\widehat{G}/\widehat{G}^S$ giving a continuous homomorphism $\varphi_S : \widehat{G}/\widehat{G}^S \to \prod_{p \in S} G(K_p)$ with a compact kernel. Also the inclusion of $G(K)$ in $\widehat{G}$ leads to compatible inclusions of $G(K)$ in the two groups above. If we denote by $B(p)$ a maximal compact subgroup of $G(K_p)$, then $\Gamma = \varphi_S^{-1}(\prod_{p \in S} B(p)) \cap G(K)$ is an S-congruence subgroup that maps isomorphically onto its image. Also Γ is dense in $\widetilde{B} = \varphi_S^{-1}(\prod_{p \in S} B(p))$ as the latter is an open subgroup of $\widehat{G}$. We conclude thus that $\widetilde{B}$ is a quotient of the profinite completion of Γ. But according to [Sury], $\widetilde{B} \to \prod_{p\in S} B(p)$ is necessarily an isomorphism. It follows that if x is in the kernel $C(G,H)$ of the map $\widehat{G} \to \widehat{G}(c)$, x belongs to the closure of the group generated by $\widehat{G}(p), p \notin S$. It follows from Proposition (2.1) that x commutes with $\widehat{G}(p)$ for $p \in S$; and since S was arbitrary, we see that x commutes with $\widehat{G}(p)$ for *every* $p \in V_f$. But the closed subgroup $\widehat{G'}$ generated by the $\widehat{G}(p), p \in V_f$ is all of $\widehat{G}$-this is seen from the fact that $\widehat{G'}$ contains the *infinite* group $\widetilde{H}(W) \cap G(K)$ and since $\widehat{G'} \cap G(K)$ is *normal* in $G(K)$ we conclude from the projective simplicity of $G(K)$ ([Tomanov]) that $\widehat{G'} \cap G(K) = G(K)$. Thus $G(K)$ and hence $\widehat{G}$ centralises x. We see therefore that $C(G,H)$ is central in $\widehat{G}$. On the other hand, since the central extension $\widehat{G} \to \widehat{G}(c)$ *splits* over $G(K)$, it follows from [Raghunathan] that $C(G,H)$ is *finite.*

(2.2) Lemma: *Let $\Omega = \{g \in G^*(K_\infty) \mid$ there exist ∞-split $x, y \in W$ such that (1) they are in the same G^*-orbit (2) $(g(x)-y)$ is W-admissible and (3) $H((gx)-y)^\perp \cap W)$ has K_∞-rank > 0 (hence K_∞-rank $= 1)\}$. Let Ω^{-1} be $\{g \in G^*(K_\infty) \mid g^{-1} \in \Omega\}$. Then $\Psi = \Omega \cap \Omega^{-1}$ is a non-empty ∞-adic open subset of $G^*(K_\infty)$. Further, $\Psi \cap G(K)$ generates $G(K)$ as a group.*

Proof. Let $x_0 \in W$ be an admissible ∞-split vector and let $E = (x_0D)^\perp \cap W$. Let $y_0 \in E$ and $z_0 \in (y_0D)^\perp \cap W$ be admissible vectors. Then for any $\epsilon \in K^*$ sufficiently close to 0 in K_∞, $x_0 + y_0\epsilon + z_0\epsilon$ is admissible and ∞-split. Let $g \in G^*(K)$ be an automorphism defined by setting $g(x_0) = x_0, g(y_0) = -y_0$ and $g \mid_{(x_0D+y_0D)^\perp}$ is a K-rational elements ρ of the unitary group U of $h_V \mid_{(x_0D+y_0D)^\perp}$ with the property that $\rho(z_0) \notin W$ -such an element ρ exists since U acts irreducibly on $(x_0D+y_0D)^\perp$ even when this last space is treated as a D-module (note that in case 2 it is a D_L-module, $D_L \neq D$). One then has $g(x_0+y_0\epsilon+z_0\epsilon) - (x_0+y_0\epsilon+z_0\epsilon) = 2y_0\epsilon + (\rho(z_0) - z_0)\epsilon$; and since $\rho(z_0) - z_0 \notin W$ and is in $(y_0D)^\perp$ one sees that $2y_0 + (\rho(z_0) - z_0)$ is W-nonsingular. If ϵ is sufficiently small in K_∞, $x = x_0+y_0\epsilon+z_0\epsilon$ is admissible and ∞-split. Taking $y = x$ with ϵ sufficiently small, it is clear that $gx - y = gx - x$ has x in its orthogonal complement. Thus $g \in \Omega$ and if we assume that $\rho^{-1}(z_0) \notin W$, as we may, it is clear that $g \in \Omega^{-1}$ and so $\Psi = \Omega \cap \Omega^{-1}$ is nonempty. That Ω, Ω^{-1} and Ψ are open in the ∞-adic topology is clear from their definition. This proves the first part. Since Ψ is open, it is immediate that if $\mathcal{G}$ is the connected component of the identity in $G^*(K_\infty)$, $\Psi \cap \mathcal{G}$ generates $\mathcal{G}(K)$. Now if we are in case 2, $\mathcal{G} = G^*(K_\infty)$ and the proof is finished. In case 1, $\mathcal{G}$ is a subgroup of index 2 in $G^*(K)$ and we need only show that $\Psi \cap G^*(K)$ contains an element of the connected component other than $\mathcal{G}$. Now as is easily seen, Ω is stable under left and right multiplication by $H'(K)$ so that the same holds for the subgroup generated by $\Psi \cap G(K)$ in $G(K)$. Hence it sufficies to show that $H'(K)$ meets every connected component of $G(K_\infty)$. This is ensured by our assumption that $\mathrm{rank}_D(W) \geq 3$ (we refer to the last part of (1.4)).

(2.3) Proof of Proposition (2.1). We will prove by induction on n the following statement which is stronger than proposition (2.1).
Let $\Psi(1) = \Psi \cap G(K)$ and $\Psi(n) = \Psi(1)\Psi(n-1)$. Then given any ∞-split admissible vector $w \in W$ and any $g \in \Psi(n)$, there is a neighbourhood $U(g)$ of identity in $G(K)$ for the ∞-adic topology such that for all $p, q \in V_f$ with $q \notin T, u \in U(g)$, we have, $\widetilde{H}(gwD)(K_p)$ commutes with $\widehat{u}(\widetilde{H}(W)(K_q))$. Assume that the assertion is proved for all $g \in \Psi(n-1)(n-1 \geq 1)$. We will then show that it holds for all $g \in \Psi(n)$. Let then $g \in \Psi(n)$ so that $g = g_1 g'$, with $g_1 \in \Psi(1)$ and $g' \in \Psi(n-1)$. By induction hypothesis, $\widetilde{H}(g'wD)(K_p) = \widehat{g'}(\widetilde{H}(wD)(K_p))$ commutes with $\widehat{g}(\widetilde{H}(W)(K_q))$ for all $u \in U(g')$. Then $\widetilde{H}(gwD)(K_p) = \widehat{g_1}(\widetilde{H}(g'w)(K_p))$ commutes with $\widehat{g_1}\widehat{u}(\widetilde{H}(W)(K_q))$ for all $u \in U(g')$. Replacing $U(g')$ by $U_1 = g_1 U(g') g_1^{-1}$, one sees that $\widehat{g_1}(\widetilde{H}(g'wD)(K_p))$ commutes with $\widehat{u}\widehat{g_1}(\widetilde{H}(W)(K_q))$ for all $u \in U_1$. Let $x, y \in W$ be admissible ∞-split vectors in the same $H^*(K)$-orbit such that $(g_1(x)-y)^{\perp} \cap W = W'$ is free, of minimal possible rank over D ($\mathrm{rank}_D(W') = m-1$ in case 1 and $m-2$ in case 2), with $H(W')$ split at ∞. Replacing x and y by close approximations in the ∞-adic topology which also approximate respectively $w \in W_p$ and a q-split vector in W_q, we may assume that x is p-adically as close as we want to w while y is q-split. Let $W'' = g_1(x)D + W'$; then W'' is admissible. Indeed let r be a reflection taking $g_1(x)$ to y and fixing $(g_1(x)-y)^{\perp}$. Then r acts trivially on $W' \subset (g_1(x)-y)^{\perp}$. Therefore $r(W'') = yD + W' \subset W$. Let U_2 be an ∞-adic neighbourhood of 1 in $G^*(K)$ such that $U_2^{-1}.U_2 \subset U_1$. Now for all $u \in U_2$ and $u' \in H(W'')(K) \cap U_2$, $\widehat{u}(\widetilde{H}(gxD)(K_p))$ commutes with $\widehat{u'}(\widetilde{H}(g_1(x)D)(K_q))$. But the group B generated by $\{\widehat{u'}(\widetilde{H}(g_1xD)(K_q)), u' \in U_2 \cap H(W'')(K)\}$ in the closure of $i_{W''}(H(W'')(K))$ is all of $\widetilde{H}(W'')(K_q)$ - this follows from the fact that B is *open* and *non-compact*. We see thus that $\widehat{u}(\widetilde{H}(gxD)(K_p))$ commutes with $\widetilde{H}(W''(K_q)$ and hence with $\widetilde{H}(W')(K_q)$. Let U_3 be an ∞-adic neighbourhood of 1 in $G^*(K)$ such that $U_3^{-1}.U_3 \subset U_2$. Then for all $v \in U_3$ we have $\widehat{v}(\widetilde{H}(gxD)(K_q))$ commutes with $\widehat{h}(\widetilde{H}(W')(K_q))$ for all $h \in \widetilde{H}(W)(K) \cap U_3$. But h can approximate any element in $\widetilde{H}(W)(K_q)$ and therefore $\widehat{v}(\widetilde{H}(gxD)(K_p))$ commutes with $\widetilde{H}(W)(K_q)$ (since $\widetilde{H}(W)(K_q)$ is generated $\widehat{h}(\widetilde{H}(W')(K_q)); h \in H(K) \cap U_3$) for all $v \in U_3$. Since x can be assumed to be as close as we want to w p-adically, in the limit $\widetilde{H}(gwD)(K_q)$ commutes with $\widehat{v^{-1}}(\widetilde{H}(W)(K_q))$ for all $v^{-1} \in U_3^{-1} = U(g)$. This leaves us to show that we can start the induction at $n = 1$. Since $\Psi \cap G^*(K)$ is open in the ∞-adic topology, it clearly suffices to show that for $g \in \Psi$ and $w \in W, \widetilde{H}(gwD)(K_q)$ commutes with $\widetilde{H}(W)(K_q)$. Choose $x, y \in W$ such that they are in the same $H^*(K)$-orbit, $g(x) - y$ is W-admissible and $W' = (g(x)-y)^{\perp} \cap W$ is split at ∞. Once again as before we may assume in addition that y is q-split and that x is as close an approximation of w p-adically as we want. Now there is an ∞-adic neighbourhood U of g in $G^*(K)$ such that all the above holds for all $u.g$ with $u \in U$ as well. Consider now the D-module $W'_u = (ugx - y)^{\perp} \cap W$ with $u \in U$. Let

$W_u'' = ugxD + W_u'$. Then $\widetilde{H}(W_u'')(K_p)$ commutes with $\widetilde{H}(W_u'')(K_q)$ and hence with $\widetilde{H}(W_u')(K_q)$. Let $z \in W$ be a q-split admissible vector with $z^\perp \cap W \neq W_u'$ and such that $z^\perp \cap W_u''$ is free of minimal D-rank among the $\{t^\perp \cap W_u''; t \in W\}$. Let $x' \in W_u''$ be a vector that approximates ugx p-adically as closely as needed and at the same time q-adically approximates an admissible vector in $z^\perp \cap W_u''$. We may assume also that x' is W-admissible. Then clearly $\widetilde{H}(x'D)(K_p)$ commutes with the group C generated by $\widetilde{H}(W_u')(K_q)$ and $\widetilde{H}(x'^\perp \cap W)(K)$. Since x' approximates q-adically a vector in $z^\perp \cap W_u''$, we see that $\widetilde{H}(x'^\perp \cap W)$ is split at q. Further $x'^\perp \cap W$ cannot be orthogonal to W_u' if this q-adic approximation is sufficiently good. It follows that C contains an open noncompact subgroup of $\widetilde{H}(W)(K_q)$, hence is all of $\widetilde{H}((W)(K_q)$. We conclude thus that $\widetilde{H}(x'D)(K_p)$ commutes with $\widetilde{H}(W)(K_q)$. Allowing x' to tend to ugx p-adically inside W_u'', we conclude that $\widetilde{H}(W)(K_q)$ commutes with $\widetilde{H}(ugxD)(K_p)$ for all $u \in U$. But once again we may let x tend to w p-adically in W leading to the conclusion that $\widetilde{H}(ugwD)(K_p)$ commutes with $\widetilde{H}(W)(K_q)$ for all $u \in U$ or equivalently that $\widetilde{H}(gwD)(K_p)$ commutes with $\widehat{u}(\widetilde{H}(W)(K_q))$ for all $u^{-1} \in U$. This completes the proof of proposition (2.1).

(2.4) Proof of Theorem 3. From the tables of [Tits] (see also [Weil]) it is clear that (unless $n = 7$ and the arithmetic structure on $SO(7,1)$ comes, in the notation of [Tits], from ${}^6D_{40}$ or ${}^3D_{40}$) the only arithmetic cocompact lattices in $H = SO(2m-1,1)(m \geq 3)$ come from those defined in (1.4). Let $G = SO(2m+1,1)$ as in case 1 of (1.5). Then from the main theorem (1.6), C is finite and by proposition (1.3), the restriction map at the H^1-level is injective. This proves Theorem 3.

(2.5) Proof of Theorem 2. We must now assume $H = SO(2m-1,1)$ with $m \geq 4$. Then $G = SU(2m-1,1)$ as in case 2 of (1.4). By the main theorem (1.6), C is finite and by proposition (1.3), Theorem 2 follows.

(2.6) Proof of Theorem 1. Kazhdan's Theorem (see [Kazhdan]) together with Theorems 2 and 3 already imply Theorem 1 if $H = SO(2m-1,1)$ with $m \geq 3$. Suppose now that $H_0 = SO(2m,1)$. Then by [Tits] or [Weil] the only cocompact arithmetic lattices come from quadratic forms over a totally real field K and hence H_0 is caught between $H = SO(2n-1,1)$ and $G = SO(2m+1,1)$ as in case 1 of (1.5), with $D = M_2(K)$. By Theorem 3 for G and H, the restriction map at H^1-level from G to H is injective and hence Res : $H^1(\Gamma \cap G : \mathbf{C}) \to \prod_{\sigma \in G^*(K)} H^1(\sigma(\Gamma) \cap H_0; \mathbf{C})$ is also injective. Since by Theorem 1 for $G, H^1(\Gamma \cap G; \mathbf{C}) \neq 0$ for some congruence subgroup Γ of $G(K)$, it follows that $H^1(\sigma(\Gamma) \cap H_0; \mathbf{C}) \neq 0$ for a suitable $\sigma \in G^*(K)$. Thus Theorem 1 holds for H_0 as well.

Acknowledgement. We are very grateful to B.Sury for helpful conversations and for his help in preparing the manuscript. We also thank Priyan for typing the manuscript excellently.

REFERENCES

[Kazhdan] Kazhdan,D.A. Some Applications of the Weil Representation, Journal d' analyse,**32** (1977)233-248.

[Li] Li,J.-S. Nonvanishing Theorems for the Cohomology of certain Arithmetic Quotients J. Reine angew. Math.**428** (1992) 177-217.

[Li-Millson] Li,J.S. and Millson,J. On the First Betti Number of a Hyperbolic Manifold with an Arithmetic Fundamental Group, Preprint.

[Millson] Millson,J. On the First Betti Number of a constant negatively curved manifold, Ann. of Math. **104**, (1976) 235-247.

[Platonov] Platonov,V.P. The Problem of Strong Approximation and the Kneser-Tits conjecture for algebraic groups Math. USSR Izv. **3** (1969) 1139-1147.

[Raghunathan] Raghunathan,M.S. Torsion in cocompact Lattices in coverings of Spin $(2, n)$, Math. Ann. **266**, 403-419 (1984).

[Sury] Sury,B. Congruence subgroup problem for anisotropic groups over semilocal rings, Proc. Indian Acad. Sci. (Math. Sci.), Vol. 101, No.**2**, (1991) pp.87-110.

[Tits] Tits,J.Classification of Algebraic semi-simple groups Proc. Symp. Pure Math. (Boulder, Colorado, 1965) p. 33-62, (AMS providence) (1966).

[Tomanov] Tomanov,G. Projective simplicity of Groups of Rational Points of simply Connected Algebraic Groups over Number Fields, Banach Cente Publ. (Topics in Algebra) **26**, (1989).

[Weil] Weil,A. Algebras with involutions and the classical groups. J. Ind. Math.Soc.**24**(1960),589-623

SCHOOL OF MATHEMATICS, TIFR, HOMI BHABHA ROAD, BOMBAY 400 005, INDIA

E-mail address: MSR@tifrvax.tifr.res.in

Contemporary Mathematics
Volume **153**, 1993

Combinatorics and Geometry of K-orbits on the Flag Manifold

R. W. RICHARDSON AND T. A. SPRINGER

Introduction

Let G be a reductive algebraic group over an algebraically closed field F of characteristic $\neq 2$ and let θ be an automorphism of G of period two. Let K be a subgroup of the fixed point subgroup G^θ which contains $(G^\theta)^0$, the identity component of G^θ. Let $\mathcal{B} = \mathcal{B}(G)$ denote the variety of Borel subgroups of G; $\mathcal{B}$ is the flag manifold of G. If $B \in \mathcal{B}$, we may identify $\mathcal{B}$ with the coset space $B\backslash G$. We consider the K–orbits on the flag manifold $\mathcal{B}$ or, equivalently, the set $V = B\backslash G/K$ of $(B \times K)$–orbits of G. It is known that V is finite and there is a natural partial order on V given by inclusion of the orbit closures.

The classification and properties of these orbits play an important role in the representation theory of real semisimple groups (see [**V1**] and [**HMSW**]) and in a number of geometric problems. If $G = H \times H$, where H is a reductive group, and if θ is given by $\theta(x, y) = (y, x)$, $\ (x, y) \in H \times H$, then $K = \{ (h, h) \mid h \in H \}$ and the K–orbits on $\mathcal{B}(G) = \mathcal{B}(H) \times \mathcal{B}(H)$ can be naturally parametrized by $W(H)$, the Weyl group of H. In this case the partial order on the set V of orbits corresponds to the usual Bruhat order on the Weyl group $W(H)$. One would like a similar description of the poset $V = B\backslash G/K$ in the general case.

In this paper we will give an informal discussion, mostly without proofs, of some recent results of the authors on the set V of orbits. In the joint paper [**RS**], we developed some techniques for the analysis of the orbits. Using this machinery, we were able to give a purely combinatorial description of the partial order on V and to generalize to the poset V a number of standard properties of the usual Bruhat order on the Weyl group $W(G)$. Although the results of [**RS**] are based on simple geometric ideas, the discussion there sometimes gets bogged down in technical detail and parts of the paper are difficult to read (even for the authors). In Sections 1-4 of this paper, we discuss the main ideas and theorems

1991 *Mathematics Subject Classification.* 14L30, 20G15.

of [**RS**]. We have tried to omit most of the technical detail and to emphasize the underlying geometric ideas.

Let G be simple (over its center) and simply connected. Then we say that the pair (G, θ) is of "Hermitian symmetric type" if the center of $K = G^\theta$ has positive dimension. Roughly speaking, pairs (G, θ) of Hermitian symmetric type correspond to Hermitian symmetric spaces. In Section 5, we discuss some recent (unpublished) results on the parametrization of the set V of orbits for pairs (G, θ) of Hermitian symmetric type. In this case, there is an elementary combinatorial model for the poset V in terms of combinatorial data involving only the Weyl group.

In Section 6, we consider the transcendental case $F = \mathbb{C}$. In this case, there is a real form $G_\mathbb{R}$ of G such that $K_\mathbb{R} = G_\mathbb{R} \cap K$ is a maximal compact subgroup of both $G_\mathbb{R}$ and K. Matsuki has shown that there is a natural duality between the K–orbits and the $G_\mathbb{R}$–orbits on $\mathcal{B}$ which reverses the natural partial order on these orbits and he has proved a number of results concerning both the K–orbits and the $G_\mathbb{R}$–orbits [**M1-M5, MO**]. It has been shown recently by one of us (RWR) that the machinery of [**RS**] can equally well be used for the analysis of the $G_\mathbb{R}$–orbits on $\mathcal{B}$. We discuss these results and indicate some extensions of Matsuki's work.

In the applications to the representation theory of real semisimple Lie groups, an important role is played by certain representations of the Hecke algebra $\mathcal{H}$ of the Weyl group W of G (see [**LV**]). In Section 7, we give a description of these representations of $\mathcal{H}$ in terms of our analysis of the set V of orbits.

§1 Some basic constructions of [RS]

1.1. Preliminaries. Our main reference for algebraic groups and algebraic geometry will be Borel's book [**B**] and we will usually follow the terminology and notation there. All algebraic varieties will be taken over an algebraically closed field F with $\operatorname{char}(F) \neq 2$.

Since we consider left, right and two sided actions of groups, we need to be careful about the notation for the sets of orbits. Let the groups H and L act on the set E, with H acting on the left and L acting on the right. Then we let $H\backslash E$ denote the set of H–orbits of E and let E/L denote the set of L–orbits. If the actions of H and L commute, so that we get an action of $H \times L$ on E, we let $H\backslash E/L$ denote the set of $(H \times L)$–orbits of E.

Throughout the paper, G, θ and K will be as in the Introduction. In order to conform with the notation of [**RS**], we will always assume that $K = G^\theta$.

REMARKS 1.1.1. (a) If G is semisimple and simply connected, then it follows from a theorem of Steinberg [**St**, Thm. 8.1] that $K = G^\theta$ is connected. For a general reductive G however, the fixed point subgroup $K = G^\theta$ is not necessarily connected, so that the K–orbits on $\mathcal{B}$ are not necessarily connected. However, the B–orbits on G/K are always connected (and irreducible), so that one can apply standard irreducibility arguments to these orbits.

(b) The assumption that $G^\theta = K$ is not an essential restriction. The argument goes as follows. Let $\theta_1 : G_1 \to G_1$ be an involutive automorphism of a reductive group G_1 and let K_1 be a subgroup of the fixed point subgroup $G_1^{\theta_1}$ containing the identity component $(G_1^{\theta_1})^0$. We note that the center $Z(G_1)$ acts trivially on $\mathcal{B}(G_1)$, so that we get an induced action of $\mathrm{Ad}(G_1)$ on $\mathcal{B}(G_1) \cong \mathcal{B}(\mathrm{Ad}(G_1))$. A straightforward argument, using the result of Steinberg mentioned in (a) above, shows that there exists a covering group G of $\mathrm{Ad}(G_1)$ and an involutive automorphism θ of G, such that, setting $K = G^\theta$, we have $\mathrm{Ad}(K) = \mathrm{Ad}(K_1)$ (as subgroups of $\mathrm{Ad}(G_1) = \mathrm{Ad}(G)$). Thus the K–orbits of $\mathcal{B}(G) \cong \mathcal{B}(\mathrm{Ad}(G_1))$ are the same as the K_1–orbits, and we are in the case with $K = G^\theta$.

The group G acts on $\mathcal{B}$ (on the left) by conjugation. In order to avoid confusing notation, such as $K \cdot B$ for the K–orbit on $\mathcal{B}$ of a Borel subgroup B, we will often let X denote the projective G-variety corresponding to $\mathcal{B}$. If $x \in X$, we let B_x denote the corresponding Borel subgroup; thus $K \cdot x$, the K–orbit of x on X, is equal to $\{\, {}^kB_x \mid k \in K \,\}$.

1.2. In order to bring the Weyl group into the picture, we need to consider maximal tori. Let $\mathcal{T} = \mathcal{T}(G)$ denote the variety of maximal tori of G. Then θ acts on $\mathcal{T}$ and we let $\mathcal{T}^\theta$ be the subvariety of θ–stable maximal tori. If $T \in \mathcal{T}$, we let $W(T) = N_G(T)/T$ denote the corresponding Weyl group. Let

$$\mathcal{C} = \{\, (B,T) \in \mathcal{B} \times \mathcal{T} \mid T \subset B \,\} \quad \text{and let} \quad \mathcal{C}_\theta = \mathcal{C} \cap (\mathcal{B} \times \mathcal{T}^\theta).$$

The group G acts on $\mathcal{C}$ (on the left) by conjugation and it is clear that $\mathcal{C}_\theta$ is K–stable. Define $\pi : \mathcal{C} \to \mathcal{B}$ by $\pi(B,T) = B$. Then the restriction of π to $\mathcal{C}_\theta$ determines a map $\gamma : K\backslash\mathcal{C}_\theta \to K\backslash\mathcal{B}$ of the sets of K–orbits.

PROPOSITION 1.2.1. *$\gamma : K\backslash\mathcal{C}_\theta \to K\backslash\mathcal{B}$ is a bijection.*

PROOF. We will give the proof, since it is quite easy. Let $B \in \mathcal{B}$. Since $B \cap \theta(B)$ is of maximal rank, it follows from another theorem of Steinberg [**St**, Thm. 7.5] that $B \cap \theta(B)$ contains a θ–stable maximal torus. Thus γ is surjective. To prove that γ is injective, we need to prove that if T and T' are θ–stable maximal tori of B, then they are conjugate by an element of $B \cap K$. Now T and T' are maximal tori of $B \cap \theta(B)$, hence they are conjugate by an element u of $R_u(B \cap \theta(B))$, the unipotent radical of $B \cap \theta(B)$. It follows from this that $u^{-1}\theta(u) \in N_G(T) \cap R_u(B \cap \theta(B))$. But $R_u(B \cap \theta(B)) \subset R_u(B)$, so that $u^{-1}\theta(u) \in N_G(T) \cap R_u(B) = \{1\}$; thus $\theta(u) = u$ and hence $u \in K$

In the transcendental case $F = \mathbb{C}$, Proposition 1.2.1 is due to Matsuki [**M1**] and Rossman [**Ros**]. The basic idea is due to Wolf [**W**], who proved a similar theorem for the $G_\mathbb{R}$–orbits on $\mathcal{B}$. (Here $G_\mathbb{R}$ is as in the Introduction.)

We say that a pair $(B,T) \in \mathcal{C}$ is a *standard pair* if both B and T are θ–stable. It follows from [**St**, Thm. 7.5] that standard pairs exist. We choose a standard pair (B_0, T_0), which will remain fixed throughout the paper. Let $W = W(T_0)$ and $N = N_G(T_0)$. Let $\Phi = \Phi(T_0, G)$ be the set of roots of G

relative to T_0, let $\Phi^+ = \Phi(T_0, B_0)$ be the set of positive roots determined by B_0 and let $\Delta = \Delta(T_0, B_0)$ be the set of simple roots corresponding to Φ^+. If $\alpha \in \Phi$, let $s_\alpha \in W$ be the corresponding reflection. Let $S = \{ s_\alpha \mid \alpha \in \Delta \}$. Then $W = (W, S)$ is a Coxeter group. Let l denote the length function on W and let $\leq$ be the Bruhat order on W.

For each subset I of Δ, let W_I be the subgroup of W generated by $\{ s_\alpha \mid \alpha \in I \}$ and let $P_I = B_0 W_I B_0$ be the "standard" parabolic subgroup of G corresponding to I. We let w_I denote the longest element of W_I and let $w_0 = w_\Delta$ be the longest element of W.

Define $\zeta_0 : G \to \mathcal{C}$ by $\zeta_0(g) = ({}^{g^{-1}}B_0, {}^{g^{-1}}T_0)$. Then ζ_0 is constant on right T_0 cosets of G and induces a isomorphism of varieties $\zeta : T_0\backslash G \to \mathcal{C}$. The group G acts on $T_0\backslash G$ on the right and we have $\zeta(x \cdot g) = g^{-1} \cdot \zeta(x),\ \ g \in G,\ x \in T_0\backslash G$. Set

$$\mathcal{V} = \{ g \in G \mid g\theta(g)^{-1} \in N \} = \{ g \in G \mid {}^{g^{-1}}T_0 \in \mathcal{T}^\theta \} = \zeta_0^{-1}(\mathcal{C}_\theta).$$

Since $N_G(B_0) = B_0$ and $B_0 \cap N = T_0$, we see that the restriction of ζ_0 to $\mathcal{V}$ induces a bijection of $T_0\backslash\mathcal{V}$ onto $\mathcal{C}_\theta$. Thus there is an induced bijection of orbit sets $T_0\backslash\mathcal{V}/K \to K\backslash\mathcal{C}_\theta$.

The map $g \mapsto {}^{g^{-1}}B_0$ of G to $\mathcal{B}$ is constant on left B_0-cosets and induces an isomorphism of K–varieties $B_0\backslash G \to \mathcal{B}$. Thus we obtain a bijection $B_0\backslash G/K \cong K\backslash\mathcal{B}$.

Combining all of these results with Proposition 1.2.1, we obtain:

PROPOSITION 1.2.2. *There exist canonical bijections between the following four sets of orbits*: (a) $T_0\backslash\mathcal{V}/K$; (b) $K\backslash\mathcal{C}_\theta$; (c) $K\backslash\mathcal{B}$; *and* (d) $B_0\backslash G/K$.

We observe that the bijection $T_0\backslash\mathcal{V}/K \cong B_0\backslash G/K$ is induced by the inclusion map $\mathcal{V} \to G$.

We let V denote the set $T_0\backslash\mathcal{V}/K$ of $(T_0 \times K)$–orbits on $\mathcal{V}$. By Proposition 1.2.2, we may identify V with each of the orbit sets given in (b), (c) and (d) of the Proposition. Occasionally we will make make such identifications without being very explicit about it.

1.3. Notation and remarks. (a) We let $x_0 \in X$ correspond to B_0, so that $B_{x_0} = B_0$. If $v \in V$, then v is a (T_0, K) double coset, $v = T_0 g K$, with $g \in \mathcal{V}$. In this case we let $B_0 v K$ denote the set-theoretic product; thus $B_0 v K = B_0(T_0 g K)K = B_0 g K$. It $v \in V$, we will let $\mathcal{O}(v)$ denote $B_0 v K$, and let $\mathcal{K}(v)$ denote the corresponding K–orbit on X. Thus, if $v = T_0 g K$ and if $x = g^{-1} \cdot x_0$ (so that $B_x = {}^{g^{-1}}B_0$), then $\mathcal{K}(v)$ is equal to $K \cdot x$, the K–orbit of x on X.

(b) The varieties $\mathcal{T}^\theta$, $\mathcal{C}_\theta$ and $\mathcal{V}$ are not connected. In fact, it follows from [**R1**, Thm. A] that K^0, the identity component of K, acts transitively on each irreducible component of $\mathcal{T}^\theta$. Using this, one can show that K^0 acts transitively on each irreducible component of $\mathcal{C}^\theta$ and that $T_0 \times K^0$ acts transitively on each irreducible component of $\mathcal{V}$. In particular, all K-orbits on $\mathcal{C}_\theta$ are closed and all

$(T_0 \times K)$–orbits on $\mathcal{V}$ are closed. Thus, one cannot obtain any information about closures of K–orbits in X or closures of $(B_0 \times K)$-orbits in G from the closures of the corresponding orbits on $\mathcal{C}_\theta$ or $\mathcal{V}$.

1.4. The map ϕ and the W–action on V. Since T_0 and N are θ–stable, we get an induced action of θ on the Weyl group W. Let $\mathcal{I} = \{ w \in W \mid \theta(w) = w^{-1} \}$. We say that the elements of $\mathcal{I}$ are *twisted involutions.* If θ is an inner automorphism of G, then θ acts trivially on W and $\mathcal{I}$ is just the set of ordinary involutions of W. We define the "twisted action" of the group W on the set W by the following rule : if w', $w \in W$, then $w' * w$, the twisted action of w' on w, is equal to $w'w\theta(w'^{-1})$. It is clear that the set $\mathcal{I}$ of twisted involutions is stable under the twisted action of W. We have the following elementary result [**S1**, §3] concerning $\mathcal{I}$:

LEMMA 1.4.1. *Let $s \in S$ and $a \in \mathcal{I}$ and assume that $s*a \neq a$. If $l(sa) > l(a)$, then $l(s*a) = l(a)+2$ and if $l(sa) < l(a)$, then $l(s*a) = l(a)-2$.*

Define $\kappa : G \to G$ by $\kappa(g) = g\theta(g)^{-1}$. Then κ is constant on left K-cosets, and induces an isomorphism of G/K onto the closed subvariety $\kappa(G)$ of G. We note that $\mathcal{V} = \kappa^{-1}(N)$. Let $\pi : N \to W$ be the canonical projection. Then the map $g \to \pi(\kappa(g))$ from $\mathcal{V}$ to W is constant on $(T_0 \times K)$–orbits and hence induces a map $\phi : V \to W$. If $g \in G$, then it is easy to check that $\theta(\kappa(g)) = \kappa(g)^{-1}$, which shows that $\phi(V) \subset \mathcal{I}$. Let $v = T_0gK \in V$ and let $\phi(v) = a$. Then $\kappa(\mathcal{O}(v)) \subset B_0aB_0$; one can use this to give an alternate definition of the map ϕ.

The map ϕ was introduced in [**S1**]. A geometric interpretation of ϕ in terms of the *canonical Weyl group* $\mathcal{W}$ is given in 1.7. The map ϕ plays an important role in the study of the set V of orbits. For example, we have:

PROPOSITION 1.4.2. *Let $v \in V$. Then the orbit $\mathcal{O}(v)$ is closed if and only if $\phi(v) = 1$.*

See [**S1**, §6.6].

We also have the following characterization of closed orbits:

PROPOSITION 1.4.3. *Let $x \in X$. Then the orbit $K \cdot x$ is closed in X if and only if the Borel subgroup B_x is θ–stable.*

Let $V_0 = \{ v \in V \mid \phi(v) = 1 \}$ denote the set of closed orbits. An easy argument shows that all closed K-orbits on X have dimension equal to the dimension of $\mathcal{B}(K^0)$, the flag manifold of K^0; this is equivalent to the statement that if $B \in \mathcal{B}^\theta$, then $K^0 \cap B$ is a Borel subgroup of K^0. If $v \in V$, we set $l(v) = \dim \mathcal{K}(v) - \dim \mathcal{B}(K^0)$. We say that $l(v)$ is the *length* of v. In the case $G = H \times H$ and $\theta(x,y) = (y,x)$, which was discussed in the Introduction, the orbits are parametrized by the Weyl group $W(H)$ and the length function l on the set V of orbits corresponds to the usual length function on the Coxeter group $W(H)$.

Since N is θ-stable, it is clear that $\mathcal{V}$ is stable under left multiplication by N. Thus we get an action of W on $V = T_0\backslash\mathcal{V}/K$. We let $w \cdot v$ denote the action of $w \in W$ on $v \in V$. A geometric interpretation of the action of W on the set V of orbits in terms of the canonical Weyl group $\mathcal{W}$ is given in 1.7.

The W action on V and the twisted W action on $\mathcal{I}$ are related by the following proposition:

PROPOSITION 1.4.4. (1) *If $w \in W$ and $v \in V$, then $\phi(w \cdot v) = w * \phi(v)$. Thus $\phi : V \to \mathcal{I}$ is W-equivariant.* (2) *If $\phi(v) = \phi(v')$, then v and v' lie in the same W-orbit.* (3) *There are canonical bijections between the following three sets of orbits:* (i) $W\backslash V$; (ii) $W\backslash\text{image}(\phi)$; *and* (iii) $K\backslash\mathcal{T}^\theta$.

1.5. Examples. (See [**RS**, §10].)

(1) Let H be reductive, let $G = H \times H$ and let θ be defined by $\theta(x, y) = (y, x)$. In this case we may identify $W(G)$, the Weyl group of G, with $W(H) \times W(H)$ and the set $\mathcal{I}$ of twisted involutions is given by $\mathcal{I} = \{ (w, w^{-1}) \mid w \in W(H) \}$. The map $\phi : V \to \mathcal{I}$ is a bijection.

(2) Let $G = GL(n, F)$ and define $\theta : G \to G$ by $\theta(g) = {}^t g^{-1}$. Thus K is equal to the orthogonal group $O(n, F)$ and W is the symmetric group S_n. Let $\mathcal{J} = \mathcal{J}_n$ be the set of all involutions in S_n. Then $\mathcal{I} = \mathcal{J} w_0$. In this case also, $\phi : V \to \mathcal{I}$ is a bijection.

(3) Let $G = SL(n, F)$ and let θ be as in (2). Then $K = SO(n, F)$. Again $W = S_n$ and $\mathcal{I} = \mathcal{J} w_0$. The map $\phi : V \to \mathcal{I}$ is surjective. Let $\mathcal{J}^0 \subset \mathcal{J}$ be the set of fixed point free involutions in S_n and let $\mathcal{J}^1 \subset \mathcal{J}$ be the set of involutions which have a fixed point. If $a \in \mathcal{J}^0$, then $|\phi^{-1}(a w_0)| = 2$ and if $a \in \mathcal{J}^1$, then $|\phi^{-1}(a w_0)| = 1$. Thus $|V| = 2|\mathcal{J}^0| + |\mathcal{J}^1|$. So, in particular, ϕ is bijective if n is odd but is not bijective if n is even.

(4) Let $G = SL(2n, F)$ and let $J \in G$ be defined by: $J(e_i) = -e_{n+i}$ and $J(e_{n+1}) = e_i$, $i = 1, \ldots, n$, where $e_1, \ldots, e_{2n}$ is the standard basis of F^{2n}. Define the involutive automorphism $\theta : G \to G$ by $\theta(g) = J {}^t g^{-1} J^{-1}$. Then the fixed point subgroup $K = G^\theta$ is the symplectic group $Sp(2n, F)$. The Weyl group W is the symmetric group S_{2n}. Let $\mathcal{J} = \mathcal{J}_{2n}$ and $\mathcal{J}^0$ be as above. Then the set $\mathcal{I}$ of twisted involutions is equal to $\mathcal{J} w_0$ and the image of ϕ is equal to $\mathcal{J}^0 w_0$.

We note that in general the map $\phi : V \to \mathcal{I}$ is not necessarily either injective or surjective. It follows from Proposition 1.4.4 that the image of ϕ is a union of twisted W-orbits of $\mathcal{I}$.

1.6. Since $B_0\backslash G/K$ is finite, there exists a unique open $(B_0 \times K)$-orbit of G. Let $v_{max} \in V$ be such that $\mathcal{O}(v_{max})$ is open in G and let $a_{max} = \phi(v_{max})$. One can describe a_{max} in terms of the Araki diagram associated to (G, θ). (See [**S2**] for the Araki diagram.) Let $J \subset S$ correspond to the set of black dots in the Araki diagram. Then a_{max} is equal to $w_J w_0$, where w_J and w_0 are as in 1.2.

We have the following criteria for the map ϕ to be surjective:

PROPOSITION 1.6. *The following four conditions are equivalent:* (i) ϕ *is surjective*; (ii) $a_{max} = w_0$; (iii) *the Araki diagram of* (G, θ) *does not contain any black dots*; *and* (iv) *there exists a* θ*-split Borel subgroup of* G.

Recall that a Borel subgroup B is θ-split if $B \cap \theta(B)$ is a maximal torus of G.

1.7. The canonical Weyl group. The map $\phi : V \to W$ and the W-action on V are defined in terms of the standard pair (B_0, T_0). If (B_1, T_1) is another standard pair, then it can be shown that T_0 and T_1 are K-conjugate (in fact K^0-conjugate), but it is not necessarily true that B_0 and B_1 are K-conjugate. We will show that ϕ and the W-action on V are canonically defined. To discuss this, we need the *canonical Weyl group* $\mathcal{W}$. As a set $\mathcal{W}$ is equal to $G\backslash(\mathcal{B} \times \mathcal{B})$, the set of G-orbits on $\mathcal{B} \times \mathcal{B}$. For each (B, T) in $\mathcal{C}$, there is a bijective map $\eta_{B,T} : W(T) \to \mathcal{W}$ defined by $\eta_{B,T}(w) = p(B, {}^wB)$, where $p : \mathcal{B} \times \mathcal{B} \to G\backslash(\mathcal{B} \times \mathcal{B}) = \mathcal{W}$ is the canonical map. Let $\eta_0 = \eta_{B_0,T_0}$. We define the group structure on $\mathcal{W}$ by requiring that η_0 be an isomorphism of groups. This implies that each $\eta_{B,T}$ is an isomorphism of groups. If $(B, B') \in \mathcal{B} \times \mathcal{B}$ and if $\overline{w} = p(B, B')$, then we say that $\overline{w} \in \mathcal{W}$ is the *relative position* of (B, B'). Let $\overline{S} = \eta_0(S)$. Then $\mathcal{W} = (\mathcal{W}, \overline{S})$ is a Coxeter group; this Coxeter group structure on $\mathcal{W}$ is canonical. As usual, we let l denote the length function on $\mathcal{W}$.

There is a canonical action of $\mathcal{W}$ on $\mathcal{C}$ defined as follows:

Let $\overline{w} \in \mathcal{W}$ and let $(B, T) \in \mathcal{C}$. Choose w in $W(T)$ such that $\eta_{B,T}(w) = \overline{w}$. Then $\overline{w} \cdot (B, T) = ({}^{w^{-1}}B, T)$. It is a straightforward exercise to prove that this defines an action of $\mathcal{W}$ on $\mathcal{C}$. The projection $\mathcal{C} \to \mathcal{T}$, $(B, T) \mapsto T$, is a Galois covering and $\mathcal{W}$ acts on $\mathcal{C}$ as the group of "deck transformations". (For the case $F = \mathbb{C}$, the projection $\mathcal{C} \to \mathcal{T}$ is a covering map in the usual sense.)

We note that $\mathcal{C}_\theta$ is stable under the $\mathcal{W}$-action and it is clear that the $\mathcal{W}$-action on $\mathcal{C}$ commutes with the action of G on $\mathcal{C}$ by conjugation. Let $\zeta : T_0\backslash G \to \mathcal{C}$ be as in 1.2. Left multiplication by N gives an action of $W = W(T_0)$ on $T_0\backslash G$. An easy argument shows that $\zeta(w \cdot x) = \eta_0(w) \cdot \zeta(x)$, for $w \in W$ and $x \in T_0\backslash G$, so that ζ is equivariant (with respect to η_0). Now ζ maps $T_0\backslash\mathcal{V}$ equivariantly onto $\mathcal{C}_\theta$. Thus the bijection of $V = T_0\backslash\mathcal{V}/K$ onto $K\backslash\mathcal{C}_\theta$ determined by ζ is equivariant with respect to η_0. This proves that the W-action on the set of orbits is canonical. (This was not done in [**RS**].)

As regards the map $\phi : V \to \mathcal{I}$, we have the following proposition.

PROPOSITION 1.7.1. *Let* $v \in V$ *and let* $x \in \mathcal{K}(v)$, *so that* $\mathcal{K}(v) = K \cdot x$. *Let* $\overline{a} = \eta_0(\phi(v))$. *Then* $\overline{a}$ *is the relative position of* $(B_x, \theta(B_x))$.

The proof is straightforward.

It follows from Proposition 1.7.1 that $\overline{a} = \eta_0(\phi(v))$ depends only on the orbit $\mathcal{K}(v) = K \cdot x$ and is independent of the choice of standard pair (B_0, T_0).

§2. The product of a minimal parabolic and an orbit

Let $s \in S$ and let $P_s = B_0 \cup B_0 s B_0$ be the corresponding standard minimal parabolic subgroup. If $v \in V$, then the product $P_s B_0 v K = P_s v K$ is a union of a finite number of $(B_0 \times K)$–orbits. In this section, we will analyze the decomposition of $P_s vK$ into $(B_0 \times K)$–orbits. In order to do this, it is easier to work with the K–orbits on X rather than the $(B_0 \times K)$–orbits.

Most of the results of this section are contained in [**LV**].

2.1. Real, complex and imaginary roots. Let T be a θ–stable maximal torus and let $\Phi(T, G)$ be the set of roots of G relative to T. If $\alpha \in \Phi(G, T)$, let G_α be the subgroup of G generated by the root subgroups U_α and $U_{-\alpha}$; the subgroup G_α is semisimple of rank one. Let $T_\alpha = T \cap G_\alpha$; then T_α is a maximal torus of G_α.

There are three cases to consider.

(a) $\theta(\alpha) = \alpha$. In this case we say that α is *imaginary* (relative to θ). If α is imaginary, the subgroup G_α is θ-stable. There are two subcases. If the restriction of θ to G_α is trivial, so that $G_\alpha \subset K$, then α is *compact imaginary*. If $G_\alpha \not\subset K$, then α is *non-compact imaginary*. In the latter case, T_α is the identity component of G_α^θ.

(b) $\theta(\alpha) = -\alpha$. Then we say that α is *real*. In this case, G_α is θ–stable and $(G_\alpha^\theta)^0$ is a maximal torus of G_α.

(c) $\theta(\alpha) \neq \pm\alpha$. Then α is *complex*. In this case G_α is not θ–stable.

REMARK. The terminology of real, complex and imaginary roots is taken from the theory of real semisimple Lie groups.

2.2. The $\mathbb{P}^1$ approach. Let $Y = \mathcal{P}_s$ denote the variety of all conjugates of P_s. Let $\pi_s : X \to Y$ denote the morphism which assigns to every $B \in \mathcal{B}$ the unique $P \in \mathcal{P}_s$ which contains B. Let $y_0 = \pi_s(x_0)$ (recall that $B_{x_0} = B_0$). Now let $y \in Y$ and let $P_y \in \mathcal{P}_s$ be the corresponding parabolic subgroup. Let X_y denote $\mathcal{B}(P_y)$, the variety of Borel subgroups of P_y; note that $X_y = \pi_s^{-1}(y)$. Then X_y is a complete subvariety of X which is isomorphic to $\mathbb{P}^1(F) = \mathbb{P}^1$. Let $A_y = \mathrm{Aut}(X_y)$, the group of automorphisms of the algebraic variety X_y and let $h : P_y \to A_y$ be the canonical homomorphism. Then A_y is an algebraic group isomorphic to $PGL(2, F)$. Let $K_y = K \cap P_y$; then K_y is the isotropy subgroup at y for the action of K on Y. The following lemma is elementary:

LEMMA 2.2.1. *Let $x \in X_y$ and $g \in G$ be such that $g \cdot x \in X_y$. Then $g \in P_y$.*

Let $x \in X_y$. Then it follows easily from Lemma 2.2.1 that $K \cdot x \cap X_y = K_y \cdot x$. Furthermore $K \cdot x$ is a homogeneous fiber bundle over $K \cdot y$ with fiber $K_y \cdot x$. In particular, we have a bijective correspondence between the K–orbits on $K \cdot X_y$ and the K_y–orbits on X_y. Since $K \backslash X$ is finite, K_y has a finite number of orbits on X_y. Since $h(K_y)$ is an algebraic subgroup of $A_y \cong PGL(2, F)$, it is easy to analyze the possibilities for the K_y–orbits on X_y. There are four cases to consider.

Case 1. $h(K_y) \neq A_y$ and $h(K_y)$ contains a non-trivial unipotent subgroup. Then there are two K_y–orbits on X_y, one of which is a fixed point.

Case 2. $h(K_y) = A_y$. Then K_y is transitive on X_y.

Case 3. $h(K_y)$ is a maximal torus of A_y. There are three orbits, two fixed points and one open dense orbit.

Case 4. $h(K_y)$ is the normalizer of a maximal torus of A_y. There are two orbits. There are two fixed points of K_y^0, which are permuted by K_y, and there is an open dense orbit.

2.3. Further analysis of the K_y–orbits on X_y. Let $x \in X_y$ and let $B = B_x$. Let T be a θ–stable maximal torus of B, and let $\Phi(T, G)$ and $\Delta(T, B)$ be defined as usual. Then there exists $\alpha \in \Delta(T, B)$ such that $P = P_y$ is equal to $P_\alpha = B \cup Bs_\alpha B$. Note that, since T is θ–stable, θ acts on $\Phi(T, G)$.

Case A. α is complex (relative to θ). In this case $h(K_y)$ is a solvable group with non-trivial unipotent radical. Thus we are in Case 1 above.

Case B. α is compact imaginary. Then $G_\alpha \subset P \cap K$ and hence $h(K_y) = A_y$. Thus we are in Case 2 above and K_y is transitive on X_y.

Case C. α is non-compact imaginary. In this case $(G_\alpha^\theta)^0 = T_\alpha$ and $h(T_\alpha) = h(K_y)^0$. Then we are in either Case 3 or Case 4 above. We have $h(K_y)^0 \cdot x = x$, and there are either two or three K_y–orbits. If $h(K_y)$ is connected, there are three K_y–orbits on X_y and if $h(K_y)$ is not connected, there are two orbits.

Case D. α is real. Then $(G_\alpha^\theta)^0$ is a maximal torus of G_α and $h((G_\alpha^\theta)^0) = h(K_y)^0$. We are in either Case 3 or 4 above. There are either two or three K_y–orbits on X_y and $K_y \cdot x$ is the unique dense open orbit in X_y.

2.4. Case analysis for the $(B_0 \times K)$–orbits on $P_s vK$. Let $\alpha \in \Delta = \Delta(T_0, B_0)$. Let $s = s_\alpha$ and let $v = T_0 gK \in V$. We wish to describe the decomposition of $P_s vK$ into $(B_0 \times K)$–orbits. Let $x = g^{-1} \cdot x_0$, let $T = {}^{g^{-1}}T_0$ and let $B = {}^{g^{-1}}B_0$. Thus $T \in \mathcal{T}^\theta$ and $B = B_x$. Let $y = \pi_s(x)$ and let $P = P_y = {}^{g^{-1}}P_s$. The map $g' \mapsto g'^{-1} \cdot x_0$ from G to X determines a bijection from $(B_0 \times K)$–orbits on $P_s vK$ to K–orbits on $K \cdot X_y$ and it follows from the discussion above that there is a bijection from K_y–orbits on X_y to K–orbits on $K \cdot X_y$. Thus we may use the case analysis of 2.3 above for the $(B_0 \times K)$–orbits on $P_s vK$.

First we need some definitions. The inner automorphism $\mathrm{Int}(g^{-1})$ of G maps T_0 to T and maps $\Phi = \Phi(T_0, G)$ onto $\Phi(T, G)$. Let $\alpha' = \mathrm{Int}(g^{-1})(\alpha)$. We say that α (or $s = s_\alpha$) is complex, compact imaginary, ... , for v if α' is complex, compact imaginary, ... , relative to θ in the sense of 2.1. These definitions are independent of the choice of $g \in v$. Let $\phi(v) = a$. Then it is easy to check that: (i) α is complex for v if $a\theta(\alpha) \neq \pm\alpha$; (ii) α is imaginary for v if $a\theta(\alpha) = \alpha$; and (iii) α is real for v if $a\theta(\alpha) = -\alpha$. We observe that if $s = s_\alpha$ is real (respectively imaginary) for v, then $l(sa) < l(a)$ (respectively $l(sa) > l(a)$).

Now we can apply the results of 2.2 and 2.3 to the $(B_0 \times K)$–orbits of $P_s vK$.

Case A. s is complex for v. Then $P_s vK = \mathcal{O}(v) \cup \mathcal{O}(s \cdot v)$ and $s \cdot v \neq v$. Thus there are two $B_0 \times K$–orbits on $P_s vK$. If $sa > a$ (respectively $sa < a$) then $\mathcal{O}(v)$

(respectively $\mathcal{O}(s \cdot v)$) is closed in $P_s vK$ and $\mathcal{O}(s \cdot v)$ (respectively $\mathcal{O}(v)$) is open and dense in $P_s vK$.

Case B. s is compact imaginary. There is only one orbit, so that $\mathcal{O}(v) = P_s vK$.

Case C. s is non-compact imaginary for v. Then there exists $v' \in V$ such that $P_s vK = \mathcal{O}(v) \cup \mathcal{O}(s \cdot v) \cup \mathcal{O}(v')$. The orbits $\mathcal{O}(v)$ and $\mathcal{O}(s \cdot v)$ are closed in $P_s vK$ and $\mathcal{O}(v')$ is open and dense in $P_s vK$. If $s \cdot v \neq v$, there are three orbits and if $s \cdot v = v$, then there are two orbits. Both cases can occur.

Case D. s is real for v. Then there exists $v' \in V$ such that $P_s vK = \mathcal{O}(v) \cup \mathcal{O}(v') \cup \mathcal{O}(s \cdot v')$. The orbits $\mathcal{O}(v')$ and $\mathcal{O}(s \cdot v')$ are closed in $P_s vK$ and $\mathcal{O}(v)$ is open and dense in $P_s vK$. If $s \cdot v' \neq v'$, there are three orbits, otherwise there are two orbits. Both cases can occur.

In Case C (respectively Case D), s is real (respectively non-compact imaginary) for v' and $P_s vK = P_s v' K$.

§3 The monoid $M(W)$

3.1. The monoid M(W). In our analysis of the orbits, an important role is played by a certain monoid $M = M(W)$ which is canonically associated to the Coxeter group $W = (W, S)$. (See [**RS**, §3].) As a set, M consists of symbols $m(w)$, one for each $w \in W$. Multiplication in M is determined by the following rule: if $w \in W$ and $s \in S$, then $m(s)m(w)$ is equal to $m(sw)$ if $l(sw) > l(w)$ and is equal to $m(w)$ if $l(sw) < l(w)$. Note that $m(1)$ is the identity element of M and $m(w_0)$ is the "final element" of M, i.e. $m(w)m(w_0) = m(w_0) = m(w_0)m(w)$ for every $w \in W$. If $w \in W$ and if $\mathbf{s} = (s_1, \ldots, s_k)$ is a reduced decomposition of w, then $m(w) = m(s_1)m(s_2) \cdots m(s_k)$. If $s \in S$, then $m(s)^2 = m(s)$.

The monoid algebra $\mathbb{Z}[M(W)]$ can be viewed as a degeneration of the Hecke algebra $\mathcal{H}$ of W (see Section 7).

There is a geometric interpretation of the multiplication in M in terms of the product of $(B_0 \times B_0)$–orbits of G. By the Bruhat Lemma, we have $G = \coprod_{w \in W} B_0 w B_0$, where the symbol $\coprod$ denotes the disjoint union. Let w, w' and $w'' \in W$. Then $m(w)m(w') = m(w'')$ if and only if $B_0 w'' B_0$ is the unique dense open $(B_0 \times B_0)$–orbit in the product $B_0 w B_0 \cdot B_0 w' B_0$.

3.2. Action of M on V. There is an action of M on the set V (or, equivalently, an action on $B_0 \backslash G / K$ or $\mathcal{K} \backslash \mathcal{B}$) defined as follows: If $v \in V$ and $w \in W$, then $m(w) \cdot v$ is the unique element $v' \in V$ such that $B_0 v' K$ is the dense open $(B_0 \times K)$–orbit in the product $B_0 w B_0 \cdot B_0 v K$. It follows from the geometric description of the multiplication in M that this defines an action of M on V. If $v \in V$, then $M \cdot v = \{ m(w) \cdot v \mid w \in W \}$ denotes the M–orbit of V.

Let $s \in S$, let $v \in V$ and let $a = \phi(v)$. Then $B_0 m(s) \cdot vK$ is the dense open $(B_0 \times K)$–orbit in $P_s vK$. If $m(s) \cdot v \neq v$, then we write $v \to m(s) \cdot v$. It follows from 2.5 that $v \to m(s) \cdot v$ if and only if one of the following two conditions holds: (i) s is complex for v and $l(sa) > l(a)$; or (ii) s is non-compact imaginary for v.

REMARK 3.2.1. If $v \to m(s) \cdot v$, then it is clear that $\dim \mathcal{O}(m(s) \cdot v) = \dim \mathcal{O}(v) + 1$ and $\dim \mathcal{K}(m(s) \cdot v) = \dim \mathcal{K}(v) + 1$.

LEMMA 3.2.2. *Let $s \in S$, let $v \in V$ and let $a = \phi(v)$. Assume that $l(sa) < l(a)$. Then there exists $v' \in V$ such that $v' \to m(s) \cdot v' = v$. Furthermore, if $v'' \in V$ is such that $v'' \to m(s) \cdot v'' = v$, then either* (i) $v'' = v'$ *or* (ii) *s is non-compact imaginary for v and $v'' = s \cdot v' \neq v'$.*

The proof follows easily from 2.4.

DEFINITION 3.2.3. Let $v \in V$. A *reduced decomposition* of v is a pair $(\mathbf{v}, \mathbf{s})$, where $\mathbf{v} = (v_0, \ldots, v_k)$ is a sequence in V and $\mathbf{s} = (s_1, \ldots, s_k)$ is a sequence in S, which satisfies the following conditions: (1) $v_0 \in V_0$ and $v_k = v$; and (2) for each $i = 0, 1, \ldots, k-1$, we have $v_i \to m(s_{i+1}) \cdot v_i = v_{i+1}$. We say that k is the *length* of the reduced decomposition $(\mathbf{v}, \mathbf{s})$.

It follows from Lemma 3.2.2 that every $v \in V$ has a reduced decomposition. It follows from Remark 3.2.1 that every reduced decomposition of v has length equal to $l(v)$. If $(\mathbf{v}, \mathbf{s})$ is a reduced decomposition of v, it is clear that v_0 and $\mathbf{s}$ determine $(\mathbf{v}, \mathbf{s})$. It is not necessarily the case that $\mathbf{v}$ determines $\mathbf{s}$.

Reduced decompositions of elements of V should be considered as the analogue of reduced decompositions of elements of the Weyl group. For the case in which $G = H \times H$ and $\theta(h, h') = (h', h)$, the orbits are parametrized by the Weyl group $W(H)$ and reduced decompositions of elements of V correspond to reduced decompositions of the corresponding elements of $W(H)$.

3.3. Action of M on $\mathcal{I}$. (See [**RS**, §3].) There is an action of the monoid M on the set $\mathcal{I}$ of twisted involutions which, in a sense, parallels the twisted action of the Weyl group W on $\mathcal{I}$. If $s \in S$ and $a \in \mathcal{I}$, we define $s \circ a \in \mathcal{I}$ as follows: if $s * a = a$, then $s \circ a = sa$; if $s * a \neq a$, then $s \circ a = s * a$. For each $s \in S$, the map $a \mapsto s \circ a$ is a fixed point free bijection of $\mathcal{I}$ of order two. We now define the twisted action of elements $m(s)$, $s \in S$, on $\mathcal{I}$. If $s \in S$ and $a \in \mathcal{I}$, we set $m(s) * a$ equal to $s \circ a$ if $l(sa) > l(a)$ and equal to a if $l(sa) < l(a)$. The proof that this extends to give an action of M on $\mathcal{I}$ is a bit tricky. If $m \in M$ and $a \in \mathcal{I}$, then $m * a$ denotes the twisted action of m on a.

LEMMA 3.3.1. *If $a \in \mathcal{I}$, then there exists $m \in M$ such that $m * 1 = a$.*

DEFINITION 3.3.2. Let $a \in \mathcal{I}$. Then $L(a)$, *the length of a as a twisted involution*, is the smallest integer k for which there exists $w \in W$ with $l(w) = k$ and $m(w) * 1 = a$.

It is clear that $L(a) \leq l(a)$, where $l(a)$ is the length of a as an element of the Coxeter group $W = (W, S)$, but it seldom happens that $L(a) = l(a)$. We can also define $L(a)$ in terms of the -1 eigenspace of the involution $a\theta$, acting on

$E = X(T) \otimes_{\mathbb{Z}} \mathbb{R}$. If $a \in \mathcal{I}$, let $E_-(a)$ denote the -1 eigenspace of $a\theta$ on E. Then $2L(a) = l(a) + \dim E_-(a) - \dim E_-(1)$.

Let $s \in S$ and $a \in \mathcal{I}$. If $a \neq m(s) * a$, then we write $a \to m(s) * a$. In this case $L(a) + 1 = L(m(s) * a)$.

The action of M on V and the twisted action of M on $\mathcal{I}$ are related by the following proposition.

PROPOSITION 3.3.3. *Let v, $v' \in V$ and let $s \in S$. (1) If $v' \to m(s) \cdot v' = v$, then $\phi(v') \to m(s) * \phi(v') = \phi(v)$. (2) Assume that $\phi(v') \to m(s) * \phi(v')$. Then $v' \to m(s) \cdot v'$ unless s is compact imaginary for v', in which case $v' = m(s) \cdot v'$.*

Note that it is not necessarily the case that the map $\phi : V \to \mathcal{I}$ is M-equivariant.

Now let $(\mathbf{v}, \mathbf{s} = (s_1, \dots, s_k))$ be a reduced decomposition of $v \in V$ and let $a = \phi(v)$. Then it follows from Proposition 3.3.3 that $a = m(s_k) * m(s_{k-1} * \cdots * m(s_1) * 1$ and that $L(a) = k$. Thus we obtain:

PROPOSITION 3.3.4. *Let $v \in V$ and let $a = \phi(v)$. Then $l(v) = L(a)$. Consequently* $\dim \mathcal{K}(v) = L(a) + \dim \mathcal{B}(K^0)$ *and* $\dim \mathcal{O}(v) = L(a) + \dim \mathcal{B}(K^0) + \dim B$.

We see from Proposition 3.3.4 that the dimensions of the orbits $\mathcal{K}(v)$ and $\mathcal{O}(v)$ are determined by $L(\phi(v))$, the length of $\phi(v)$ as a twisted involution.

We define the *weak order* on $\mathcal{I}$, denoted by $\vdash$, as follows: Let a, $b \in \mathcal{I}$. Then $a \vdash b$ if $b \in M * a$, where $M * a$ denotes the (twisted) M-orbit of a.

The next proposition describes the image of the map ϕ.

PROPOSITION 3.3.5. *The following three conditions on $a \in \mathcal{I}$ are equivalent:* (i) *$a \in$* image(ϕ); (ii) *$a \vdash a_{max}$; and* (iii) *there exists $b \in W * a$ such that $E_-(b) \subset E_-(a_{max})$.*

3.4. $(P \times K)$**–orbits on** G. Let I be a subset of Δ and let $P = P_I$ be the corresponding standard parabolic subgroup. Then each $(B_0 \times K)$–orbit of G is contained in a unique $(P \times K)$–orbit, so that we have a surjective map $f_I : B_0 \backslash G/K \to P \backslash G/K$ of the orbit sets. It is of interest to describe the fibres of the map f_I. We shall how indicate how to handle this problem in terms of the $M(W)$ formalism. Since this problem is not discussed in [**RS**], we shall give proofs in this subsection.

PROPOSITION 3.4.1. *Let v', $v \in V$. Then the following conditions are equivalent*:

(1) $Pv'K = PvK$.
(2) $m(w_I) \cdot v' = m(w_I) \cdot v$.

PROOF. (1) $\Rightarrow$ (2). Since $B_0 w_I B_0$ is a dense open subset of P, it is clear that the product $B_0 w_I B_0 \cdot B_0 v' K$ (respectively $B_0 w_I B_0 \cdot B_0 v K$) is open and dense in $Pv'K$ (respectively PvK). Consequently we see that the sets $B_0 w_I B_0 \cdot B_0 v' K$ and $B_0 w_I B_0 \cdot B_0 v K$ intersect in a dense open subset of $Pv'K = PvK$, which implies that (2) holds.

(2) $\Rightarrow$ (1). Let $v'' = m(w_I) \cdot v'$. Then $B_0 v'' K$ is a dense open subset of $B_0 w_I B_0 \cdot B_0 v' K$, hence a dense open subset of $Pv'K$. Similarly it follows that $B_0 v'' K$ is a dense open subset of PvK. Thus we see that $Pv'K = Pv''K = PvK$.

It is clear that each $(P \times K)$–orbit of G contains a (unique) open dense $(B_0 \times K)$–orbit. Let V_I be the set of all $v \in V$ such that $B_0 v K$ is an open dense subset of PvK. Thus the map $v \mapsto PvK$ is a bijection from V_I to $P\backslash G/K$.

PROPOSITION 3.4.2. *The following conditions on $v \in V$ are equivalent*:

(1) $v \in V_I$.
(2) $m(s_\alpha) \cdot v = v$ *for every* $\alpha \in I$.
(3) $m(w_I) \cdot v = v$.
(4) $v = m(w_I) \cdot v'$ *for some* $v' \in V$.

PROOF. (1) $\Rightarrow$ (2). If (1) holds, then $B_0 v K$ is dense in PvK. If $\alpha \in I$ and $s = s_\alpha$, then $P_s \subset P$, so that $P_s v K \subset PvK \subset \overline{B_0 v K}$. Thus (2) holds.

(2) $\Leftrightarrow$ (3) If (2) holds, then (3) holds, since $m(w_I)$ is a product of terms of the form $m(s_\alpha)$, $\alpha \in I$. Assume (3) holds. If $\alpha \in I$, then $s_\alpha w_I < w_I$, so that $m(s_\alpha) m(w_I) = m(w_I)$. Hence (2) holds.

(3) $\Rightarrow$ (4). Trivial

(4) $\Rightarrow$ (1). If (4) holds, then $B_0 v K$ is dense in $B_0 w_I B_0 \cdot B_0 v' K$, and hence is dense in $Pv'K$, so that $v \in V_I$.

COROLLARY 3.4.3. *For each $v \in V_I$, let $\Gamma(v) = \{ v' \in V \mid m(w_I) \cdot v' = v \}$. Then we have*:

(1) $G = \coprod_{v \in V_I} PvK$.
(2) *For each $v \in V_I$, we have* $PvK = \coprod_{v' \in \Gamma(v)} B_0 v' K$.

We note that Proposition 3.4.2 and Corollary 3.4.3 give a description of the fibres of the map f_I in terms of the action of M on V.

Let $\mathcal{P}_I$ be the variety of all conjugates of the parabolic subgroup P. Then there is an obvious correspondence between the K–orbits on $\mathcal{P}_I$ and the $(P \times K)$–orbits on G. Thus all of the results of this subsection have an interpretation in terms of the K–orbits on $\mathcal{P}_I$.

In a number of concrete examples, one can give quite precise descriptions of the sets V_I and $\Gamma(v)$. See the comments in §5.4.

4. The partial order on the orbits

(See [**RS**, §7].) We define a partial order on the set V of orbits as follows: $v' \leq v$ if and only if $\mathcal{O}(v') \subset \overline{\mathcal{O}(v)}$. We call this the Bruhat order on V because of the analogy with the usual Bruhat order on W. It is also useful to define another partial order on V, the *weak order*, denoted by $\vdash$. If v', $v \in V$, we write $v' \vdash v$ if v belongs to $M \cdot v'$, the M-orbit of v'. This is equivalent to the condition that there exists a reduced decomposition $(\mathbf{v} = (v_0, \dots, v_k), \mathbf{s})$ of v and an integer j, $0 \leq j \leq k$, such that $v' = v_j$. It is clear that $\vdash$ is a partial order on V and that $v' \vdash v$ implies that $v' \leq v$.

The following result is the key to the combinatorial description of the Bruhat order on V.

PROPOSITION 4.1. *Let $s \in S$ and $v \in V$ and assume that $v \to m(s) \cdot v$. Then*

$$\overline{\mathcal{O}(m(s) \cdot v)} = \bigcup_{v' \leq v} P_s \mathcal{O}(v').$$

The proof of Proposition 4.1 is by an easy induction on $l(v)$.

It will be convenient to translate Proposition 4.1 into a combinatorial framework involving the action of M on V. If $s \in S$ and $v \in V$, we let

$$p(s,v) = \{ v' \in V \mid \mathcal{O}(v') \subset P_s \mathcal{O}(v) \} = \{ v' \in V \mid m(s) \cdot v' = m(s) \cdot v \}.$$

It follows from the case analysis of 2.5 that $p(s,v)$ contains either one, two or three elements. If s is compact imaginary for v, then $p(s,v) = \{v\}$. If s is complex for v, then $p(s,v) = \{v, s \cdot v\}$ and $s \cdot v \neq v$. If s is either real or noncompact imaginary for v, then $p(s,v)$ contains either two or three elements. We observe that $\{v, m(s) \cdot v\} \subset p(s,v)$ and that, if $v' \in p(s,v)$, then $p(s,v) = p(s,v')$.

The following result is a reformulation of Proposition 4.1:

PROPOSITION 4.2. *Let $s \in S$ and $v \in V$ and assume that $v \to m(s) \cdot v$. Then*

$$\{ v' \in V \mid v' \leq m(s) \cdot v \} = \bigcup_{v' \leq v} p(s, v').$$

Thus, for $y \in V$, we have $y \leq m(s) \cdot v$ if and only if there exists $v' \leq v$ such that $m(s) \cdot y = m(s) \cdot v'$.

We note that Proposition 4.2 gives an elementary inductive description of the partial order on V in terms of the M-action on V. In [**RS**], the property of the partial order $\leq$ on V given by Proposition 4.2 is called the "one-step property". It describes how the partial order behaves as we go up one step from v to $m(s) \cdot v$. (The description of the set $p(s,v)$ in [**RS**] is slightly different from the one we have given here.)

The Bruhat order on V has a number of properties which are generalizations of standard properties of the Bruhat order on the Coxeter group W = (W,S). We list below a number of these properties. The proofs are in [**RS**, §5-§7] and involve a considerable amount of combinatorial formalism.

DEFINITION 4.3. Let $(\mathbf{v} = (v_0, \ldots, v_k), \mathbf{s} = (s_1, \ldots, s_k))$ be a reduced decomposition of v. A sequence $\mathbf{u} = (u_0, \ldots, u_k)$ in V is a *subexpression* of $(\mathbf{v}, \mathbf{s})$ if $u_0 = v_0$ and, for every $i = 0, \ldots, k-1$, one of the following three alternatives holds: (a) $u_{i+1} = u_i$; (b) $u_i \to m(s_{i+1}) \cdot u_i = u_{i+1}$; or

(c)
$$u_i \to m(s_{i+1}) \cdot u_i, \qquad u_{i+1} \to m(s_{i+1}) \cdot u_{i+1}$$
$$\text{and} \quad m(s_{i+1}) \cdot u_i = m(s_{i+1}) \cdot u_{i+1}.$$

If $\mathbf{u} = (u_0, \ldots, u_k)$ is a subexpression of $(\mathbf{v}, \mathbf{s})$, then u_k is the *final term* of $\mathbf{u}$.

In Definition 4.3, if alternative (c) holds and if $u_i \neq u_{i+1}$, then s_{i+1} is non-compact imaginary for u_i and $u_{i+1} = s_{i+1} \cdot u_i$.

PROPOSITION 4.4. *Let v', $v \in V$ and let $(\mathbf{v}, \mathbf{s})$ be a reduced decomposition of v. Then $v' \leq v$ if and only if there is a subexpression $\mathbf{u}$ of $(\mathbf{v}, \mathbf{s})$ with final term v'.*

PROPOSITION 4.5 (The exchange property). *Let $(\mathbf{v}, \mathbf{s} = (s_1, \ldots, s_k))$ be a reduced decomposition of v. Let s, v' be such that $v' \to m(s) \cdot v' = v$. Then there exists $i \in [1, k]$ and a reduced decomposition $(\mathbf{u} = (u_0, \ldots, u_k), \mathbf{s}')$ of v such that $u_{k-1} = v'$ and $\mathbf{s}' = (s_1, \ldots, \widehat{s}_i, \ldots, s_k, s)$.*

PROPOSITION 4.6 (Property $Z(s, u, v)$). *Let u, $v \in V$ and $s \in S$ be such that $u \to m(s) \cdot u$ and $v \to m(s) \cdot v$. Then the following three properties are equivalent:* (i) *either $u \leq v$ or there exists u' with $u' \to m(s) \cdot u' = m(s) \cdot u$ such that $u' \leq v$;* (ii) *$m(s) \cdot u \leq m(s) \cdot v$; and* (iii) *$u \leq m(s) \cdot v$.*

PROPOSITION 4.7. *Let u, $v \in V$ be such that $v \to m(s) \cdot v$ and $u \leq m(s) \cdot v$. Then one of the following three conditions holds:* (i) *$u \leq v$;* (ii) *$u \to m(s) \cdot u$ and there exists u' such that $u' \to m(s) \cdot u' = m(s) \cdot u$ and $u' \leq v$; and* (iii) *there exists $u'' \leq v$ such that $u'' \to m(s) \cdot u'' = u$.*

PROPOSITION 4.8 (The chain condition). *If $u \leq v$, then there exists a sequence $u = u_0 < u_1 < \cdots < u_k = v$ with $l(u_{i+1}) = l(u_i) + 1$ for $i = 0, \ldots, k-1$.*

DEFINITION 4.9. A partial order $\preceq$ on V is *compatible* with the M-action on V if the following three conditions are satisfied for all u, $v \in V$ and $s \in S$: (i) $v \preceq m(s) \cdot v$; (ii) if $u \preceq v$, then $m(s) \cdot u \preceq m(s) \cdot v$; and (iii) if $u \preceq v$ and $l(v) \leq l(u)$, then $u = v$.

PROPOSITION 4.10. *The Bruhat order on V is the weakest partial order on V which is compatible with the M-action on V.*

A direct description of the weakest partial order on V compatible with the M-action on V is given in [**RS**, §5.2].

4.11. The Bruhat order on $\mathcal{I}$. Using the M-action on $\mathcal{I}$ and the length function L on $\mathcal{I}$, we can define precise analogues of all of the earlier results of this section for the set $\mathcal{I}$ of twisted involutions. The definition of a reduced decomposition of an element of $\mathcal{I}$ is essentially the same as that given in 4.3 for reduced decompositions of elements of V. We also have an obvious definition of a partial order $\preceq$ on $\mathcal{I}$ being compatible with the M-action. The Bruhat order on $\mathcal{I}$ is defined to be the weakest partial order on $\mathcal{I}$ which is compatible with the M-action. It is a surprising fact (at least to the authors) that this Bruhat order on $\mathcal{I}$ agrees with the restriction to $\mathcal{I}$ of the usual Bruhat order on W. In [**RS**], it was incorrectly stated that these two partial orders on $\mathcal{I}$ were not necessarily equal. A proof that they are equal appears in [**RS1**]. We refer the reader to [**RS**, §5 and §8] for the precise formulation of analogues for the Bruhat order on $\mathcal{I}$ of 4.3-4.10 above. We will only formulate the appropriate exchange condition:

PROPOSITION 4.11.1 (The exchange condition). *Let $a \in \mathcal{I}$, let $L(a) = k$ and let $\mathbf{s} = (s_1, \ldots, s_k)$ be a sequence in S such that $a = s_k \circ s_{k-1} \circ \cdots \circ s_1 \circ 1$. Let $s \in S$ be such that $s \circ a < a$ (or, equivalently, such that $sa < a$). Then there exists $i \in 1, \ldots, k$ such that $s \circ a = s_k \circ \cdots \circ \widehat{s_i} \circ \cdots \circ s_1 \circ 1$.*

It seems to be a non-trivial exercise to give a direct proof of 4.11.1.

§5 Combinatorial parametrization of the orbits. The Hermitian symmetric case

It follows from the Bruhat Lemma that the $(B_0 \times B_0)$–orbits on G (or, equivalently, the G–orbits on $\mathcal{B} \times \mathcal{B}$) can be canonically parametrized by the Weyl group W. We would like a similar parametrization for the K–orbits on $\mathcal{B}$. For one class of pairs (G, θ), those of "Hermitian symmetric type", there exists an elementary parametrization which is very satisfactory. We shall describe this parametrization in this section. A detailed exposition will appear in a paper now in preparation by one of us (RWR).

5.1. The hermitian symmetric type. We will assume in §5 that G is simple (over its center) and simply connected. We consider first the case $F = \mathbb{C}$. In this case, G is a simple complex Lie group and has an underlying structure of a simple real Lie group. By a Cartan involution of G we mean a Cartan involution of the underlying simple real Lie group. It is known that there exists a Cartan involution τ of G which commutes with θ. Let $\sigma = \theta\tau = \tau\theta$. Then the fixed point subgroups $U = G^\tau$ and $G_\mathbb{R} = G^\sigma$ are real forms of G and U is a maximal compact subgroup of G. Let $K_\mathbb{R} = U^\theta = U \cap K = (G_\mathbb{R})^\theta = G_\mathbb{R} \cap U$. Then $K_\mathbb{R}$ is a maximal compact subgroup of both K and $G_\mathbb{R}$, and the coset spaces $U/K_\mathbb{R}$ and $G_\mathbb{R}/K_\mathbb{R}$ are dual irreducible Riemannian symmetric spaces. We say that the pair (G, θ) is of Hermitian symmetric type if $G_\mathbb{R}/K_\mathbb{R}$ is a Hermitian symmetric space; this is equivalent to the condition that $U/K_\mathbb{R}$ be a Hermitian symmetric space. It is known that $G_\mathbb{R}/K_\mathbb{R}$ is Hermitian symmetric if and only if the center of $K_\mathbb{R}$ (or of K) has positive dimension (see [**H**, Chap. 8]).

We return to the case where F is algebraically closed of characteristic $\neq 2$.

DEFINITION 5.1.1. Let G be simple and simply connected and let θ be an involutive automorphism of G. Then we say that (G, θ) is of *Hermitian symmetric type* if the center of $G^\theta = K$ has positive dimension.

Roughly speaking, involutions θ of G such that (G, θ) is of Hermitian symmetric type correspond to parabolic subgroups of G with abelian unipotent radical. A precise statement is given in Theorem 5.1.2 below.

In Theorem 5.1.2, G is simple and simply connected and we do not assume that we are given in advance an involution θ and a standard pair (B_0, T_0).

THEOREM 5.1.2. (1) *Let P be a parabolic subgroup of G with abelian unipotent radical. Let B_0 be a Borel subgroup of P and let T_0 be a maximal torus of B_0. Let $\Phi = \Phi(G, T_0)$, let $\Phi^+ = \Phi(T_0, B_0)$ and let $\Delta = \Delta(B_0, T_0)$. Let $\widetilde{\alpha} =$*

$\sum_{\alpha\in\Delta} n_\alpha(\widetilde{\alpha})\alpha$ denote the highest root. Let $J \subset \Delta$ be such that $P = P_J$. Then J is equal to $\Delta \setminus \{\gamma\}$, where γ is an element of Δ such that $n_\gamma(\widetilde{\alpha}) = 1$. In particular, P is a maximal (proper) parabolic subgroup of G. Let $c_\gamma \in T_0$ be such that $\gamma(c_\gamma) = -1$ and $\alpha(c_\gamma) = 1$ for $\alpha \in J$ and let $\theta = \mathrm{Int}(c_\gamma)$ denote the inner automorphism induced by c_γ. Then θ is an involutive automorphism of G and $G^\theta = K$ is the unique Levi subgroup of P containing T_0. In particular, K has a center of positive dimension, so that (G, θ) is of Hermitian symmetric type.

(2) *Let B_0 be a Borel subgroup of G and let T_0 be a maximal torus of B_0. Let Φ, Δ and $\widetilde{\alpha}$ be as in* (1) *above. Let $\gamma \in \Delta$ be such that $n_\gamma(\widetilde{\alpha}) = 1$ and let $J = \Delta \setminus \{\gamma\}$. Then the standard parabolic subgroup P_J has abelian unipotent radical.*

(3) *Let θ be an involutive automorphism of G such that the center of $K = G^\theta$ has positive dimension. Then there exist P, B_0, T_0, c_γ, $\dots$ as in* (1) *above such that $\theta = \mathrm{Int}(c_\gamma)$ and $K = G^\theta$ is the unique Levi subgroup of P containing T_0.*

See [**RRoSt**] for (1) and (2).

Using the classification of involutions [**S2**], one can show that the classification (over F) of pairs (G, θ) of Hermitian symmetric type corresponds exactly to the classification of simply connected, irreducible Hermitian symmetric spaces of compact type. See [**H**, p. 518] for this classification.

REMARK 5.1.3. Let $\Delta = \{\alpha_1, \dots, \alpha_n\}$ and let the simple roots be indexed as in [**Bou**, Planches I-IX]. We list below all simple roots γ such that $n_\gamma(\widetilde{\alpha}) = 1$. The corresponding (maximal) parabolic subgroups P_J, where $J = \Delta \setminus \{\gamma\}$, give all standard parabolic subgroups with abelian unipotent radical.

(1) Type A_n. $\gamma = \alpha_1, \alpha_2, \dots, \alpha_n$.
(2) Type B_n. $\gamma = \alpha_1$.
(3) Type C_n. $\gamma = \alpha_n$.
(4) Type D_n. $\gamma = \alpha_1, \alpha_{n-1}, \alpha_n$.
(5) Type E_6. $\gamma = \alpha_1, \alpha_6$.
(6) Type E_7. $\gamma = \alpha_7$.

For types E_8, F_4 and G_2, there are no proper parabolic subgroups with abelian unipotent radical.

5.2. The parameter set $\mathcal{E}$. For the rest of §5, we will assume that (G, θ) is of hermitian symmetric type. Let P, B_0, T_0, Δ, $J = \Delta \setminus \{\gamma\}$, $\dots$, be as in Theorem 5.1.2(1). We note that, since θ is an inner automorphism, $\mathcal{I}$ is the set of involutions of W. For each $v \in V$, the involution $\phi(v)$ is an invariant associated to the orbit $\mathcal{O}(v)$. In the case at hand, there is a second natural invariant which one can associate to $\mathcal{O}(v)$. Since $K \subset P$, each $(B_0 \times K)$–orbit of G is contained in a unique $(B_0 \times P)$–orbit and it is known that the set of $(B_0 \times P)$–orbits is canonically parametrized by W/W_J. Thus we obtain a surjective map $\nu : V \to W/W_J$ defined by: $\nu(v) = dW_J$ if $B_0vK \subset B_0dP$. Let $D = D_J = \{d \in W \mid d(J) \subset \Phi^+\}$ be the set of minimal left coset representatives for W/W_J. For $d \in D$, let $V(d) = \{v \in V \mid \nu(v) = dW_J\}$. Thus, for $v \in V$

and $d \in D$, we have $v \in V(d)$ if and only if $B_0vK \subset B_0dP$. Consequently, $B_0dP = \coprod_{v \in V(d)} \mathcal{O}(v)$.

If $sd < d$, then $sd \in D$ and we have

$$P_sB_0sdP = B_0sdP \cup B_0sB_0sdP = B_0sdP \cup B_0dP.$$

The following theorem gives detailed information on the corresponding relationship between the sets of orbits $V(sd)$ and $V(d)$.

THEOREM 5.2.1. *Let $s \in S$ and $d \in D$, with $sd < d$.*

(1) *Let $v \in V(sd)$. Then:* (i) $v \to m(s) \cdot v$ *and* $v \neq s \cdot v$; (ii) $BsB\mathcal{O}(v) = \mathcal{O}(s \cdot v) \cup \mathcal{O}(m(s) \cdot v) \subset BdP$; *and* (iii) $P_s\mathcal{O}(v) = \mathcal{O}(v) \cup \mathcal{O}(s \cdot v) \cup \mathcal{O}(m(s) \cdot v)$. *Furthermore s is either complex for v, in which case $s \cdot v = m(s) \cdot v$, or noncompact imaginary for v, in which case $s \cdot v \neq m(s) \cdot v$.*

(2) *Let v, $v' \in V(sd)$, with $v \neq v'$. Then $P_s\mathcal{O}(v)$ and $P_s\mathcal{O}(v')$ are disjoint.*

(3) $V(d) = s \cdot V(sd) \cup m(s) \cdot V(sd) = \coprod_{v \in V(sd)} \{s \cdot v, m(s) \cdot v\}$.

We now have the following key theorem:

THEOREM 5.2.2. *Define $\eta : V \to \mathcal{I} \times W/W_J$ by $\eta(v) = (\phi(v), \nu(v))$. Then η is injective.*

SKETCH OF PROOF. In order to prove Theorem 5.2.2, it suffices to prove that the restriction of ϕ to each subset $V(d)$, $d \in D$, is injective. The proof is by induction on $l(d)$. Let $sd < d$. In the inductive step of the proof, one needs to analyze the passage from $V(sd)$ to $V(d)$, and the necessary information for doing this is given by Theorem 5.2.1.

It follows from Theorem 5.2.2 that the subset set $\mathcal{E} = \text{image}(\eta)$ of $\mathcal{I} \times W/W_J$ is a parameter set for the set V of orbits. If $d \in D$, let $\mathcal{I}(d) = \{ a \in \mathcal{I} \mid (a, d) \in \mathcal{E} \}$. Thus $\mathcal{E} = \{(a, d) \in \mathcal{I} \times W/W_J \mid a \in \mathcal{I}(d) \}$. Note that $1 \in D$. Since $B_0K = B_0P = P$, it is easy to see that $\mathcal{I}(1) = \{1\}$. The following lemma gives an elementary inductive description of the sets $\mathcal{I}(d)$, and hence of the parameter set $\mathcal{E}$.

LEMMA 5.2.3. *Let $d \in D$ and $s \in S$ be such that $sd < d$. Then*

$$\mathcal{I}(d) = s * \mathcal{I}(sd) \cup m(s) * \mathcal{I}(sd).$$

The proof follows from 1.4.4, 3.3.3 and 5.2.1(3).

Next we describe the image of η. Let Ψ denote $\Phi(T_0, R_u(P))$, the support of $R_u(P)$.

THEOREM 5.2.4. *Let $(a, dW_J) \in \mathcal{I} \times W/W_J$, with $d \in D$. Then $(a, dW_J) \in$ image(η) if and only if there exists a sequence $(\beta_1, \dots, \beta_k)$ of mutually orthogonal roots in $\Phi^+ \cap d(-\Psi)$ such that $a = s_{\beta_1}s_{\beta_2}\cdots s_{\beta_k}$.*

Briefly, the proof goes as follows. For each $d \in D$, let $\mathcal{A}(d)$ be the set of all $a \in \mathcal{I}$ such that $a = s_{\beta_1}s_{\beta_2}\cdots s_{\beta_k}$, where $(\beta_1, \dots, \beta_k)$ is a sequence of mutually orthogonal roots in $\Phi^+ \cap d(-\Psi)$. Let $sd < d$. Then one proves that

$$\mathcal{A}(d) = s * \mathcal{A}(sd) \cup m(s) * \mathcal{A}(sd).$$

Since it is clear that $\mathcal{A}(1) = \{1\}$, it follows from 5.2.3 that $\mathcal{A}(d) = \mathcal{I}(d)$, which proves 5.2.4.

5.3. An example. Let $G = SL(n, F)$ and let $\gamma = \alpha_k$ (we follow the notation of Remark 5.1.3). In this example, the coset space G/P is the Grassmann variety of k-planes in F^n and

$$K = \{ (g, h) \in GL(k, F) \times GL(n-k, F) \mid \det(g)\det(h) = 1 \},$$

where we embed $GL(k, F) \times GL(n-k, F)$ in $GL(n, F)$ in the obvious way. Assume that $k \leq n-k$.

We will give an elementary description of the parameter set $\mathcal{E}$.

We identify the Weyl group with the symmetric group S_n. Then W_J is the stabilizer of $[1, k]$ in S_n. Each involution a in S_n is a product of disjoint two-cycles:

$$a = (i_1, j_1)(i_2, j_2) \cdots (i_r, j_r) \quad \text{with} \quad i_p < j_p, \quad p = 1, \ldots, r.$$

We set $\mathrm{Hi}(a) = \{j_1, \ldots, j_r\} = \{j \in [1, n] \mid a(j) < j\}$ and $\mathrm{Lo}(a) = \{i_1, \ldots, i_r\} = \{i \in [1, n] \mid a(i) > i\}$; we say that $\mathrm{Hi}(a)$ (resp. $\mathrm{Lo}(a)$) is the set of *high points* (resp. *low points*) of a. Let $\Sigma = \Sigma(k, n)$ denote the set of all k-element subsets of $[1, n]$. If $w \in W$, let $\underline{w} \in \Sigma$ denote $\{w(1), \ldots, w(k)\}$. Then the map $w \to \underline{w}$, $W \to \Sigma$, is constant on left coset of W_J and induces a bijection from W/W_J to Σ.

The set D of minimal left coset representatives is given by

$$D = \{ d \in W \mid d(1) < d(2) \cdots < d(k) \text{ and } d(k+1) < d(k+2) \cdots < d(n) \}.$$

We have $\Sigma = \{ \underline{d} \mid d \in D \}$. The subset Ψ (the support of $R_u(P)$) is equal to $\{ \varepsilon_i - \varepsilon_j \mid 1 \leq i \leq k < j \leq n \}$. (We follow the notation of [**Bou**, Planche I].) Let $d \in D$. Then an easy argument shows that

$$\Phi^+ \cap d(-\Psi) = \{ \varepsilon_i - \varepsilon_j \mid i < j,\ j \in \underline{d} \text{ and } i \notin \underline{d} \}.$$

It follows from this that

$$\mathcal{I}(d) = \{ a \in \mathcal{I} \mid \mathrm{Hi}(a) \subset \underline{d} \text{ and } \mathrm{Lo}(a) \cap \underline{d} = \emptyset \}.$$

Thus the set

$$\mathcal{M} = \{ (a, \underline{d}) \in \mathcal{I} \times \Sigma \mid \mathrm{Hi}(a) \subset \underline{d} \text{ and } \mathrm{Lo}(a) \cap \underline{d} = \emptyset \}$$

is a parameter set for the set V of orbits.

We discuss in more detail the case $n = 4$, $k = 2$. In this case, G/P is the set of 2-planes in F^4 and Σ is the set of two-element subsets of $[1, 4]$. For each $\underline{d} \in \Sigma$, we list below the involutions in $\mathcal{I}(d)$.

$\underline{d}$	$\mathcal{I}(d)$
$\{1,2\}$	1
$\{1,3\}$	$1,\ (2,3)$
$\{1,4\}$	$1,\ (2,4),\ (3,4)$
$\{2,3\}$	$1,\ (1,2),\ (1,3)$
$\{2,4\}$	$1,\ (1,2),\ (1,4),\ (3,4),\ (1,2)(3,4)$
$\{3,4\}$	$1,\ (1,3),\ (1,4),\ (2,3),\ (2,4),\ (1,3)(2,4),\ (1,4)(2,3)$

Thus we see that there are 21 orbits. Note that this list agrees with the list in **[C]**.

These examples for type A_n were worked out by P. D. Ryan **[Ry]** in his M.Sc. thesis at the Australian National University. Ryan also gives similar concrete models of the parameter sets for the other pairs (G, θ) of Hermitian symmetric type whenever G is a classical group (i.e., for the examples of type B_n, C_n and D_n in Remark 5.1.3).

5.4. The actions of M and W on the parameter set $\mathcal{E}$. We define actions of M and W on the parameter set $\mathcal{E} = \text{image}(\eta)$ by requiring that $\eta : V \to \mathcal{E}$ be an isomorphism of M-sets and of W-sets. Thus, if $w \in W$ and $v \in V$, then we have $\eta(m(w) \cdot v) = m(w) \cdot \eta(v)$ and $\eta(w \cdot v) = w \cdot \eta(v)$.

In order to describe the action of M on $\mathcal{E}$, we first need to define an action of M on the set W/W_J. Let $s \in S$ and let $d \in D$. Then we define $m(s) \cdot dW_J$ by the following rules: (i) if $sd < d$, then $m(s) \cdot dW_J = dW_J$; (ii) if $sd > d$, then $m(s) \cdot dW_J = sdW_J$. This determines an action of M on W/W_J.

REMARK. In the definition above, if $sd > d$, then $sd \in D$ if and only if $sdW_J \neq dW_J$. If $sdW_J = dW_J$, then we have $m(s) \cdot dW_J = dW_J$.

The following theorem determines the actions of M and W on $\mathcal{E}$:

THEOREM 5.4.1. *Let $s \in S$ and $(a, dW_J) \in \mathcal{E}$.*

(1) *$m(s) \cdot (a, dW_J)$ is equal to $(m(s) * a, m(s) \cdot dW_J)$ if $(m(s) * a, m(s) \cdot dW_J) \in \mathcal{E}$ and is equal to (a, dW_J) if $(m(s) * a, m(s) \cdot dW_J) \notin \mathcal{E}$. The latter case can only occur if s is imaginary for a and $sdW_J = dW_J$.*
(2) *$s \cdot (a, dW_J)$ is equal to (sas, sdW_J) if $(sas, sdW_J) \in \mathcal{E}$ and is equal to (a, dW_J) if $(sas, sdW_J) \notin \mathcal{E}$. The latter case occurs if and only if s is real for a and $sd < d$.*

5.5. The opposite parabolic P^-. Let P^- be the unique parabolic subgroup opposite to P which contains T_0. Then $K = P \cap P^-$. If $J' = -w_0(J)$, then $P^- = w_0 P_{J'} w_0$. In particular, P^- is conjugate to P if and only if $J = J'$.

LEMMA 5.5.1. (1) *The map $c \to BcP^-$ is a bijection from D to $B_0 \backslash G/P^-$.* (2) *Let c, $d \in D$. Then $BdP^- \subset \overline{BcP^-}$ if and only if $c \leq d$.* (3) *Let c, $d \in D$.*

Then $BdP \cap BcP^- \neq \emptyset$ if and only if $c \leq d$. If $c \leq d$, then $BdP \cap BcP^-$ is a smooth irreducible variety of dimension $l(d) - l(c) + \dim P$.

PROOF. The proof of (1) follows from [**RRoSt**, §5]. The proof of (2) and (3) follows from [**R2**, §3] (see, in particular, 3.6, 3.7 and 3.9.1).

Note the reversal of order in (2) of Lemma 5.6.1.

Since $K \subset P^-$, every $(B_0 \times K)$–orbit is contained in a unique $(B_0 \times P^-)$–orbit. It follows from Lemma 5.6.1 that the $(B_0 \times P^-)$–orbits are also parametrized by W/W_J. We define a map $\nu' : V \to W/W_J$ as follows: $\nu'(v) = cW_J$ if $B_0vK \subset B_0cP^-$. We let $\rho : V \to \mathcal{I} \times W/W_J \times W/W_J$ denote the map $v \mapsto (\phi(v), \nu(v), \nu'(v))$. We let $\mathcal{E}^*$ denote the subset image(ρ) of $\mathcal{I} \times W/W_J \times W/W_J$. Thus ρ maps V bijectively onto $\mathcal{E}^*$ and we may also use $\mathcal{E}^*$ as a parameter set for the orbits.

The relation between the maps ν and ν' is given by the following lemma:

LEMMA 5.5.2. *Let $v \in V$ and let $\rho(v) = (a, dW_J, cW_J)$. Then $dW_J = acW_J = m(a) \cdot cW_J$.*

5.6 Partial order on the orbits. The following lemma is easy to prove:

LEMMA 5.6.1. *Let v', $v \in V$, let $\rho(v') = (a', d'W_J, c'W_J)$ and let $\rho(v) = (a, dW_J, cW_J)$, with c', d', c, $d \in D$. If $v' \leq v$, then $a' \leq a$ and $c \leq c' \leq d' \leq d$.*

It has been conjectured by P. D. Ryan that the converse of Lemma 5.6.1 holds.

CONJECTURE 5.6.2 (P. D. Ryan). *Let v', $v \in V$, let $\rho(v') = (a', d'W_j, c'W_J)$ and let $\rho(v) = (a, dW_J, cW_J)$, with c', d', c, $d \in D$. Then $v' \leq v$ if and only if $a' \leq a$ and $c \leq c' \leq d' \leq d$.*

This conjecture is implicit in Ryan's M.Sc. thesis [**Ry**], although it is not explicitly stated there. In loc. cit., Ryan gives considerable evidence for this conjecture. We (PDR and RWR) now have a proof of the following slightly weaker result:

THEOREM 5.6.3. *Let the notation be as in Conjecture 5.6.2 and assume further that $l(v) = l(v') + 1$. Then $v' \leq v$ if and only if $a' \leq a$ and $c \leq c' \leq d' \leq d$.*

Since the partial order on V satisfies the chain condition (see Proposition 4.8), we see that Theorem 5.6.3 gives an elementary description of the partial order on the set V of orbits in terms of the parameter set $\mathcal{E}^*$.

If $d \in D$, it is not difficult to see that B_0dK is a closed subset of G. Let $v(d)$ be the corresponding element of V, so that $B_0v(d)K = B_0dK$. Then V_0, the set of closed orbits, is equal to $\{\, v(d) \mid d \in D \,\}$ and $v(d) \neq v(d')$ if $d \neq d'$. If $v \in V$, let $\mathcal{C}_0(v) = \{\, v_0 \in V_0 \mid v_0 \leq v \,\}$. We can also prove the following special case of Ryan's conjecture (which is not covered by Theorem 5.6.3).

THEOREM 5.6.4. *Let $v \in V$ and let $\rho(v) = (a, dW_J, cW_J)$, with c, $d \in D$. Then $\mathcal{C}_0(v) = \{\, v(d') \mid c \leq d' \leq d \,\}$.*

Here is an equivalent form of Ryan's conjecture:

CONJECTURE 5.6.5. *Let v', $v \in V$, let $a' = \phi(v)$ and let $a = \phi(v)$. Then $v' \leq v$ if and only if $a' \leq a$ and $\mathcal{C}_0(v') \subset \mathcal{C}_0(v)$.*

§6. Orbits of real forms of G and duality of orbits

In §6 the base field F will be the field $\mathbb{C}$ of complex numbers. Other than this, the notation is as in §1–§4. We will have occasion to consider two distinct topologies on algebraic varieties and their subsets, the classical (Hausdorff) topology and the Zariski topology. Unless explicitly indicated otherwise, in §6 all references to topological concepts will refer to the classical topology. Thus a closed set is closed in the classical topology and a Zariski-closed subset is closed in the Zariski topology. By a torus, we mean a compact Lie group isomorphic to a product of circle groups, as opposed to an algebraic torus, which is an algebraic group isomorphic to a product of multiplicative groups $\mathbb{C}^*$.

Let τ be a Cartan involution of (the underlying real Lie group of) G which commutes with θ. Let $\sigma = \theta\tau = \tau\theta$ and let $G_{\mathbb{R}} = G^{\sigma}$ and $U = G^{\tau}$ be the fixed point subgroups. Then U and $G_{\mathbb{R}}$ are real forms of G and U is a maximal compact subgroup of G. We wish to study the orbits of the real Lie group $G_{\mathbb{R}}$ on the flag manifold $\mathcal{B}$ or, equivalently, the $(B_0 \times G_{\mathbb{R}})$–orbits on G. In general, the $G_{\mathbb{R}}$–orbits on $\mathcal{B}$ will not be complex submanifolds of $\mathcal{B}$. We also study orbits of $G_{\mathbb{R}}$ on generalized flag manifolds $\mathcal{P}_I \cong G/P_I$. Matsuki has shown that there is a natural duality between K–orbits and $G_{\mathbb{R}}$–orbits on $\mathcal{B}$ (or on $\mathcal{P}_I$) which reverses the natural partial order on these orbits. It turns out that all of the machinery of the paper [**RS**] applies to the $G_{\mathbb{R}}$ orbits in a natural way. In §6, we will review (and sometimes reformulate) a number of Matsuki's results and indicate how to apply the methods of [**RS**] to these problems. We will only sketch the results here. A detailed exposition will appear in a paper by one of us (RWR).

6.1. Duality of orbits. Let $K_{\mathbb{R}} = G_{\mathbb{R}} \cap K$. Then $K_{\mathbb{R}}$ is a maximal compact subgroup of both $G_{\mathbb{R}}$ and K. In particular every connected component of $G_{\mathbb{R}}$ and of K meets $K_{\mathbb{R}}$. Let $\Gamma = \{\mathrm{Id}_G, \theta, \tau, \sigma\}$. Then Γ is a subgroup of the group of all Lie group automorphisms of G. It is clear that $K_{\mathbb{R}} = G^{\Gamma}$. We note that Γ acts on $\mathcal{T}$ and $\mathcal{B}$; recall that $\mathcal{T}$ denotes the set of all maximal algebraic tori of G. We let $\mathcal{T}^{\Gamma}$ denote the set of all Γ–stable maximal algebraic tori.

PROPOSITION 6.1.1. *Each K–orbit on $\mathcal{T}^{\theta}$ meets $\mathcal{T}^{\Gamma}$ in a unique $K_{\mathbb{R}}$–orbit and each $G_{\mathbb{R}}$–orbit on $\mathcal{T}^{\sigma}$ meets $\mathcal{T}^{\Gamma}$ in a unique $K_{\mathbb{R}}$-orbit. Thus we have canonical bijections between the three following sets of orbits:* (i) $K\backslash\mathcal{T}^{\theta}$; (ii) $G_{\mathbb{R}}\backslash\mathcal{T}^{\sigma}$; *and* (iii) $K_{\mathbb{R}}\backslash\mathcal{T}^{\Gamma}$.

Note that Proposition 6.1.1 also gives a bijective correspondence between the following sets: (i) $G_{\mathbb{R}}$ conjugacy classes of Cartan subalgebras of Lie($G_{\mathbb{R}}$); (ii) K conjugacy classes of θ–stable Cartan subalgebras of Lie(G); and (iii) $K_{\mathbb{R}}$ conjugacy classes of θ-stable Cartan subalgebras of Lie($G_{\mathbb{R}}$).

For the rest of §6, we assume that the standard pair (B_0, T_0) of 1.2 is chosen such that $T_0 \in \mathcal{T}^{\Gamma}$. Let $C = U \cap T_0$. Then C is the unique maximum compact subgroup of T_0 and is a maximal torus of the compact connected Lie group U. Let $W_K(T_0)$ (respectively $W_{G_{\mathbb{R}}}(T_0)$), denote the image of $N_K(T_0)$ (respectively $N_{G_{\mathbb{R}}}(T_0)$) in $W = W(T_0)$. Similarly, let $W_{K_{\mathbb{R}}}(C)$ denote the image of $N_{K_{\mathbb{R}}}(C)$ in $W(C) = N_U(C)/C$.

LEMMA 6.1.2. (1) *The inclusion map* $U \hookrightarrow G$ *maps* $N_U(C)$ *into* $N_G(T_0)$ *and induces an isomorphism of* $W(C)$ *onto* $W(T_0)$. (2) *The inclusion map* $K_{\mathbb{R}} \hookrightarrow K$ *induces an isomorphism of* $W_{K_{\mathbb{R}}}(C)$ *onto* $W_K(T_0)$. (3) *The inclusion map* $K_{\mathbb{R}} \hookrightarrow G_{\mathbb{R}}$ *induces an isomorphism of* $W_{K_{\mathbb{R}}}(C)$ *onto* $W_{G_{\mathbb{R}}}(T_0)$.

COROLLARY 6.1.3. $W_K(T_0) = W_{G_{\mathbb{R}}}(T_0)$.

We define subsets $\mathcal{V}_\theta$, $\mathcal{V}_\sigma$ and $\mathcal{V}_\Gamma$ of G by:

(6-1) $\quad \mathcal{V}_\theta = \{\, g \in G \mid g\theta(g^{-1}) \in N_G(T_0) \,\}.$

(6-2) $\quad \mathcal{V}_\sigma = \{\, g \in G \mid g\sigma(g^{-1}) \in N_G(T_0) \,\}.$

(6-3) $\quad \mathcal{V}_\Gamma = \{\, u \in U \mid u\theta(u^{-1}) \in N_U(C) \,\}.$

We note that $g \in \mathcal{V}_\sigma$ if and only if the maximal algebraic torus ${}^{g^{-1}}T_0$ is σ–stable. Similarly, if $u \in U$, then $u \in \mathcal{V}_\Gamma$ if and only if the maximal torus ${}^{u^{-1}}C$ of U is θ–stable. Note also that $\mathcal{V}_\Gamma = \mathcal{V}_\theta \cap U = \mathcal{V}_\sigma \cap U$. We observe that $\mathcal{V}_\Gamma$ (respectively $\mathcal{V}_\sigma$) is stable under the action of $C \times K_{\mathbb{R}}$ (respectively $T_0 \times G_{\mathbb{R}}$). We have the following orbit sets:

(6-4) $\quad V_\Gamma = C\backslash\mathcal{V}_\Gamma/K_{\mathbb{R}}.$

(6-5) $\quad V_\theta = T_o\backslash\mathcal{V}_\theta/K.$

(6-6) $\quad V_\sigma = T_0\backslash\mathcal{V}_\sigma/G_{\mathbb{R}}.$

REMARK. In the earlier sections, the set $\mathcal{V}_\theta$ was denoted by $\mathcal{V}$ and V_θ was denoted by V.

THEOREM 6.1.4. (1) *Each* $(T_0 \times K)$*–orbit on* $\mathcal{V}_\theta$ *meets* $\mathcal{V}_\Gamma$ *in a unique* $(C \times K_{\mathbb{R}})$*–orbit. Thus the inclusion* $\mathcal{V}_\Gamma \hookrightarrow \mathcal{V}_\theta$ *induces a bijection of orbit sets* $V_\Gamma \to V_\theta$. (2) *Each* $(T_0 \times G_{\mathbb{R}})$*–orbit on* $\mathcal{V}_\sigma$ *meets* $\mathcal{V}_\Gamma$ *in a unique* $K_{\mathbb{R}}$*–orbit. Consequently the inclusion map* $\mathcal{V}_\Gamma \hookrightarrow \mathcal{V}_\sigma$ *induces a bijection from* V_Γ *to* V_σ. (3) *The inclusion* $\mathcal{V}_\Gamma \hookrightarrow G$ *induces bijections from* V_Γ *onto* $B_0\backslash G/K$ *and* $B_0\backslash G/G_{\mathbb{R}}$.

It follows from Theorem 6.1.4 that the set V_Γ naturally parametrizes the following sets of orbits: (a) $B_0\backslash G/K$; (b) $B_0\backslash G/G_{\mathbb{R}}$; (c) $K\backslash\mathcal{B}$; and (d) $G_{\mathbb{R}}\backslash\mathcal{B}$.

If $v \in V_\Gamma$, we let $\mathcal{O}(v) = B_0vK$ and let $\mathcal{R}(v) = B_0vG_{\mathbb{R}}$. We let $\mathcal{K}(v)$ (respectively $\mathcal{G}(v)$) denote the K–orbit on X (respectively the $G_{\mathbb{R}}$–orbit on X) corresponding to v. Thus, if $v = CuK_0 \in V_\Gamma$ and if $x = u^{-1} \cdot x_0$ (where $x_0 \in X$ corresponds to $B_0 \in \mathcal{B}$), then $\mathcal{K}(v) = K \cdot x$ and $\mathcal{G}(v) = G_{\mathbb{R}} \cdot x$. We have:

(6-7) $\quad B_0\backslash G/K = \{\, \mathcal{O}(v) \mid v \in V_\Gamma \,\}.$

(6-8) $\quad B_0\backslash G/G_{\mathbb{R}} = \{\, \mathcal{R}(v) \mid v \in V_\Gamma \,\}.$

(6-9) $\quad K\backslash X = \{\, \mathcal{K}(v) \mid v \in V_\Gamma \,\}.$

(6-10) $\quad G_{\mathbb{R}}\backslash X = \{\, \mathcal{G}(v) \mid v \in V_\Gamma \,\}.$

It is well-known that U acts transitively on $\mathcal{B}$ and that $U \cap B_0 = U \cap T_0 = C$. It follows easily that, for every $B \in \mathcal{B}$, the intersection $U \cap B$ is a maximal torus of U. We also note that $\sigma(B_0) = \tau(B_0) = {}^{w_0}B_0$. The intersection $B \cap \tau(B)$ is a (τ–stable) maximal algebraic torus of G and $B \cap U$ is the unique maximal compact subgroup of $B \cap \tau(B)$. We let $\mathcal{B}^*$ denote the set of all $B \in \mathcal{B}$ such that $B \cap U$ is θ–stable, or, equivalently, such that $B \cap \tau(B)$ is θ–stable. We observe further that $B \in \mathcal{B}^*$ if and only if $B \cap \theta(B) \cap \tau(B) \cap \sigma(B)$ is of maximal rank.

THEOREM 6.1.5. (1) *Each K–orbit on $\mathcal{B}$ meets $\mathcal{B}^*$ in a unique $K_{\mathbb{R}}$–orbit. Thus the inclusion map $\mathcal{B}^* \hookrightarrow \mathcal{B}$ induces a bijection of orbit sets $K_{\mathbb{R}}\backslash\mathcal{B}^* \to K\backslash\mathcal{B}$.* (2) *Each $G_{\mathbb{R}}$–orbit on $\mathcal{B}$ meets $\mathcal{B}^*$ in a unique $K_{\mathbb{R}}$–orbit, so that the inclusion $\mathcal{B}^* \hookrightarrow \mathcal{B}$ induces a bijection $K_{\mathbb{R}}\backslash\mathcal{B}^* \to K\backslash\mathcal{B}$.*

The proof follows easily from Theorem 6.1.4.

We let $X^* = \{ x \in X \mid B_x \in \mathcal{B}^* \}$. Then X^* is a $K_{\mathbb{R}}$–stable, closed differentiable submanifold of the projective variety X.

PROPOSITION 6.1.6. *Let $x \in X^*$. Then $G_{\mathbb{R}} \cdot x \cap K \cdot x = K_{\mathbb{R}} \cdot x$.*

DEFINITION 6.1.7. (1) Let x, $y \in X$. Then the orbits $K \cdot x$ and $G_{\mathbb{R}} \cdot y$ are *dual orbits* if $K \cdot x \cap G_{\mathbb{R}} \cdot y$ meets X^*. If this happens, then there exists $z \in X^*$ such that $K \cdot x \cap G_{\mathbb{R}} \cdot y = K_{\mathbb{R}} \cdot z$. (2) Let g, $g' \in G$. Then the $(B_0 \times K)$–orbit B_0gK and the $(B_0 \times G_{\mathbb{R}})$–orbit $B_0g'G_{\mathbb{R}}$ are *dual orbits* if there exists $h \in G$ such that $B_0gK \cap B_0g'G_{\mathbb{R}} = B_0hK_{\mathbb{R}}$.

PROPOSITION 6.1.8. *Let v, $v' \in V_\Gamma$. Then the orbits $\mathcal{K}(v')$ and $\mathcal{G}(v)$ (respectively $\mathcal{O}(v')$ and $\mathcal{R}(v)$) are dual orbits if and only if $v = v'$.*

6.2. The map ϕ and the action of M and W on the $G_{\mathbb{R}}$-orbits. For the moment, we let f denote the bijection $V_\Gamma \to V_\theta$ given by Theorem 6.1.4. We use the bijection f to transfer the actions of M and W on V_θ to actions on V_Γ. Similarly, we may consider ϕ as a map from V_Γ to $\mathcal{I}$. Thus let $v \in V_\Gamma$, let $v_1 = f(v)$ and let $w \in W$. Then we set $\phi(v) = \phi(v_1)$, $w \cdot v = f^{-1}(w \cdot v_1)$ and $m(w) \cdot v = f^{-1}(m(w) \cdot v_1)$. Since we also have a bijection from V_Γ to V_σ, and since V_σ parametrizes the orbit sets $B_0\backslash G/G_{\mathbb{R}}$ and $G_{\mathbb{R}}\backslash\mathcal{B}$, we obtain actions of W and M on these orbit sets; we may also consider ϕ as a map from these orbits sets.

For $w \in W$, we let $\overline{w} \in \mathcal{W}$ denote $\eta_0(w)$, where $\eta_0 = \eta_{B_0,T_0} : W \to \mathcal{W}$ is as in Remark 1.7.

The following proposition gives a geometric interpretation of the map ϕ in terms of the action of σ on $\mathcal{B}$.

PROPOSITION 6.2.1. *Let $v \in V_\Gamma$, let $a = \phi(v)$ and let $x \in \mathcal{G}(v)$ (so that $\mathcal{G}(v) = G_{\mathbb{R}} \cdot x$). Then $\overline{a}\,\overline{w}_0 = \eta_0(aw_0)$ is the relative position of $(B_x, \sigma(B_x))$.*

We define a partial order on V_Γ by requiring that the bijection $V_\Gamma \to V$ be an isomorphism of posets.

6.3. Case analysis for the $(B_0 \times G_{\mathbb{R}})$–orbits on $P_svG_{\mathbb{R}}$. We follow the notation of the earlier sections as regards real, complex and imaginary roots.

Thus, let $v \in V_\Gamma$, let $v_1 = f(v)$ and let $s \in S$. Then we say that s is real (respectively compact imaginary, ...) for v if s is real (respectively compact imaginary, ...) for v_1.

Let $s \in S$ and $v \in V_\Gamma$. We can carry out an analysis of the $(B_0 \times G_\mathbb{R})$–orbits on $P_s v G_\mathbb{R}$ as in §2. For the $(B_0 \times G_\mathbb{R})$–orbits, the analysis of the corresponding orbits on $\mathbb{P}^1$ is a bit different from that of §2 and we will discuss it first. Let $Y = \mathcal{P}_s$, let $y \in Y$ and let $P = P_y$. Then $G_\mathbb{R} \cap P = (P \cap \sigma(P))^\sigma$. As before, let $X_y = \mathcal{B}(P) \cong \mathbb{P}^1$. Let H_y denote the image of $G_\mathbb{R} \cap P$ in $A_y = \mathrm{Aut}(X_y)$; note that H_y is a (real) Lie subgroup of A_y. There are essentially three cases:

Case 1. s is complex for v. In this case, there are two H_y–orbits on X_y, one of which is a fixed point.

Case 2. s is compact imaginary for v. In this case H_y is a maximum compact subgroup of A_y and acts transitively on X_y.

Case 3. s is either real or non-compact imaginary for v. Then H_y^0, the identity component of H_y, is isomorphic to $PSL(2, \mathbb{R})$. There are three H_y^0–orbits on X_y, the "equator", the "upper hemisphere" and the "lower hemisphere". The equator is a circle and the two hemispheres are open orbits. If H_y is not connected, then it permutes the two hemispheres, and there are only two H_y orbits.

We have the following analysis for the $(B_0 \times G_\mathbb{R})$–orbits on $P_s v G_\mathbb{R}$.

Case A. s is complex for v. Then $P_s v G_\mathbb{R} = \mathcal{R}(v) \cup \mathcal{R}(s \cdot v)$ and $s \cdot v \neq v$. If $sa > a$ (respectively $sa < a$), then $\mathcal{R}(v)$ (respectively $\mathcal{R}(s \cdot v)$) is open and dense in $P_s v G_\mathbb{R}$ and $\mathcal{R}(s \cdot v)$ (respectively $\mathcal{R}(v)$) is closed of codimension two in $P_s v G_\mathbb{R}$.

Case B. s is compact imaginary for v. In this case there is only one orbit, so that $P_s v G_\mathbb{R} = \mathcal{R}(v)$.

Case C. s is non-compact imaginary for v. Then there exists $v' \in V_\Gamma$ such that $\mathcal{R}(v')$ is closed and of codimension one in $P_s v G_\mathbb{R}$. We have $P_s v G_\mathbb{R} = \mathcal{R}(v') \cup \mathcal{R}(v) \cup \mathcal{R}(s \cdot v)$. The orbits $\mathcal{R}(v)$ and $\mathcal{R}(s \cdot v)$ are open in $P_s v G_\mathbb{R}$. There are either two or three orbits, depending on whether or not $v = s \cdot v$.

Case D. s is real for v. The orbit $\mathcal{R}(v)$ is closed of codimension one in $P_s v G_\mathbb{R}$. There exists $v' \in V_\Gamma$ such that $P_s v G_\mathbb{R} = \mathcal{R}(v) \cup \mathcal{R}(v') \cup \mathcal{R}(s \cdot v')$. The orbits $\mathcal{R}(v')$ and $\mathcal{R}(s \cdot v')$ are open in $P_s v G_\mathbb{R}$ and there are either two or three orbits, depending on whether or not $s \cdot v' = v'$.

We note that the case analysis for the $(B_0 \times G_\mathbb{R})$–orbits on $P_s v G_\mathbb{R}$ is a bit different from the corresponding analysis for the $B_0 \times K$–orbits. In the first place, all of the closure relations get reversed. Furthermore, the dimensions of the orbits behave somewhat differently.

6.4. Further results. Using the case analysis of 6.3, we can now prove a number of theorems for the $(B_0 \times G_\mathbb{R})$–orbits on G or, equivalently, for the $G_\mathbb{R}$–orbits on X. We can also obtain a number of results relating to the duality of orbits. We list below some of these results. In the results below, all dimensions referred to will be the real dimensions.

LEMMA 6.4.1. *Let $v \in V_\Gamma$ and $s \in S$. Then $\mathcal{R}(m(s) \cdot v)$ is the unique closed $(B_0 \times G_\mathbb{R})$–orbit in $P_s v G_\mathbb{R}$.*

THEOREM 6.4.2. *Let $v \in V_\Gamma$ and let $a = \phi(v)$. Then the codimension of $\mathcal{R}(v)$ in G is equal to $l(a)$. Equivalently, the codimension of $\mathcal{G}(v)$ in X is equal to $l(a)$.*

COROLLARY 6.4.3. *Let $x \in X$ and let $\bar{b} \in \mathcal{W}$ be the relative position of $(B_x, \sigma(B_x))$. Then $\dim G_\mathbb{R} \cdot x = |\Phi^+| + l(\bar{b})$.*

Note that the dimension of the orbit $G_\mathbb{R} \cdot x$ is explicitly determined by the relative position of $(B_x, \sigma(B_x))$.

THEOREM 6.4.4. *Let v', $v \in V_\Gamma$, let $a' = \phi(v')$ and let $a = \phi(v)$. Then the orbits $\mathcal{G}(v')$ and $\mathcal{K}(v)$ intersect transversally in X. In particular, each connected component of $\mathcal{G}(v') \cap \mathcal{K}(v)$ is a smooth locally closed submanifold of X whose dimension is $2L(a) + \dim \mathcal{B}(K^0) - l(a')$.*

THEOREM 6.4.5. *Let v', $v \in V_\Gamma$. Then the following conditions are equivalent:*

(1) $v' \leq v$.
(2) $\mathcal{K}(v') \subset \overline{\mathcal{K}(v)}$.
(3) $\mathcal{G}(v) \subset \overline{\mathcal{G}(v')}$.
(4) $\mathcal{G}(v') \cap \mathcal{K}(v) \neq \emptyset$.

THEOREM 6.4.6. *Let v', $v \in V_\Gamma$ with $v' \leq v$. Then*

$$\begin{aligned}\overline{\mathcal{G}(v') \cap \mathcal{K}(v)} &= \overline{\mathcal{G}(v')} \cap \overline{\mathcal{K}(v)} \\ &= \coprod_{v' \leq v_2 \leq v_1 \leq v} \mathcal{G}(v_2) \cap \mathcal{K}(v_1).\end{aligned}$$

All of the results of §4 concerning the partial order on the $(B_0 \times K)$–orbits of G (or the K–orbits of X) apply equally well to the $(B_0 \times G_\mathbb{R})$–orbits of G (or the $G_\mathbb{R}$–orbits of X). However all of the inclusion relations are reversed. In this case the length function on the set V_Γ (carried over from the length function on V_θ via the bijection $V_\Gamma \to V_\theta$) does not relate quite so directly to the dimension of the corresponding orbit $\mathcal{R}(v)$ (or $\mathcal{G}(v)$). We give below the analogue for $(B_0 \times G_\mathbb{R})$–orbits of Proposition 4.1. First we need more notation. If $E \subset G$, we set $E^\frown = G \setminus (\overline{E} \setminus E)$.

PROPOSITION 6.4.7. *Let $s \in S$ and $v \in V_\Gamma$ and assume that $v \to m(s) \cdot v$. Then*

$$\mathcal{R}(m(s) \cdot v)^\frown = \bigcup_{v' \leq v} P_s \mathcal{R}(v').$$

6.5. The Hermitian symmetric case. Assume that (G, θ) is of Hermitian symmetric type, as defined in §5. This is equivalent to the condition that $G_{\mathbb{R}}/K_{\mathbb{R}}$ be a Hermitian symmetric space. The combinatorial classification of $(B_0 \times K)$–orbits on G given in §5 also gives a combinatorial classification of the $(B_0 \times G_{\mathbb{R}})$–orbits on G via the duality between these two sets of orbits. Note that, by Theorem 6.3.2, one can easily read off the dimensions of the $(B_0 \times G_{\mathbb{R}})$–orbits from the parameter set $\mathcal{E}$. Note also that we have a complete description of the closure relations between the $G_{\mathbb{R}}$–orbits in terms of $\mathcal{E}$. There does not seem to be any easy geometric interpretation of the map $\nu : V \to W/W_J$ of §5 in terms of the $G_{\mathbb{R}}$–orbits.

6.6. Duality of orbits on generalized flag manifolds. Let $I \subset \Delta$ and let $\mathcal{P} = \mathcal{P}_I$ be the variety of all conjugates of the standard parabolic subgroup P_I. We will briefly indicate how to set up a correspondence between K–orbits and $G_{\mathbb{R}}$-orbits on $\mathcal{P}$. If $Q \in \mathcal{P}_I$, set

$$Q_\Gamma = Q \cap \tau(Q) \cap \theta(Q) \cap \sigma(Q).$$

One can show that Q_Γ is a reductive algebraic subgroup of G. Let $\mathcal{P}^*$ denote the set of all $Q \in \mathcal{P}$ such that Q_Γ is of maximal rank. We note that $\mathcal{P}^*$ is stable under conjugation by elements of $K_{\mathbb{R}}$.

THEOREM 6.6.1. (1) *Each K–orbit on $\mathcal{P}$ meets $\mathcal{P}^*$ in a unique $K_{\mathbb{R}}$–orbit. Thus the inclusion $\mathcal{P}^* \hookrightarrow \mathcal{P}$ induces a bijection $K_{\mathbb{R}}\backslash\mathcal{P}^* \to K\backslash\mathcal{P}$ of orbit sets.* (2) *Each $G_{\mathbb{R}}$ orbit on $\mathcal{P}$ meets $\mathcal{P}^*$ in a unique $K_{\mathbb{R}}$–orbit. Hence the injection $\mathcal{P}^* \to \mathcal{P}$ induces a bijection $K_{\mathbb{R}}\backslash\mathcal{P}^* \to G_{\mathbb{R}}\backslash\mathcal{P}$.*

Let $Z = K_{\mathbb{R}}\backslash\mathcal{P}^*$. By Theorem 6.5.1, Z naturally parametrizes the sets of K–orbits and $G_{\mathbb{R}}$–orbits on $\mathcal{P}$, so that again we get a duality between these orbits. For each $z \in Z$, let $\mathcal{K}(z)$ (respectively $\mathcal{G}(z)$) denote the corresponding K-orbit (respectively $G_{\mathbb{R}}$–orbit) on $\mathcal{P}$.

THEOREM 6.6.2. *Let z', $z \in Z$. Then $\mathcal{K}(z') \subset \overline{\mathcal{K}(z)}$ if and only if $\mathcal{G}(z) \subset \overline{\mathcal{G}(z')}$.*

We note that, if (G, θ) is of Hermitian symmetric type, one can get quite explicit information on the $G_{\mathbb{R}}$-orbits on $\mathcal{P}$ if we combine the results of §5 and of 3.4. In particular, if G is a classical group, then **[Ry]** gives elementary and explicit combinatorial models for the set $K\backslash\mathcal{P}$ of K–orbits on $\mathcal{P}$ (and hence for the $G_{\mathbb{R}}$–orbits on $\mathcal{P}$).

6.7. Comments. (1) Most of the results of §6 are due to Matsuki. The work of Matsuki related to K–orbits and $G_{\mathbb{R}}$–orbits on flag manifolds (and generalized flag manifolds) appears in a series of six papers **[M1-M5, MO]**. Matsuki considers a somewhat more general problem, namely the orbits of minimal parabolic subgroups on (real) affine symmetric spaces. In addition, there is a considerable amount of cross referencing in this series of papers. Perhaps for these reasons, we have sometimes had difficulty in understanding the precise statements of his theorems. The proof that there is a natural bijective correspondence between the K-orbits and the $G_{\mathbb{R}}$-orbits is in Matsuki's original paper [**M1**], although the proof that this bijection is order reversing does not seem to appear until a later paper. As nearly as we can tell, most of the results of §6.1 are essentially due to Matsuki. The results of Theorem 6.4.2, Corollary 6.4.3 and Theorem 6.4.4 on the orbit dimensions and the transversality of intersections seem to be new. Theorems 6.4.5 and 6.4.6 are due to Matsuki. Our proofs of these theorems, which rely on Theorem 6.4.4, are quite different. The duality of K-orbits and $G_{\mathbb{R}}$-orbits on generalized flag manifolds is due to Matsuki.The application of the techniques of [**RS**] to the $G_{\mathbb{R}}$-orbits is new, as are the more explicit results for the Hermitian symmetric case.

(2) Throughout §6, we have assumed that $G_{\mathbb{R}} = G^{\sigma}$. This is not an essential assumption, and all of the results of this section go through if one only assumes that $G_{\mathbb{R}}$ is a subgroup of G^{σ} containing $(G^{\sigma})^0$. In this case, one lets $K_{\mathbb{R}} = G_{\mathbb{R}} \cap U$ and lets K denote the Zariski closure of $K_{\mathbb{R}}$ in G. Then we have $(G^{\theta})^0 \subset K \subset G^{\theta}$, and the arguments of Remark 1.1 apply.

§7. A Hecke algebra representation

7.1. Introduction. We keep the notations of the preceding sections. Denote by $\mathcal{H}$ the Hecke algebra of the Weyl group (W, S). Recall that $\mathcal{H}$ is an algebra over the ring of Laurent polynomials $\mathbb{Z}[q, q^{-1}]$ with a basis e_w $(w \in W)$. The multiplication is defined by the following rules, where $w \in W$, $s \in S$,

$$e_s e_w = \begin{cases} e_s e_w & \text{if } l(sw) > l(w), \\ (q-1)e_w + q e_{sw} & \text{if } l(sw) < l(w). \end{cases}$$

See [**Hu**, Ch. 7]

To our data G, θ, B_0, T_0 one can associate a representation of $\mathcal{H}$, i.e. an $\mathcal{H}$-module $\mathcal{M}$. This representation was found by Lusztig and Vogan. It is described in geometric terms in [**LV**]. We shall describe it here in purely combinatorial terms, following [**MS**]. We shall not go into the details of the construction, which invokes the powers of l-adic cohomology. But in order to describe the module $\mathcal{M}$ we need to say a bit about the fine structure of the $(B_0 \times K)$-orbits $\mathcal{O}(v)$.

Namely, one ingredient of the construction is a finer geometric analysis of the product morphism $P_s \times \mathcal{O}(v) \to P_s vK$. The analysis leads one to consider sheaves on $\mathcal{O}(v)$ which are locally constant for the étale topology (briefly: local systems), which moreover are $(B_0 \times K)$-equivariant. Let $x \in \mathcal{O}(v)$ and let I_x be its isotropy group in $(B_0 \times K)$. The local systems in question are classified by the characters of the finite group I_x/I_x^0. Therefore we need some information about these groups.

7.2. Component groups. Let $x \in \mathcal{V}$, let $v = T_0 x K$ and write write $n = x(\theta x)^{-1}$. Let $a = \phi(v)$, so that a is the image of n in W. Write $T_v = \{t \in T_0 \mid a\theta(t) = t\}$ and denote by A_v the component group T_v/T_v^0. Let $U = R_u(B_0)$.

LEMMA 7.2.1. (1) *I_x is isomorphic to the semi-direct product of T_v and a connected subgroup of U. (2) A_v is an elementary abelian 2-group.*

The proof is easy.

Let $y = tuxz \in \mathcal{O}(v)$, where $t \in T_0$, $u \in U$, $z \in K$. Then the coset tT_v depends only on y. We define a morphism $\mu_x : \mathcal{O}(v) \to T_0/T_v$ by $\mu_x(y) = tT_v$. We have an isogeny of tori $T_0/T_v^0 \to T_0/T_v$ which is a Galois covering with group A_v. A character χ of that group defines a local system of rank one $\mathcal{L}_\chi$ on T_0/T_v. The inverse image $\mathcal{L}_{x,\chi} = \mu_x{}^*(\mathcal{L}_\chi)$ is a $(B_0 \times K)$–equivariant local system on $\mathcal{O}(v)$, whose isomorpism class does not depend on the choice of the representative x.

The Hecke algebra module $\mathcal{M}$ which we are going to describe has a basis indexed by such isomorphism classes. In the geometric analysis alluded to before one has to relate the groups A_v and $A_{m(s)\cdot v}$. In relating these groups, we have to distinguish cases (see §2).

LEMMA 7.2.2. *Assume that s is complex for v. Then s induces an isomorphism $A_v \cong A_{m(s)\cdot v}$.*

This is straightforward. Recall that in this case $m(s) \cdot v = s \cdot v$.

We assume (as we may) that $m(s) \cdot v > v$. If s is not complex for v then s is non-compact imaginary for v. We then have $\phi(m(s) \cdot v) = sa = a\theta(s)$. Let α be the simple root with $s = s_\alpha$. Then $sa\theta(\alpha) = -\alpha$, from which one sees that $\alpha(T_{m(s)\cdot v}) \subset \{1, -1\}$. It follows that α induces a character $\chi(\alpha)$ of $A_{m(s)\cdot v}$. Denote by $\alpha^\vee$ the co-root associated to α. Then $\alpha^\vee(-1) \in T_{m(s)\cdot v}$ and $\eta(\alpha^\vee) = \alpha^\vee(-1)T^0_{m(s)\cdot v}$ defines an element of $A_{m(s)\cdot v}$. From $\langle \alpha, \alpha^\vee \rangle = 2$ it follows that $\eta(\alpha^\vee)$ lies in the kernel of $\chi(\alpha)$.

LEMMA 7.2.3. *Assume that s is non-compact imaginary for v. Then*

$$A_v \cong \mathrm{Ker}\chi(\alpha)/\{1, \eta(\alpha^\vee)\}.$$

PROOF. One shows that

$$\begin{aligned} T_v &= \mathrm{Im}(\alpha^\vee)\left(T_{m(s)\cdot v} \cap \mathrm{Ker}(\alpha)\right), \quad \text{and} \\ T_v^0 &= \mathrm{Im}(\alpha^\vee)(T^0_{m(s)\cdot v}), \end{aligned}$$

whence

$$\begin{aligned} \mathrm{Ker}(\chi(\alpha)) &= \left(T_{m(s)\cdot v} \cap \mathrm{Ker}(\alpha)\right)/T^0_{m(s)\cdot v} \to \\ &\to \left(\mathrm{Im}(\alpha^\vee)(T_{m(s)\cdot v} \cap \mathrm{Ker}(\alpha)\right) / \left(\mathrm{Im}(\alpha^\vee)T^0_{m(s)\cdot v}\right) = T_v/T_v^0, \end{aligned}$$

from which the lemma follows.

This lemma shows that the quotient of the order of $A_{m(s)\cdot v}$ by the order of A_v equals 1, 2 or 4. Hence the order of A_v is increasing with respect to the weak order on V.

Denote by Γ_v the character group of A_v. The lemma implies that Γ_v is isomorphic to a quotient of the annihilator $\Gamma'_{m(s)\cdot v}$ of $\eta(\alpha^\vee)$ in $\Gamma_{m(s)\cdot v}$. We denote by ψ or $\psi_{m(s)\cdot v,s}$ the induced homomorphism $\Gamma'_{m(s)\cdot v} \to \Gamma_v$. Notice ψ is bijective if and only if $\chi(\alpha) = 0$.

The following criterion for the vanishing of $\chi(\alpha)$ follows readily from the definitions.

LEMMA 7.2.4. *$\chi(\alpha) = 0$ if and only if $s \cdot v \neq v$.*

Notice that in this case there is an obvious isomorphism $s : \Gamma_v \to \Gamma_{s\cdot v}$.

If s is real for v, then there exists $v' \in V$ such that s is non-compact imaginary for v' and $v = m(s)\cdot v'$. The preceding results then apply with v, $m(s)\cdot v$ replaced by v', v.

7.3. The Hecke algebra module. Denote by $\mathcal{M}$ the free $\mathbb{Z}[q, q^{-1}]$–module with a basis $f_{v,\chi}$, where v runs through V and $\chi \in \Gamma_v$. In **[LV]** $\mathcal{M}$ is endowed with a structure of $\mathcal{H}$-module. In order to describe this module structure, it suffices to describe the products $e_s f_{v,\chi}$, where $s \in S$. We have to distinguish several cases, to be described now.

(a) s is complex for v and $m(s) \cdot v = s \cdot v > v$. Then

$$e_s f_{v,\chi} = f_{m(s)\cdot v, s\cdot\chi}\,.$$

The notation $s.\chi$ is explained by Lemma 7.2.2.

(b) s is complex for v and $v = m(s) \cdot (s \cdot v) > s \cdot v$. Then

$$e_s f_{v,\chi} = (q-1) f_{v,\chi} + q f_{s\cdot v, s\cdot\chi}\,.$$

(c) s is compact imaginary for v. Then

$$e_s f_{v,\chi} = q f_{v,\chi}\,.$$

(d) s is non-compact imaginary for v and $s \cdot v \neq v$. Using Lemma 7.2.4 we see that now the homomorphism ψ of the preceding section is an isomorphism of a subgroup of $\Gamma_{m(s)\cdot v}$ onto Γ_v. We have

$$e_s f_{v,\chi} = f_{m(s)\cdot v, \psi^{-1}(\chi)} + f_{s\cdot v, s\cdot\chi}\,.$$

(e) s is non-compact imaginary for v and $s\cdot v = v$. Then here are two elements χ' and χ'' in $\Gamma'_{m(s)\cdot v}$ whose image under ψ is χ. We have

$$e_s f_{v,\chi} = f_{m(s)\cdot v,\chi'} + f_{m(s)\cdot v,\chi''} + f_{v,\chi}\,.$$

In the remaining cases s is real for v. Then there is $v' \in V$ such that $v = m(s) \cdot v' > v'$ and s is non-compact imaginary for v'. We denote by ψ' the homomorphism $\psi_{v,s}$ of of the preceding section. Let $v \in V$, $\chi \in \Gamma_v$.

(f) If $\chi \notin \Gamma'_v$ then

$$e_s f_{v,\chi} = -f_{v,\chi}\,.$$

(g) If $\chi \in \Gamma'_v$ and $s \cdot v' \neq v'$ then

$$e_s f_{v,\chi} = (q-2) f_{v,\chi} + (q-1)(f_{v',\nu} + f_{s\cdot v', s\cdot\nu}),$$

where $\nu = \psi'(\chi)$.

(h) If $\chi \in \Gamma'_v$ and $s \cdot v' = v'$ then

$$e_s f_{v,\chi} = (q-1) f_{v,\chi} - f_{v,\chi'} + (q-1) f_{v',\nu}\,,$$

where $\chi' \neq \chi$, $\psi'(\chi') = \psi'(\chi)$ and where ν is as in case (g).

That the preceding formulas do indeed define an $\mathcal{H}$-module is proved in [**LV**] by geometric means. Of course, a direct algebraic proof ought to be possible, but would be cumbersome. Our formulas look somewhat different from those of [**LV**, pp. 371-372] or [**V1**, p. 403], as we have formulated everything in combinatorial terms. Our formulation follows [**MS**], where a more general situation is considered. Our cases (a), ... ,(h) correspond to the cases labeled (b1),(b2),(a),(d1),(c1),(e),(d2),(c2) in [**LV,V1**]. In these references, Kazhdan-Lusztig polynomials are defined for this situation. We shall not go into this. See also [**MS**].

7.4. Specializations. The formulas describing the module structure on $\mathcal{M}$ show that they define, in fact, a representation of the $\mathbb{Z}[q]$-subalgebra $\mathcal{H}_0$ of $\mathcal{H}$ with basis e_w $(w \in W)$. We can therefore specialize q to some element of F, obtaining a representation of the specialization of the Hecke algebra.

(a) $q \to 1$. It is well-known that now $\mathcal{H}_0$ specializes to the group ring $\mathbb{Z}[W]$. Hence we obtain a representation of W in a free $\mathbb{Z}$-module with basis $(f_{v,\chi;1})$, the index set being as before. The action of a simple reflection $s \in W$ is read off from the formulas for the action of e_s, with $q = 1$.

(b) $q \to 0$. Now $\mathcal{H}_0$ specializes to the ring with a free $\mathbb{Z}$-basis $m(w)$ $(w \in W)$, the multiplication of the basis elements being as in the monoid $M = M(W)$. The element e_w specializes to $(-1)^{l(w)} m(w)$. We have a representation of the monoid M in a free module with basis $f_{v,\chi;0}$. The action of $m(s)$ $(s \in S)$, is obtained from the formulas for the action of e_s by changing the sign of the right-hand sides and specializing q to 0.

(c) $q \to \infty$. For $w \in W$, $v \in V$, and $\chi \in \Gamma_v$, put

$$\widetilde{e}_w = q^{-l(w)} e_w, \quad \text{and} \quad \tilde{f}_{v,\chi} = q^{-d(v)} f_{v,\chi},$$

where $d(v)$ is the dimension of the orbit $\mathcal{O}_v$. One checks that the e_w span a $\mathbb{Z}[q^{-1}]$-subalgebra of $\mathcal{H}$ and that its elements stabilize the $\mathbb{Z}[q^{-1}]$-submodule of $\mathcal{M}$ with basis $\widetilde{f}_{v,\chi}$ of $\mathcal{M}$. We can specialize q to ∞, obtaining a representation of $M(W)$ in a free module with basis $f_{v,\chi;\infty}$. We obtain that in cases (b),(c),(g),(h)

$$m(s) \cdot f_{v,\chi;\infty} = f_{v,\chi;\infty}\,.$$

In case (f) the left-hand side is 0, and it equals $f_{s\cdot v,s.\chi;\infty}$ in case (a),

$$f_{m(s)\cdot v,\psi^{-1}(\chi);\infty}$$

in case (d) and

$$f_{m(s)v,\chi';\infty}+f_{m(s)v,\chi'';\infty}$$

in case (e). The notations are as before.

7.5. Examples. We now give a few examples of the Hecke algebra representations in low rank. In the first two examples the groups Γ_v have order ≤ 2. We shall write $f_{v,\chi}=f_{di}$, where d is the dimension of the orbit $\mathcal{O}(v)$ and $i=0$ (respectively 1) designates the trivial (respectively the non-trivial) element of Γ_v. If Γ_v is trivial we write $f_{d0}=f_d$. If there are several orbits with the same dimension we write $f_{di}, f'_{di}, \dots$ for the corresponding basis elements. In these examples there will be at most two such orbits. Working out the details is a straightforward matter.

(a) $G=SL(2,\mathbb{C})$, $\theta(g)=cgc^{-1}$, where $c=\operatorname{diag}(1,-1)$. We write e for the only generator of $\mathcal{H}$. The representation is as follows.

$$\begin{aligned} ef_0&=f'_0+f_{10}\,,\\ ef'_0&=f_0+f_{10}\,,\\ ef_{10}&=(q-2)f_{10}+(q-1)(f_0+f'_0)\,,\\ ef_{11}&=-f_{11}\,. \end{aligned}$$

(b) $G=PSL(2,\mathbb{C})$ and θ is induced by the automorphism of the previous example. The formulas are different.

$$\begin{aligned} ef_0&=f_0+f_{10}+f_{11}\,,\\ ef_{10}&=(q-1)f_{10}-f_{11}+(q-1)f_0\,,\\ ef_{11}&=(q-1)f_{11}-f_{10}+(q-1)f_0\,. \end{aligned}$$

Write

$$\widetilde{f_0}=-f_{10}^{\star}\,,\quad \widetilde{f'_0}=-f_{11}^{\star}\,,\quad\text{and}\quad \widetilde{f_{10}}=f_0^{\star}\,,$$

where the star denotes the dual basis of the dual module. One checks that the in the dual representation $(q-1)-e$ acts on the elements $\widetilde{f_0}$, $\widetilde{f'_0}$, and $\widetilde{f_{10}}$ as e acts on the corresponding elements without a tilde in example (a). This is an example of a general duality phenomenon, studied at length in [**V2**].

(c) $G=SL(3,\mathbb{C})$, $\theta(g)={}^tg^{-1}$. The Hecke algebra $\mathcal{H}$ has two generators, e_1 and e_2, corresponding to the generators (12) resp. (23) of the Weyl group S_3. The corresponding simple roots are α_1 and α_2.

There is one orbit of length 0 and 2, and there are two of length 1. See [**RS**, pp. 432-433], where the case of $SL(n,\mathbb{C})$ is dealt with. One shows that the groups Γ_v are trivial, except when $\mathcal{O}(v)$ is the maximal orbit, in which case it has order 4. In fact, the analysis of [loc. cit.] gives in that case an isomorphism of Γ_v onto the diagonal subgroup of $SL(3,\mathbb{C})$ with (diagonal) entries ± 1. The roots α_1, α_2, and $\alpha_1 + \alpha_2$ define characters of Γ_v. The corresponding basis elements of $\mathcal{M}$ are denoted by f_{21}, f_{22} and f_{23}, respectively. The other notations are as before. The representation is as follows.

$$e_1 f_0 = f_1', \quad e_2 f_0 = f_1, \quad e_1 f_1 = f_{20} + f_{21} + f_1,$$
$$e_2 f_1 = (q-1) f_1 + q f_0, \quad e_1 f_1' = (q-1) f_1' + q f_0,$$
$$e_2 f_1' = f_{20} + f_{22} + f_1, \quad e_1 f_{20} = (q-1) f_{20} - f_{21} + (q-1) f_1,$$
$$e_2 f_{20} = (q-1) f_{20} - f_{22} + (q-1) f_1', \quad e_1 f_{21} = (q-1) f_{21} - f_{20} + (q-1) f_1,$$
$$e_2 f_{21} = -f_{21}, \quad e_1 f_{22} = -f_{22}, \quad e_2 f_{22} = (q-1) f_{22} - f_{20} + (q-1) f_1',$$
$$e_1 f_{23} = e_2 f_{23} = -f_{23}.$$

Editors' note. In the final stages of the production of this volume the editors learned that one of the contributors, Roger W. Richardson, died in Canberra, Australia, on June 15, 1993. Roger was a valued colleague and a close friend of many of the contributors to this volume, as well as of many others who participated in the conference. His sudden death at a time when he was actively engaged in mathematical work is a sad loss.

References

[B] A. Borel, *Linear Algebraic groups, 2nd ed.*, Springer-Verlag, New York, Berlin, Heidelberg, 1991.

[Bou] N. Bourbaki, *Groupes et algèbres de Lie, Chapitres 4, 5 et 6*, Hermann, Paris, 1968.

[C] D. Collingwood, *Orbits and characters associated to highest weight representations*, Proc. Amer. Math. Soc. **114** (1992), 1157-1165.

[H] S. Helgason, *Differential Geometry, Lie Groups and Symmetric Spaces*, Academic Press, New York, San Francisco, London, 1978.

[HMSW] H. Hecht, D. Miličić, W. Schmid and J. A. Wolf, *Localization and standard modules for real semisimple Lie groups*, Invent. Math. **90**, 297-332.

[Hu] J. Humphreys, *Reflection Groups and Coxeter Groups*, Cambridge Univ. Press, Cambridge, 1990.

[LV] G. Lusztig and D. A. Vogan, Jr., *Singularities of closures of K–orbits on flag manifolds*, Invent. Math **71** (1983), 365-379.

[MS] J. G. M. Mars and T. A. Springer, (to appear).

[M1] T. Matsuki, *The orbits of affine symmetric spaces under the action of minimal parabolic subgroups*, J. Math. Soc. Japan **31** (1979), 331-356.

[M2] T. Matsuki, *Orbits on affine symmetric spaces under the action of parabolic subgroups*, Hiroshima Math. J. **12** (1982), 307-320.

[M3] T. Matsuki, *Closure relations for orbits on affine symmetric spaces under the action of minimal parabolic subgroups*, Advanced Studies in Pure Mathematics. Representations of Lie Groups, vol. 14, Kinokuniya–North Holland, 1988, pp. 541-559.

[M4] T. Matsuki, *Closure relations for orbits on affine symmetric spaces under the actions of parabolic subgroups. Intersections of associated orbits*, Hiroshima Math. J. **18** (1988), 59-67.

[M5] T. Matsuki, *Orbits on flag manifolds*, Proceedings of the International Congress of Mathematicians, Kyoto, 1990, Springer-Verlag, Tokyo, 1991, pp. 807-813.

[MO] T. Matsuki and T. Oshima, *Embeddings of discrete series into principal series*, The Orbit Method in Representation Theory, Proceedings of the Copenhagen Conference, 1988, ed. M. Duflo, N. V. Pederson and M. Vergne, Birkhäuser, Boston, pp. 147-175.

[R1] R. W. Richardson, *On orbits of Lie groups and algebraic groups*, Bull. Austral. Math. Soc. **25** (1982), 1-28.

[R2] R. W. Richardson, *Intersections of double cosets in algebraic groups*, Indag. Math. (new series) **3** (1992), 69-78.

[RRoSt] R. Richardson, G. Röhrle and R. Steinberg, *Parabolic subgroups with abelian unipotent radical*, Invent. Math. (to appear).

[RS] R. W. Richardson and T. Springer, *The Bruhat order on symmetric varieties*, Geom. Dedicata **35** (1990), 389-436.

[RS1] R. Richardson and T. A. Springer, *Corrections and additions to "The Bruhat order on symmetric varieties"*, Geom. Dedicata (to appear).

[Ros] W. Rossman, *The structure of semisimple symmetric spaces*, Canad. J. Math. **31** (1979), 157-180.

[Ry] P. D. Ryan, *Some examples in the Bruhat order on symmetric varieties*, M.Sc. thesis at the Australian National University, Canberra, 1991.

[S1] T. A. Springer, *Some results on algebraic groups with involutions*, Advanced Studies in Pure Mathematics, vol 6, Kinokuyina–North Holland, 1985, pp. 525-543.

[S2] T. Springer, *The classification of involutions of simple algebraic groups*, J. Fac. Sci. Univ. Tokyo Sect. 1A Math. **34** (1987), 655-670.

[St] R. Steinberg, *Endomorphisms of linear algebraic groups*, Mem. Amer. Math. Soc. **80** (1968).

[V1] D. A. Vogan, Jr., *Irreducible characters of semisimple Lie groups III. Proof of the Kazdhan-Lusztig conjecture in the integral case,* Invent. Math **71** (1983), 381-417.

[V2] D. A. Vogan, Jr., *Irreducible characters of semisimple Lie groups IV. Character Multiplicity Duality*, Duke Math. J. **49** (1982), 943-1073.

[W] J. A. Wolf, *The action of a real semisimple group on a complex flag manifold, I: Orbit structure and holomorphic arc components*, Bull. Amer. Math. Soc. **75** (1969), 1121-1237.

R. W. RICHARDSON, CENTRE FOR MATHEMATICS AND ITS APPLICATIONS, AUSTRALIAN NATIONAL UNIVERSITY, CANBERRA, ACT 0200, AUSTRALIA

T. A. SPRINGER, MATHEMATISCH INSTITUUT, RIJKSUNIVERSITEIT UTRECHT, 3508 TA UTRECHT, THE NETHERLANDS

Contemporary Mathematics
Volume **153**, 1993

ON EXTRASPECIAL PARABOLIC SUBGROUPS

GERHARD E. RÖHRLE

0. INTRODUCTION

In this note we extend a result from joint work with R. Richardson and R. Steinberg [RRS] regarding the structure of abelian radicals of parabolic subgroups in Chevalley Groups to the so called 'extraspecial' case.

Let G be a simple, connected algebraic group over an algebraically closed field of characteristic $p > 2$ or 0. Let T be a maximal torus of G and Ψ the corresponding root system of G. We fix a Borel subgroup $B \supset T$ and denote the set of simple roots of Ψ with respect to B by Δ. Let ρ (ρ_s) denote the highest (short) root in Ψ. Assume that G is not of type A_n. There is a unique simple root α satisfying $\langle\alpha, \rho\rangle = 1$. (Here $\langle\beta, \gamma\rangle = (\beta, \gamma^\vee)$, where $\gamma^\vee = 2\gamma/(\gamma, \gamma)$ is the coroot of γ, for $\beta, \gamma \in \Psi$.) Let $P = P_\alpha = LV$ be the corresponding maximal standard parabolic subgroup of G with $V = R_u(P)$, the unipotent radical of P. The Levi complement of V in P, L, is generated by T and all root subgroups whose roots are orthogonal to ρ. There is a unique parabolic subgroup P^- of G satisfying $P^- \cap P = L$ with Levi decomposition $P^- = LV^-$, where $V^- = R_u(P^-)$. The derived subgroup of V is U_ρ, the root subgroup corresponding to the highest root ρ. Let V_1, respectively V_1^-, denote the commutator quotient of V, respectively V^-. Our goal is to study the conjugation action of L on V_1 or V_1^-.

By Ψ, Ψ_L, Ψ_V, etc. we denote the root system of the groups G, L, V, etc. We also write Ψ_{V_1} for the roots in the support of V_1, i.e., $\Psi_{V_1} = \Psi_V \setminus \{\rho\}$. The Weyl groups of G and L are W and W_L, respectively. We call a subset of roots of Ψ an *orthogonal set of roots* provided these roots are mutually orthogonal. Let $\mathcal{A}$ denote the set of all orthogonal sets of long roots in Ψ_{V_1} including the empty set. Note that W_L acts on $\mathcal{A}$.

It is known that L acts on V_1 with a finite number of orbits. This is a special case of a result due to R. W. Richardson [Ri2, Theorem E]. Our main theorem describes a connection between the L-orbits on V_1, the (P, P) double cosets of G, and the W_L-orbits on $\mathcal{A}$. This is the precise analogue for the case of an abelian radical [RRS]. Let $u_\beta \in U_\beta$ and $u_\beta \neq 1$.

Main Theorem. *Let G be a simple, connected algebraic group over an algebraically closed field of characteristic $p > 2$ or 0, not of type A_n. Let $P_\alpha = LV$ be*

1991 *Mathematics Subject Classification.* 20G15.

the unique maximal standard parabolic subgroup of G corresponding to the simple root α satisfying $\langle \alpha, \rho \rangle = 1$, V_1 the commutator quotient of V. Let $\mathcal{A}$ denote the set of all orthogonal sets of long roots in Ψ_{V_1}. Assume that α is long and $\operatorname{rank} G \geq 4$. *Then there exist natural bijections among:*

(a) *The L-orbits of V_1 (or V_1^-).*
(b) *The $N_L(T)$-orbits of elements in V_1 of the form $\prod u_{\beta_i} U_\rho$ with $\{\beta_1, \dots, \beta_r\}$ in $\mathcal{A}$.*
(c) *The W_L-orbits on $\mathcal{A}$.*
(d) *The W_L-orbits of involutions in W of the form $\prod w_{\beta_i}$ with $\{\beta_1, \dots, \beta_r\}$ in $\mathcal{A}$.*
(e) *The (W_L, W_L) double cosets of W.*
(f) *The (P, P) double cosets of G.*

The maximal number of mutually orthogonal long roots in Ψ_{V_1} is 4 *and the cardinalities in* (a) *through* (f) *are all equal to the number of components of $\Psi_L + 4$.*

Remarks. The bijections in the main theorem between (b), (c), (d) and (e) are the obvious ones and the one between (e) and (f) is a well known consequence of the Bruhat lemma.

The situation is somewhat different in type C_n, G_2, and B_3. For G_2 and B_3 the statements (a)–(f) in the main theorem hold for a modified version of $\mathcal{A}$ whose tuples may contain roots of different length. While for C_n (the only instance when α is short) the correspondence between the L-orbits on V_1 (there are always two) and the (P, P) double cosets of G (there are always three) fails. The methods used to prove the main theorem are similar to the ones in [RRS].

After dealing with some basic lemmas, we show in section 2 that the number of W_L-orbits on $\mathcal{A}$ only depends on the components of Ψ_L and moreover, W_L acts transitively on tuples in $\mathcal{A}$ of the same cardinality except for pairs. These results are then used to establish bijections between the set of W_L-orbits on $\mathcal{A}$, the (W_L, W_L) double cosets of W, and the L-orbits of V_1 in the following two sections. We close with some results on the dimension of the various L-orbits of V_1.

I would like to thank Robert Steinberg and Nicolas Spaltenstein for their advice and their helpful comments.

1. Preliminary Lemmas

Lemma 1.1. *Let ρ be the highest root in Ψ. Then $C_W(\rho) = W_L$.*

Proof. Clearly, $W_L \subseteq C_W(\rho)$, since ρ is the unique root in Ψ with coefficient 2 at α. Suppose there is an x in $C_W(\rho)$ but not in W_L. Then w_α appears in each reduced expression for x. After conjugating by a suitable element in W_L we may assume that x has a reduced expression ending in w_α. Thus $x(\alpha) < 0$ and $x(\rho - \alpha) = \rho - x(\alpha) > \rho$ which is impossible. Thus $C_W(\rho) = W_L$, as desired.

The α*-height* of a root $\beta \in \Psi$ is the coefficient of α in β.

Lemma 1.2. *For $\beta \in \Psi_{V_1}$ we have $\langle \beta, \rho \rangle = 1$. Thus $\rho - \beta \in \Psi_{V_1}$.*

Proof. Since $\beta \in \Psi_{V_1}$, we have $\beta = \alpha + \alpha_J$ with $\alpha_J \in \mathbb{Z}\Psi_L$ and $\langle \alpha_J, \rho \rangle = 0$.

Therefore, $\langle \beta, \rho \rangle = \langle \alpha, \rho \rangle = 1$ and so $-w_\rho(\beta) = \rho - \beta \in \Psi_{V_1}$, because the α–height of $\rho - \beta$ is 1.

Lemma 1.3. *Let $\{\beta_1, \dots, \beta_s\}$ be in $\mathcal{A}$ and set $w_s = \prod_{i=1}^{s} w_{\beta_i}$. Then the α–height of $w_s(\rho)$ is $2 - s$ and $w_s(\rho)$ is in $\{\rho\}$, Ψ_{V_1}, Ψ_L, $-\Psi_{V_1}$, or $\{-\rho\}$, according to $s = 0, 1, \dots, 4$ respectively. In particular, s is at most 4.*

Proof. By 1.2 we have $w_s(\rho) = \rho - \sum_{i=1}^{s} \beta_i$. Thus the α–height of $w_s(\rho)$ is $2 - s$ as required. Further s is at most 4, since the α–height of any root in Ψ is at least -2.

Corollary 1.4. *If $\{\beta_1, \dots, \beta_4\}$ is a maximal sequence in $\mathcal{A}$, then*

$$\beta_1 + \beta_2 + \beta_3 + \beta_4 = 2\rho.$$

Proof. For $w_4 = \prod_{i=1}^{4} w_{\beta_i}$ we have $-\rho = w_4(\rho) = \rho - (\beta_1 + \beta_2 + \beta_3 + \beta_4)$ by 1.3.

Remarks. The fact that the maximal number of pairwise orthogonal long roots in Ψ_{V_1} is at most 4 and the identity in 1.4 can be interpreted in terms of a subsystem of Ψ of type D_4. Namely, it follows from 1.2 that together with $-\rho$ any three mutually orthogonal long roots in Ψ_{V_1} span a D_4 subsystem in Ψ, and the only additional orthogonal root may be identified with the negative of the highest root of that D_4 by 1.4.

From this it is also evident that a sequence of length 4 occurs in $\mathcal{A}$ precisely when one of length 3 does, and in that case $\operatorname{rank} G \geq 4$, as mutually orthogonal roots are linearly independent characters on T. One easily exhibits three pairwise orthogonal long roots in Ψ_{V_1} provided α is long and $\operatorname{rank} G \geq 4$. Thus in our situation tuples of length 4 do occur in $\mathcal{A}$.

Lemma 1.5. *Let $\beta, \gamma \in \Psi_{V_1}$, $\beta \neq \gamma \neq \rho - \beta$, where at least one of the roots is long. Then β is orthogonal to γ if and only if β is not orthogonal to $\rho - \gamma$ if and only if γ is not orthogonal to $\rho - \beta$.*

Proof. Assume without loss that β is long. If $\langle \gamma, \beta \rangle = 0$, then $\langle \rho - \gamma, \beta \rangle = \langle \rho, \beta \rangle = 1$, and likewise $\langle \gamma, \rho - \beta \rangle = \langle \gamma, \rho \rangle = 1$. On the other hand, if $\langle \gamma, \beta \rangle \neq 0$, then $\langle \gamma, \beta \rangle = 1$, since $\beta \neq \gamma \neq \rho - \beta$ and β is long. Thus $\langle \rho - \gamma, \beta \rangle = \langle \rho, \beta \rangle - \langle \gamma, \beta \rangle = 0$ and also $\langle \gamma, \rho - \beta \rangle = \langle \gamma, \rho \rangle - \langle \gamma, \beta \rangle = 0$.

Lemma 1.6. *Assume Ψ admits two root lengths, α is long, and* $\operatorname{rank} G \geq 4$. *Let $\beta \in \Psi_{V_1}$ be short and orthogonal to α. Then:*

(i) *$\gamma = (\rho - 2\beta) + (\rho - \alpha)$ is the only root in Ψ_{V_1} which is orthogonal to both α and β.*

(ii) *There exists a (unique) short root $\eta \in \Psi_L^+$ which is orthogonal to α, β, and γ (γ as in (i)) such that $\{\alpha, \beta - \eta, \beta + \eta, \gamma\}$ is a maximal element in $\mathcal{A}$. In particular, $\beta - \eta, \beta + \eta$, and γ are long. We list all possible pairs (β, η): for B_n ($n \geq 4$) there is only the case (ρ_s, α_n) and for F_4 we have $(1221, \alpha_4)$, $(1231, \alpha_3 + \alpha_4)$, and (ρ_s, α_3).*

Proof. Here and elsewhere we refer to the usual labelling of the simple roots as in [B]. We leave it to the reader to check the details of part (ii). For part (i) assume that γ is a root in Ψ_{V_1} orthogonal to both α and β. It follows from 1.2 that $\rho - 2\beta$ and $\rho - 2\beta - \alpha - \gamma$ are roots, the latter, since α, β, and γ are mutually orthogonal.

Considering α–heights, we have $\rho - 2\beta - \alpha - \gamma = -\rho$. The result follows.

2. The W_L–Orbits on $\mathcal{A}$

Throughout the remaining sections we assume the hypotheses of the main theorem, namely that α is long and $\operatorname{rank} G \geq 4$.

Lemma 2.1. *W_L is transitive on roots in Ψ_{V_1} of the same length.*

Proof. This is a special case of a more general result. For instance see [ABS,§2 Lemma 1].

Lemma 2.2. *W_L is transitive on triples in $\mathcal{A}$.*

Proof. In view of 2.1 we may consider triples of the form $\{\beta_1, \beta_2, \beta_3\}$, $\{\beta_1, \gamma_2, \gamma_3\}$ in $\mathcal{A}$. In case $\{\beta_2, \beta_3\} = \{\gamma_2, \gamma_3\}$ there is nothing to show.

Case 1. Assume that, e.g., $\gamma_2 = \beta_2$ and $\gamma = \gamma_3 \neq \beta_3$. If $\langle \beta_3, \gamma \rangle = 0$, then $\{\beta_1, \beta_2, \beta_3, \gamma\}$ is a maximal tuple in $\mathcal{A}$ by 1.3 and from 1.4

$$\gamma = -\left(\beta_1 + \beta_2 + \beta_3 - 2\rho\right). \tag{2.2.1}$$

Set $\delta_{ij} = \beta_i + \beta_j - \rho \in \Psi_L$ for $1 \leq i < j \leq 3$. Using (2.2.1) one checks that

$$w_{\delta_{23}} w_{\delta_{13}} \cdot \{\beta_1, \beta_2, \beta_3\} = \{\beta_2, \beta_1, \gamma\}.$$

If $\langle \beta_3, \gamma \rangle = 1$, then $\delta = \beta_3 - \gamma \in \Psi_L$. Noting that $\langle \delta, \beta_1 \rangle = \langle \delta, \beta_2 \rangle = 0$ we see that $w_\delta \cdot \{\beta_1, \beta_2, \beta_3\} = \{\beta_1, \beta_2, \gamma\}$.

Finally, $\langle \beta_3, \gamma \rangle = -1$ does not occur, as otherwise $\gamma = \rho - \beta_3$ and then β_1 is orthogonal to both β_3 and $\rho - \beta_3$ contradicting 1.5.

Case 2. Now we assume that $\beta_i \neq \gamma_j$ for $2 \leq i, j \leq 3$. As $\langle \beta_1, \beta_i \rangle = \langle \beta_1, \gamma_j \rangle = 0$, we have $\langle \beta_i, \gamma_j \rangle \geq 0$ for $2 \leq i, j \leq 3$ by 1.5. Moreover, $\langle \beta_i, \gamma_j \rangle = 0$ for all $2 \leq i, j \leq 3$ is impossible, because of 1.3. Therefore, without loss of generality, we may assume that $\langle \gamma_2, \beta_2 \rangle = 1$. Set $\delta = \beta_2 - \gamma_2 \in \Psi_L$. We then have $w_\delta \cdot \{\beta_1, \beta_2, \beta_3\} = \{\beta_1, \gamma_2, w_\delta(\beta_3)\}$, and it follows from case 1 that the original triples are W_L–conjugate.

Corollary 2.3. *W_L is transitive on quadruples in $\mathcal{A}$.*

Proof. It follows from the previous lemma that any two quadruples in $\mathcal{A}$ are W_L–conjugate to ones of the form $\{\beta_1, \beta_2, \beta_3, \beta\}$ and $\{\beta_1, \beta_2, \beta_3, \gamma\}$ respectively. Moreover, from 1.4 the roots β and γ are uniquely determined by $\{\beta_1, \beta_2, \beta_3\}$ and thus agree. Whence the original quadruples are conjugate.

Lemma 2.4. *The number of W_L–orbits on pairs in $\mathcal{A}$ equals the number of components of Ψ_L.*

Proof. Assume that Ψ_L is connected. Then α is an end-node of the Dynkin-diagram of Ψ. Let α' be the unique simple root of Ψ_L satisfying $\langle \alpha', \alpha \rangle < 0$. We need to show that any two pairs of the form $\{\alpha, \beta\}$ and $\{\alpha, \gamma\}$ in $\mathcal{A}$ are W_L–conjugate. Since $\langle \beta, \alpha \rangle = \langle \gamma, \alpha \rangle = 0$, the coefficient of α' in β and γ is 2. This implies that β and γ are conjugate under $W_{\Delta_L \setminus \{\alpha'\}}$ (Δ_L denotes the set of simple roots of Ψ_L). Therefore, $\{\alpha, \beta\}$ and $\{\alpha, \gamma\}$ are $W_{\Delta_L \setminus \{\alpha'\}}$–conjugate, as α is fixed by $W_{\Delta_L \setminus \{\alpha'\}}$.

If Ψ_L has two components, then G is of type B_n, $n \geq 4$ ($\operatorname{rank} G \geq 4$), or D_n, $n \geq 5$ and $\alpha = \alpha_2$. Write $\Psi_L = \Psi_{L_1} \cup \Psi_{L_2}$, where $\Psi_{L_1} = \{\pm\alpha_1\}$ is of type A_1. Let ρ_2 be the highest long root in Ψ_{L_2}. Since $\operatorname{rank} G \geq 4$, the roots $\alpha_1 + \alpha + \alpha_3$ and $\alpha + \alpha_3 + \rho_2$ are both in Ψ_{V_1}, long, and orthogonal to α. Moreover, $\{\alpha, \alpha_1 + \alpha + \alpha_3\}$ and $\{\alpha, \alpha + \alpha_3 + \rho_2\}$ are not W_L-conjugate. Let γ be a long root in Ψ_{V_1} orthogonal to α. It follows from case 1 that $\{\alpha, \gamma\}$ is conjugate to either $\{\alpha, \alpha_1 + \alpha + \alpha_3\}$ or $\{\alpha, \alpha + \alpha_3 + \rho_2\}$ depending on whether the support of γ involves only Δ_{L_2} or else all of Δ_L.

Finally, there are three components of Ψ_L in case Ψ is of type D_4. One easily checks that there are three W_L-orbits on pairs in $\mathcal{A}$ with representatives $\{\alpha, \alpha_1 + \alpha + \alpha_3\}$, $\{\alpha, \alpha_1 + \alpha + \alpha_4\}$, and $\{\alpha, \alpha_3 + \alpha + \alpha_4\}$.

Corollary 2.5. *If Ψ_L is connected, then W_L is transitive on tuples in $\mathcal{A}$ of the same cardinality.*

Proof. This follows from 1.3 and 2.1–2.4.

Theorem 2.6. *The number of W_L-orbits on $\mathcal{A}$ equals the number of components of Ψ_L + 4.*

Proof. Since we also consider the empty tuple in $\mathcal{A}$, the assertion follows from 1.3 and 2.1–2.4.

3. The (W_L, W_L) Double Cosets of W

Let $\mathcal{A}/W_L$ denote the set of W_L-orbits on $\mathcal{A}$. We define a map

$$\varphi_1 : \mathcal{A}/W_L \to W_L\backslash W/W_L$$

via $W_L \cdot \{\beta_1, \dots, \beta_r\} \mapsto W_L \prod w_{\beta_i} W_L$. Our aim is to show that φ_1 is bijective.

Lemma 3.1. *Let $\{\beta_1, \dots, \beta_r\}$ and $\{\gamma_1, \dots, \gamma_s\}$ be in $\mathcal{A}$. Set $x = \prod_{i=1}^r w_{\beta_i}$ and $y = \prod_{j=1}^s w_{\gamma_j}$. If x and y are in the same (W_L, W_L) double coset of W, then $r = s$.*

Proof. If x and y are in the same (W_L, W_L) double coset of W, we may assume, after possibly replacing $\{\beta_1, \dots, \beta_r\}$ by a W_L-conjugate, that (*) $u \cdot x = y$ for some $u \in W_L$. By induction we may further assume that $\beta_i \neq \gamma_j$ for all i and j. Applying (*) to ρ and using 1.1 and 1.2 yields

$$\rho - \sum_{i=1}^{r} u(\beta_i) = \rho - \sum_{j=1}^{s} \gamma_j.$$

Comparing α-heights implies the desired result.

We have seen in 2.4 that in type B_n and D_n the Weyl group W_L is not transitive on pairs. However, we have the following result.

Lemma 3.2. *Let $\{\alpha, \beta\}$, $\{\gamma, \delta\} \in \mathcal{A}$. Then $w_\alpha w_\beta$ and $w_\gamma w_\delta$ are in the same (W_L, W_L) double coset of W if and only if $\{\alpha, \beta\}$ and $\{\gamma, \delta\}$ are conjugate under W_L.*

Proof. One direction is obvious. For the converse assume that $w_\alpha w_\beta$ and $w_\gamma w_\delta$ with $\{\alpha, \beta\}$, $\{\gamma, \delta\}$ in $\mathcal{A}$ are in the same (W_L, W_L) double coset of W. After possibly replacing $\{\alpha, \beta\}$ by a W_L-conjugate, we may assume that

$$u \cdot w_\alpha w_\beta = w_\gamma w_\delta \tag{3.2.1}$$

for some $u \in W_L$. If $\{\alpha, \beta\} \cap \{\gamma, \delta\}$ is non-empty, we may cancel a reflection on both sides of (3.2.1) implying $u = 1$ and $\{\alpha, \beta\} = \{\gamma, \delta\}$. So we may assume that α, β, γ, and δ are all distinct. Applying (3.2.1) to α and β implies that $\langle \alpha, \gamma \rangle = \langle \alpha, \delta \rangle = 1$ and likewise $\langle \beta, \gamma \rangle = \langle \beta, \delta \rangle = 1$. Further, applying (3.2.1) to $\alpha + \beta - \rho \in \Psi_L$ yields $\delta = \alpha + \beta - \gamma$. Then $w_{\alpha - \gamma} \cdot \{\alpha, \beta\} = \{\gamma, \alpha + \beta - \gamma\} = \{\gamma, \delta\}$.

Corollary 3.3. *The tuples $\{\beta_1, \dots, \beta_r\}$ and $\{\gamma_1, \dots, \gamma_s\}$ in $\mathcal{A}$ are W_L-conjugate if and only if $\prod_{i=1}^{r} w_{\beta_i}$ and $\prod_{j=1}^{s} w_{\gamma_j}$ are in the same (W_L, W_L) double coset of W.*

Proof. One direction is obvious. The other follows from 3.1 together with the results in the previous section, and 3.2.

Corollary 3.4. *The map $\varphi_1 : \mathcal{A}/W_L \to W_L \backslash W / W_L$ is injective.*

Theorem 3.5. *φ_1 is a bijection.*

Proof. In view of 3.4 we need to show that every (W_L, W_L) double coset of W has a representative of the form

$$\prod_{i=1}^{s} w_{\beta_i} \quad \text{with } \{\beta_1, \dots, \beta_s\} \in \mathcal{A}. \tag{3.5.1}$$

Let x be in W. We express x as a product of reflections relative to simple roots. All reflections with support in Δ_L may be moved to the left by conjugating each occuring w_α in the product for x into a reflection w_γ for some $\gamma \in \Psi_{V_1}$ long (since α is long). Then we drop the reflections of the first kind, while remaining in the original (W_L, W_L) double coset of x. Thus we have a representative of the form

$$x = w_{\gamma_1} w_{\gamma_2} \cdots w_{\gamma_l} \qquad \text{with } \gamma_i \in \Psi_{V_1} \text{ and } \gamma_i \text{ long for each } 1 \le i \le l. \tag{3.5.2}$$

We show by induction on l that the (W_L, W_L) double coset containing x has a representative of the desired form:

If $l = 0$ or 1, then x itself satisfies (3.5.1) for $s = 0$ respectively $s = 1$.

Now let $l \ge 2$. We may assume that $\gamma_i \ne \gamma_{i+1}$ for all $1 \le i \le l - 1$, as otherwise the corresponding reflections in the product in (3.5.2) cancel each other, that way shortening the product so that by induction the double coset of x has a representative of desired type.

If there is an i $(1 \le i \le l - 1)$ with $\langle \gamma_i, \gamma_{i+1} \rangle = 1$, then $\delta = w_{\gamma_i}(\gamma_{i+1}) = \gamma_{i+1} - \gamma_i \in \Psi_L$ and we replace $w_{\gamma_i} w_{\gamma_{i+1}}$ in the expression (3.5.2) for x by $w_{\gamma_i} w_{\gamma_{i+1}} w_{\gamma_i}^{-1} w_{\gamma_i} = w_\delta w_{\gamma_i}$. As before w_δ is moved to the left and dropped and the product for x has been shortened. We now may assume $\langle \gamma_i, \gamma_{i+1} \rangle \le 0$ for all $1 \le i \le l - 1$.

Suppose $\langle \gamma_i, \gamma_{i+1} \rangle = 0$ for all $1 \le i \le l - 1$. If $l = 2$, then x satisfies (3.5.1). For $l \ge 3$ it follows from 1.5 that $\langle \gamma_1, \gamma_3 \rangle \ge 0$. If $\langle \gamma_1, \gamma_3 \rangle = 1$, then we can shorten the product as above (after interchanging w_{γ_2} and w_{γ_3}). Thus we may assume γ_1, γ_2, and γ_3 are pairwise orthogonal. By iterating this argument for γ_4 etc., it is clear that we can either shorten the product for x and the result follows by induction or all γ_i's involved in x are mutually orthogonal, whence x itself is of desired type (3.5.1).

Finally, we consider the case $\langle \gamma_i, \gamma_{i+1} \rangle = -1$ for some $1 \le i \le l - 1$. Then $\gamma_{i+1} = \rho - \gamma_i$. Without loss of generality we may assume $i = 1$, i.e., $x = w_\gamma w_{\rho - \gamma} w_{\gamma_3} \cdots$ for some $\gamma \in \Psi_{V_1}$. If $\langle \rho - \gamma, \gamma_3 \rangle = 0$, then $\langle \gamma, \gamma_3 \rangle = 1$ by 1.5 and (after interchanging $w_{\rho-\gamma}$ and w_{γ_3}) we again may shorten the product. This leaves the case $\langle \rho - \gamma, \gamma_3 \rangle = -1$ which implies $\gamma_3 = \gamma$. Iterating the same argument for γ_4 etc., we see that we can either shorten the product for x, or else x is an alternating product of w_γ and $w_{\rho-\gamma}$, i.e.,

$$x = w_\gamma w_{\rho-\gamma} w_\gamma w_{\rho-\gamma} \cdots . \tag{3.5.3}$$

We examine several possibilities. For that fix a maximal element $\{\beta_1, \ldots, \beta_4\} \in \mathcal{A}$ and set $w_s = \prod_{i=1}^{s} w_{\beta_i}$ for $0 \le s \le 4$. Note that each w_s satisfies (3.5.1).

If $l = 2$, then $x = w_\gamma w_{\rho-\gamma}$ and $x(\rho) = -\gamma$. By 1.3 we have $w_3(\rho) \in -\Psi_{V_1}$. Since both $x(\rho) = -\gamma$ and $w_3(\rho)$ are in $-\Psi_{V_1}$ and long, there exists a u in W_L such that $x(\rho) = u \cdot w_3(\rho)$ by 2.1. Hence $x^{-1}uw_3(\rho) = \rho$ and so $x^{-1}uw_3$ is in W_L by 1.1. Therefore, $x = w_\gamma w_{\rho-\gamma}$ and w_3 are in the same double coset (3.5.4).

For $l = 3$ we have $x = w_\gamma w_{\rho-\gamma} w_\gamma = w_\rho$. It follows from 1.3 that $w_4(\rho) = -\rho = x(\rho)$. Hence $x^{-1}w_4(\rho) = \rho$ and $x^{-1}w_4 \in W_L$ by 1.1; whence $x = w_\rho$ and w_4 are in the same (W_L, W_L) double coset of W (3.5.5).

If $l \ge 4$, then

$$\begin{aligned} x = (w_\gamma w_{\rho-\gamma})^2 \cdots &= w_\gamma w_\rho \cdots \qquad \text{since } w_{\rho-\gamma}(\gamma) = \rho \\ &= w_\gamma w_\rho (w_\gamma^{-1} w_\gamma) \cdots \\ &= w_{\rho-\gamma} w_\gamma \cdots . \end{aligned}$$

Hence the product in (3.5.3) has been shortened. This completes the proof of 3.5.

4. The L–Orbits on V_1

Define a map $\varphi_2 : \mathcal{A}/W_L \to L$-orbits on V_1 via $W_L \cdot \{\beta_1, \ldots, \beta_r\} \mapsto L \cdot \prod u_{\beta_i} U_\rho$, where $u_{\beta_i} \in U_{\beta_i}$ and $u_{\beta_i} \ne 1$. The goal of this section is to show that φ_2 is bijective.

Lemma 4.1. *Let $\{\beta_1, \ldots, \beta_r\}$ and $\{\gamma_1, \ldots, \gamma_s\}$ be in $\mathcal{A}$. Set $\mathcal{O}_1 = L \cdot \prod_{i=1}^{r} u_{\beta_i} U_\rho$ and $\mathcal{O}_2 = L \cdot \prod_{j=1}^{s} u_{\gamma_j} U_\rho$. If $\mathcal{O}_1 = \mathcal{O}_2$, then $r = s$.*

Proof. Suppose that $\mathcal{O}_1 = \mathcal{O}_2$ and $r < s$. In view of the results of section 2 we may conjugate $\prod u_{\gamma_j} U_\rho$ by a suitable element in $N_L(T)$ such that (after possibly reordering the γ_i's)

$$\prod_{j=1}^{s} u_{\gamma_j} U_\rho = \prod_{i=1}^{r} u_{\beta_i} \cdot \prod_{j=r+1}^{s} u_{\gamma_j} U_\rho .$$

Since $\{\gamma_1, \ldots, \gamma_s\}$ is a set of orthogonal long roots the corresponding SL_2's commute elementwise. Therefore, $r < s$ implies that $\mathcal{O}_1$ is in the closure of $\mathcal{O}_2$ but not the other way around which is a contradiction. Hence $r = s$, as desired.

Proposition 4.2. *The map $\varphi_2 : \mathcal{A}/W_L \to L$-orbits on V_1 is injective.*

Proof. If Ψ_L is connected, this follows from 4.1 and 2.5. Otherwise W_L is not transitive on pairs in $\mathcal{A}$ (see 2.4). It is apparent from 4.1 that φ_2 is injective unless the L–orbits corresponding to non-conjugate pairs coincide. Using the representatives of the W_L–orbits on pairs in $\mathcal{A}$ given in 2.4 it is easy to see that the centralizers of elements corresponding to non-conjugate pairs (via φ_2) are not conjugate in L. Thus the L–orbits themselves are distinct. For instance in case Ψ

is of type D_4 there are three distinct W_L–orbits on pairs in $\mathcal{A}$. The corresponding L–orbit representatives in V_1 are $x_1 = u_\alpha u_{\alpha_1+\alpha+\alpha_3} U_\rho$, $x_2 = u_\alpha u_{\alpha_1+\alpha+\alpha_4} U_\rho$, and $x_3 = u_\alpha u_{\alpha_3+\alpha+\alpha_4} U_\rho$. Clearly, $1 \neq u_{-\alpha_4} \in C_L(x_1)$. However, no L–conjugate of $u_{-\alpha_4}$ is contained in either $C_L(x_2)$ or $C_L(x_3)$. Therefore, the centralizers are not conjugate. Similar arguments may be used for the remaining cases. We will see however in 5.3 that for D_n with $n \geq 5$ and B_n the L–orbits corresponding to non-conjugate pairs have different dimensions.

The proof that φ_2 is bijective will follow from a series of lemmas. Let m_α denote the number of roots in Ψ_L^+ that are not orthogonal to α.

Lemma 4.3. $|\Psi_{V_1}| = 2m_\alpha + 2$.

Proof. Fix β in Ψ_L^+ not orthogonal to α. Then $\langle \alpha, \beta \rangle < 0$ and so $\alpha + \beta \in \Psi_{V_1}$, since $\alpha - \beta$ is not a root. Further, since $\langle \rho - \alpha, \beta \rangle = -\langle \alpha, \beta \rangle > 0$, we see that $(\rho - \alpha) - \beta \in \Psi_{V_1}$. If β and γ are two distinct roots in Ψ_L^+ both not orthogonal to α, then α, $\alpha + \beta$, $\alpha + \gamma$, $(\rho - \alpha) - \beta$, $(\rho - \alpha) - \gamma$, and $\rho - \alpha$ are all distinct roots in Ψ_{V_1}. For if $\alpha + \beta = \rho - \alpha$, then $2\alpha + \beta = \rho$ which is impossible, since α is long. If $\alpha + \beta = (\rho - \alpha) - \beta$, then $2(\alpha + \beta) = \rho$, again a contradiction, as twice a root is not another root. Finally, consider the possibility that $\alpha + \beta = (\rho - \alpha) - \gamma$. So $\alpha + \beta + \gamma = \rho - \alpha$. But we have $\langle \rho - \alpha, \alpha \rangle = -1$, while $\langle \alpha + \beta + \gamma, \alpha \rangle = 0$ which is impossible. It follows that there are at least $2m_\alpha + 2$ roots in Ψ_{V_1}.

To complete the proof it will suffice to show that each root δ in Ψ_{V_1} besides α and $\rho - \alpha$ is of the form $\alpha + \beta$ or $(\rho - \alpha) - \beta$ for some β in Ψ_L^+ not orthogonal to α. If $\langle \delta, \alpha \rangle \neq 0$, then $\beta = \delta - \alpha \in \Psi_L^+$ $(\delta \neq \rho - \alpha)$. If $\langle \delta, \alpha \rangle = 0$, then $\langle \delta, \rho - \alpha \rangle = 1$ $(\delta \neq \alpha)$ from 1.5 and therefore $\delta = (\rho - \alpha) - \beta$ for some β in Ψ_L^+. In both cases β is not orthogonal to α.

Let X be a smooth G–variety. We say that a set of one-parameter subgroups of G acts *infinitesimally independently* on $x \in X$ provided the corresponding one-dimensional tangent spaces are linearly independent in $T_x(X)$, the tangent space of X at x.

Corollary 4.4. *The L–orbit of $u_\alpha u_{\rho-\alpha} U_\rho$ is dense in V_1.*

Proof. It suffices to show that $L \cdot u_\alpha u_{\rho-\alpha} U_\rho$ is of maximal dimension $|\Psi_{V_1}| = \dim V_1$. It follows from the proof of lemma 4.3 that for each $\beta \in \Psi_L^+$ not orthogonal to α the root subgroups U_β and $U_{-\beta}$ act non-trivially on $u_\alpha u_{\rho-\alpha} U_\rho$. We consider two additional actions as follows. Fix a root δ in Ψ_{V_1} orthogonal to α and consider the actions of the one-dimensional tori T_δ and $T_{\rho-\delta}$ on $u_\alpha u_{\rho-\alpha} U_\rho$. We observe that the described actions on $u_\alpha u_{\rho-\alpha} U_\rho$ are infinitesimally independent. Therefore, the dimension of the tangent space of $L \cdot u_\alpha u_{\rho-\alpha} U_\rho$ is at least $2m_\alpha + 2$ which equals $|\Psi_{V_1}|$ by 4.3. Hence the dimension of $L \cdot u_\alpha u_{\rho-\alpha} U_\rho$ is $|\Psi_{V_1}|$ as desired. Concerning specific information on the commutator relations among root subgroups of G, used here and later on, we refer the reader to [S].

Proposition 4.5. *Every L–orbit on V_1 has a representative of the form*

$$u_\alpha u_{\beta_2} \cdots u_{\beta_l} (u_{\rho-\alpha})^c U_\rho, \tag{4.5.1}$$

where $\{\alpha = \beta_1, \beta_2, \beta_3, \beta_4\}$ *is a maximal sequence in* $\mathcal{A}$, $0 \leq l \leq 4$, *and* $c \in \{0, 1\}$.

Proof. Let $x = uU_\rho$ be in V_1 with $u = \prod u_\beta$ $(\beta \in \Psi_{V_1})$. If $u = 1$, then x is as in (4.5.1) for $l = 0$ and $c = 0$. Assume $1 \neq u = \prod u_\beta$, i.e., $u_\beta \neq 1$ for some $\beta \in \Psi_{V_1}$. If β is long, then by 2.1 we may assume $\beta = \alpha$ and in case β is short it follows from 1.5 and 2.1 that u_β is L–conjugate to an element which has α in its support. We may therefore assume that the original orbit has a representative of the form

$$x = u_\alpha(\Pi u_\beta)U_\rho \quad \text{with each } \beta \in \Psi_{V_1},\ \beta \neq \alpha.$$

We remove all factors u_β with $\langle \beta, \alpha \rangle = 1$ arguing by induction on $\mathrm{ht}(\beta)$. Suppose γ occurs in the support of x satisfying $\langle \gamma, \alpha \rangle = 1$ and all u_β's with $\langle \beta, \alpha \rangle = 1$ and $\mathrm{ht}(\beta) < \mathrm{ht}(\gamma)$ have already been removed. Then conjugating x by a suitable $u_{\gamma-\alpha}$ $(\gamma - \alpha \in \Psi_L^+)$ removes the u_γ factor from x. Additional roots introduced this way are of the form $\beta + \gamma - \alpha$, $\beta + 2(\gamma - \alpha)$, or $2\beta + \gamma - \alpha = \rho$ which are of larger height than γ. The latter roots only occur in case $\gamma - \alpha$ respectively β is short. Proceeding in this manner, we obtain a representative of the form

$$y = u_\alpha(\Pi u_\beta)(u_{\rho-\alpha})^c U_\rho \quad \text{with } \beta \in \Psi_{V_1} \text{ orthogonal to } \alpha \text{ and } c \in \{0, 1\}. \tag{4.5.2}$$

Now suppose $\delta \neq \alpha$ is a root of minimal height in the support of y such that there is a factor u_β in (4.5.2) $(\beta \neq \rho - \alpha)$ with β not orthogonal to δ. Arguing as above by induction on height we can remove all these u_β's from y as follows. Let γ be in the support of y not orthogonal to δ such that all u_β's occuring in (4.5.2) with β not orthogonal to δ and β of smaller height than γ have already been removed. We conjugate y by a suitable $u_{\gamma-\delta}$ to remove the u_γ term from y ($\langle \gamma, \delta \rangle > 0$ by 1.5, as both roots are orthogonal to α). Note that conjugation by $u_{\gamma-\delta}$ fixes the term u_α. We show that the only new roots introduced in this process are again orthogonal to α and of larger height than γ. Let $\beta(\neq \delta)$ be a root in the support of y such that $\beta + \gamma - \delta$ is a root. In particular, $0 \geq \langle \gamma - \delta, \beta \rangle = \langle \gamma, \beta \rangle - \langle \delta, \beta \rangle$. Since β, γ, and δ are orthogonal to α, so is $\beta + \gamma - \delta$ (and $\beta + 2(\gamma - \delta)$ etc.). By 1.5 $\beta \neq \rho - \gamma, \rho - \delta$. Thus (*) $\langle \gamma, \beta \rangle, \langle \delta, \beta \rangle \geq 0$. If $\langle \gamma - \delta, \beta \rangle = 0$, then both β and $\gamma - \delta$ are short and $\langle \gamma, \beta \rangle = \langle \delta, \beta \rangle \geq 0$. It follows that at least one of γ and δ must be short, say δ. If both of these inner products are 0, then $\{\alpha, \beta, \delta\}$ is an orthogonal sequence with β and δ short contradicting 1.6. Therefore, $\langle \gamma, \beta \rangle = \langle \delta, \beta \rangle > 0$ with $\mathrm{ht}(\beta) \geq \mathrm{ht}(\delta)$ by choice of δ. But, since $\beta - \delta \in \Psi_L^+$, we must have $\mathrm{ht}(\beta) > \mathrm{ht}(\delta)$ which implies $\mathrm{ht}(\beta + \gamma - \delta) > \mathrm{ht}(\gamma)$, as required. If $\langle \gamma - \delta, \beta \rangle < 0$, then $\langle \delta, \beta \rangle > 0$ from (*) and again $\mathrm{ht}(\beta + \gamma - \delta) > \mathrm{ht}(\gamma)$. Completing and repeating this process eventually leads to an orbit representative of the form

$$z = u_\alpha u_{\beta_2} \cdots u_{\beta_l}(u_{\rho-\alpha})^c U_\rho \quad \text{with } c \in \{0, 1\}, \tag{4.5.3}$$

where $\{\alpha, \beta_2, \dots, \beta_l\}$ is an orthogonal sequence in Ψ_{V_1}. If all the β_i's are long then z is of desired type (4.5.1). Assume there are short roots occuring in (4.5.3). By 1.6 only one β_i is short and $l \leq 3$. Using the notation in 1.6 we have a representative of the form

$$z = u_\alpha u_\beta (u_\gamma)^d (u_{\rho-\alpha})^c U_\rho \quad \text{with } c, d \in \{0, 1\},$$

where β is short and γ is as in 1.6(i). There exists a short root $\eta \in \Psi_L^+$ orthogonal to α, γ, and $\rho-\alpha$ such that $\beta-\eta$ and $\beta+\eta$ are roots in Ψ_{V_1} and these are necessarily orthogonal to α and γ (see 1.6(ii)). First we conjugate z by some u_η and then by a suitable $u_{-\eta}$ in order to remove the u_β term in z. This yields

$$u_\alpha u_{\beta-\eta} u_{\beta+\eta} (u_\gamma)^d (u_{\rho-\alpha})^c U_\rho \text{ with } c, d \in \{0, 1\},$$

where $\{\alpha, \beta-\eta, \beta+\eta, \gamma\}$ is a maximal element in $\mathcal{A}$. Thus the element obtained is of desired type (4.5.1). This completes the proof of 4.5.

Lemma 4.6. *Let $\{\beta_1, \beta_2, \beta_3, \beta_4\}$ be in $\mathcal{A}$. Then the L-orbits in V_1 given by $x_l = u_{\beta_1} \cdots u_{\beta_l} u_{\rho-\beta_1} U_\rho$ for $1 \leq l \leq 3$ all coincide with the unique dense one.*

Proof. The fact that $L \cdot x_1$ is dense in V_1 is 4.4. For $1 \leq i < j \leq 4$ let $\delta_{ij} = \rho - \beta_i - \beta_j \in \Psi_L$. For $l = 2$ conjugating by a suitable element $u_{-\delta_{12}}$ removes the u_{β_2} factor from x_2. Hence x_1 and x_2 are L-conjugate. In particular, $L \cdot x_2$ is dense in V_1. For $l = 3$ we conjugate x_3 suitably by $u_{-\delta_{13}}$ to remove the u_{β_3} factor. Since $\beta_2 - \delta_{13} = \rho - \beta_4$ by 1.4, we obtain $u_{\beta_1} u_{\rho-\beta_4} u_{\beta_2} u_{\rho-\beta_1} U_\rho$. Then conjugating by a suitable $u_{\delta_{14}}$ removes the $u_{\rho-\beta_4}$ term without introducing any additional terms. Hence x_3 is conjugate to x_2 and thus to x_1.

Lemma 4.7. *Let $\{\beta_1, \beta_2, \beta_3, \beta_4\}$ be in $\mathcal{A}$. Then the L-orbits in V_1 given by $x = u_{\beta_1} \cdots u_{\beta_4} U_\rho$ or $x = u_{\beta_1} \cdots u_{\beta_4} u_{\rho-\beta_1} U_\rho$ both coincide with the unique dense one.*

Proof. It suffices to show that the dimension of $L \cdot x$ is maximal. We describe $|\Psi_{V_1}|$ distinct infinitesimally independent one-dimensional actions on x. Partition Ψ_{V_1} into six disjoint sets: $S_j = \{\beta \in \Psi_{V_1} \mid \langle \beta, \beta_i \rangle = 0 \text{ for } i < j, \langle \beta, \beta_j \rangle = 1\}$ for $1 \leq j \leq 4$, $S_5 = \{\beta_1, \beta_2, \beta_3, \beta_4\}$ and $S_6 = \{\rho - \beta_1\}$. It follows from 1.5 and the maximality of $\{\beta_1, \beta_2, \beta_3, \beta_4\}$ that Ψ_{V_1} is the disjoint union of the S_j's. For $\beta \in S_i$ $(1 \leq i \leq 4)$ $\beta - \beta_i$ is a root in Ψ_L and $U_{\beta-\beta_i}$ acts non-trivially on x. For $\beta \in S_i$, $\gamma \in S_j$ with $\beta \neq \gamma$ $(1 \leq i, j \leq 4)$ it follows from the definition of the S_j's that $\beta - \beta_i$ and $\gamma - \beta_j$ are distinct and $(\beta - \beta_i) + (\rho - \beta_1)$ is not a root. For $\beta_i \in S_5$ we let T_{β_i} act on x by scalars. Finally, let $\gamma = \rho - \beta_1 - \beta_2 \in \Psi_L$ and observe that $u_{-\gamma} \cdot x = u_{\rho-\beta_3} u_{\rho-\beta_4} x$ or $u'_{\beta_2} u_{\rho-\beta_3} u_{\rho-\beta_4} x$, because of 1.4. It follows from the definition of the S_j's that the actions of the various $U_{\beta-\beta_i}$'s, T_{β_i}'s, and $U_{-\gamma}$ are infinitesimally independent provided the characteristic is not 2. The result follows.

Theorem 4.8. *The map $\varphi_2 : \mathcal{A}/W_L \to L$-orbits on V_1 is bijective.*

Proof. This follows from 4.2 and 4.4–4.7.

This completes the proof of the main theorem.

Remark 4.9. We define a partial order on $\mathcal{A}/W_L$ as follows: set $W_L \cdot \{\beta_1, \ldots, \beta_r\} \leq W_L \cdot \{\gamma_1, \ldots, \gamma_s\}$, if either $r < s$ or $r = s$ and the two orbits coincide. We also consider the Bruhat–Chevalley orders on the set of L-orbits on V_1 and on the set of (P, P) double cosets of G given by the Zariski closure of orbits, respectively the Zariski closure of double cosets. It follows from the results in §3 and §4 that the bijections in our main theorem preserve these partial orders. Hence the poset structure on $\mathcal{A}/W_L$ gives a combinatorial description of the Bruhat–Chevalley order on the L-orbits of V_1.

5. The Dimension of the L–Orbits on V_1

Fix a maximal element $\{\beta_1, \ldots, \beta_4\}$ in $\mathcal{A}$ and set $w_s = \prod_{i=1}^{s} w_{\beta_i}$, $u_s = \prod_{i=1}^{s} u_{-\beta_i}$, and $x_s = u_s U_{-\rho}$ for $0 \le s \le 4$. Further let $\pi : V^- \to V_1^-$ be the canonical map. Then for $w \in W$ the set $\mathcal{V}(w) = \pi(PwP \cap V^-)$ is a non-empty, L–equivariant irreducible subvariety of V_1^- and non-trivial provided $w \notin W_L$ (see [R, Prop. 2.8]).

Lemma 5.1. $L \cdot x_s \subseteq \mathcal{V}(w_s)$ *for* $0 \le s \le 4$.

Proof. This is clear for $s = 0$. For $\beta \in \Psi^+$ let G_β be the rank 1 subgroup of G corresponding to β and $B_\beta = B \cap G_\beta$, the Borel subgroup of G_β. It follows from the Bruhat decomposition that $U_{-\beta} \setminus \{1\} \subset B_\beta w_\beta B_\beta$. Since $\{\beta_1, \ldots, \beta_4\}$ is an orthogonal sequence of long roots, the groups G_{β_i} and G_{β_j} $(i \ne j)$ commute elementwise, whence

$$u_s = \prod_{i=1}^{s} u_{-\beta_i} \in \prod_{i=1}^{s} B_{\beta_i} w_{\beta_i} B_{\beta_i} \subseteq B \prod_{i=1}^{s} w_{\beta_i} B = B w_s B \subseteq P w_s P.$$

The result follows.

Let $d = |\Psi_{V_1}| = \dim V_1$ and for $\{\alpha, \beta\}$ in $\mathcal{A}$ let $d(\alpha, \beta)$ be the number of roots γ in Ψ_{V_1} that are orthogonal to (both) α and β.

Lemma 5.2. $\dim P w_s P \cap V^- = 0$, $d/2$, $d - d(\beta_1, \beta_2) - 1$, d, *and* $d + 1 = |\Psi_V|$, *according to* $s = 0, \ldots, 4$.

Proof. The result is clear for $s = 0$ and $s = 4$, as $P \cap V^- = \{1\}$ and $P w_4 P$ is open dense in G. Observe that $\dim(PwP \cap V^- P) = \dim PwP$, since $V^- P$ is open dense in G. Thus $\dim PwP \cap V^- = \dim(PwP \cap V^-)P/P = \dim PwP/P = \dim P^w P/{}^w P = \dim P/P \cap {}^w P$ which equals the number of roots $\gamma \in \Psi_V$ such that $w^{-1}(\gamma) \in -\Psi_P$ (see [RRS, 2.16]).

For $w = w_1$ we may assume that $\beta_1 = \alpha$. The roots in Ψ_{V_1} that are not orthogonal to α are either α, $\rho - \alpha$ or of the form $\alpha + \delta$ with $\delta \in \Psi_L^+$ (δ not orthogonal to α). Since $w(\alpha) = -\alpha$ and $w(\alpha + \delta) = \delta$ are in $-\Psi_P$, while $w(\rho - \alpha) = \rho$ and $w(\rho) = \rho - \alpha$ are not, we have $\dim P w_\alpha P \cap V^- = m_\alpha + 1 = d/2$ by 4.3.

Next for $s = 2$ set $w = w_2 = w_{\beta_1} w_{\beta_2}$. Clearly, for any $\gamma \in \Psi_{V_1}$ which is orthogonal to both β_1 and β_2, we have $w(\gamma) = \gamma \notin -\Psi_P$. Further $w(\rho - \beta_1) = \rho - \beta_2$, $w(\rho - \beta_2) = \rho - \beta_1 \notin -\Psi_P$, while $w(\beta_i) = -\beta_i$ and $w(\rho) = \rho - \beta_1 - \beta_2$ are in $-\Psi_P$. For any root γ in Ψ_{V_1} unaccounted for as of yet $\langle \gamma, \beta_i \rangle \ge 0$ for $i = 1, 2$, where at least one of the inner products is positive. Thus $w(\gamma) \in -\Psi_P$ and the result follows in this case.

Finally, let $w = w_3$. By 1.4 the only root orthogonal to each of the β_i $(i = 1, 2, 3)$ is β_4. For $i = 1, 2, 3$ we have $w(\beta_i) = -\beta_i \in -\Psi_P$, $w(\rho - \beta_i) \in -\Psi_P$ by 1.5. And $w(\rho) \in -\Psi_P$ by 1.3. For any other root γ in Ψ_{V_1} not accounted for we have $\langle \gamma, \beta_i \rangle \ge 0$, where at least one of the inner products is positive. Thus $w(\gamma) \in -\Psi_P$. The result now follows.

It is easy to compute the values of $d(\alpha, \beta)$ in each case. One obtains the following from 5.2.

Corollary 5.3. *Let $\{\beta_1, \beta_2\}$ be in $\mathcal{A}$. Then* $\dim Pw_{\beta_1}w_{\beta_2}P \cap V^- = d/2+1$ *or* $d-3$ *for* B_n $(n \geq 4)$ *and* D_n $(n \geq 5)$*; or is* 5, 5, 25, 45, *or* 10, *for* D_4, E_6, E_7, E_8 *or* F_4 *respectively.*

Proposition 5.4. $\dim L \cdot x_s \geq \dim Pw_sP \cap V^-$ *for* $s = 0, 1, 2$ *and* $\dim L \cdot x_s \geq \dim(Pw_sP \cap V^-) - 1$ *for* $s = 3$ *and* 4.

Proof. This is clear for $s = 0$ and $s = 4$. For $s = 1$ we may take $\beta_1 = \alpha$. If $\delta \in \Psi_L^+$ is not orthogonal to α, then $U_{-\delta}$ acts non-trivially on $x_1 = u_{-\alpha}U_{-\rho}$. Moreover, the action of the various $U_{-\delta}$'s and the one of $T_{-\alpha}$ on x_1 are infinitesimally independent. The result follows for $s = 1$ from 5.2. The cases $s = 2$ and $s = 3$ can be worked out using arguments similar to the ones in 4.7. For $s = 3$ we have to assume that the characteristic is not two. We omit the details.

Theorem 5.5. *We have* $\dim L \cdot x_s = 0$, $d/2$, $d - d(\beta_1, \beta_2) - 1$, $d - 1$, *and* d, *for* $s = 0, \ldots, 4$ *respectively. In particular,* $L \cdot x_3$ *has codimension one in* V_1^-.

Proof. This is clear for $s = 4$, since $L \cdot x_4$ is open dense in V_1^-. Further for $s = 0, 1, 2$ we have $\dim L \cdot x_s \leq \dim \mathcal{V}(w_s) \leq \dim Pw_sP \cap V^- \leq \dim L \cdot x_s$ by 5.1 and 5.4. Hence equality holds at each step and the result follows in these cases from 5.2. Finally, $\dim L \cdot x_3 \geq d-1$ by 5.2 and 5.4, and, since $L \cdot x_3$ is not dense in V_1^-, equality holds.

Corollary 5.6. *The stabilizer of an element in the dense orbit is reductive.*

Proof. Since $L \cdot x_3$ has codimension one in V_1^-, its Zariski closure is a hypersurface whose complement in V_1^-, $L \cdot x_4$, is therefore an affine open subvariety in V_1^-, whence $C_L(x_4)$ is reductive by [Ri1].

Remarks 5.7. Let $x = u_\alpha u_{\rho-\alpha}U_\rho$. Then $L \cdot x$ is the dense orbit (4.4). An easy dimension argument shows that the connected semisimple part of $C_L(x)$ is generated by the root groups U_δ, where $\delta \in \Psi_L$ and δ is orthogonal to α. In particular, $C_L(x)^0$ is contained in a parabolic subgroup of L, while $C_L(x)$ itself is not. Namely, the involution in W_L that interchanges α and $\rho - \alpha$ leaves x invariant but is not in any proper parabolic subgroup of L.

In case of an abelian unipotent radical, one of the main results asserts that each (P, P) double coset intersects the opposite radical in a unique L-orbit. This gives rise to a natural bijection between the double cosets and the L-orbits on the opposite radical [RRS, Theorem 1.1]. The observations 5.8 and 5.10 below show that in the non-abelian situation the relative (P, P) double cosets in V_1^- do not separate the various L-orbits.

Lemma 5.8. *Both* $L \cdot x_3$ *and* $L \cdot x_4$ *are in* $\mathcal{V}(w_3)$.

Proof. This is clear for $L \cdot x_3$ by 5.1. Let G' be the simple subgroup of G corresponding to the subsystem of type A_2 spanned by α and $\rho - \alpha$ and let $B' = B \cap G'$ be the corresponding Borel subgroup. Then $u_{-\alpha}u_{-(\rho-\alpha)} \in B'w_\alpha B' \cdot B'w_{\rho-\alpha}B' = B'w_\alpha w_{\rho-\alpha}B' \subset Bw_\alpha w_{\rho-\alpha}B \subset Pw_\alpha w_{\rho-\alpha}P = Pw_3P$, the last equation by (3.5.4). The result follows from 4.4 and 4.7.

Corollary 5.9. $\dim \mathcal{V}(w_s) = \dim Pw_sP \cap V^-$ *for* $0 \leq s \leq 3$ *and* $\dim \mathcal{V}(w_4) = \dim(Pw_4P \cap V^-) - 1 = \dim V_1^-$.

Proof. Again this is clear for $s = 4$, as $L \cdot x_4 \subseteq \mathcal{V}(w_4)$ is dense in V_1^-. For $0 \leq s \leq 3$ this follows from 5.4 and 5.8.

More surprising than 5.8 is the following result which aserts that π is surjective when restricted to $Pw_4P \cap V^-$ in V^-.

Proposition 5.10. $\mathcal{V}(w_4) = V_1^-$. *In particular,* $\mathcal{V}(w_s) \subset \mathcal{V}(w_4)$ *for* $0 \leq s \leq 3$.

Proof. Since Pw_4P is the largest (P, P) double coset in G, $Pw_4P \cap V^-$ is open dense in V^-, and since π is an open map, $\mathcal{V}(w_4)$ is open dense in V_1^- and the complement of $\mathcal{V}(w_4)$ in V_1^- is L–invariant and closed. Thus, if the complement is not empty, then it contains a closed orbit. The only closed L–orbit in V_1^- is the trivial one, and since $1 \neq u_{-\rho} \in Bw_\rho B \subset Pw_\rho P = Pw_4P$ by (3.5.5), it is in $\mathcal{V}(w_4)$. But this implies that $\mathcal{V}(w_4)$ is all of V_1^-.

Lemma 5.11. $L \cdot x_s = \mathcal{V}(w_s)$ *for* $s = 0, 1, 2$.

Proof. This is clear for $s = 0$. By 5.1 and 5.4 $L \cdot x_s$ is open dense in $\mathcal{V}(w_s)$ for $s = 1, 2$. Since $L \cdot x_1$ is the unique non-trivial orbit of minimal dimension, the result follows for $s = 1$ and for $s = 2$ unless $L \cdot x_1 \subset \mathcal{V}(w_2)$. We have $1 \neq u_{-\rho} \in Pw_4P$, as seen in the proof of 5.10. Whence $u_{-\alpha}u_{-\rho} = u_{\rho-\alpha} \cdot u_{-\rho} \cdot u_{\rho-\alpha}^{-1} \in Pw_4P$ for any $u_{-\rho} \neq 1$. But this says that no representative of $L \cdot x_1$ is in $\mathcal{V}(w_2)$. Thus $L \cdot x_2 = \mathcal{V}(w_2)$, as desired.

Type of G	$\lvert\mathcal{A}/W_L\rvert$
B_n $(n \geq 4)$	6
D_4	7
D_n $(n \geq 5)$	6
F_4	5
E_n $(n = 6, 7, 8)$	5

The number of W_L–orbits on $\mathcal{A}$

References

[ABS] H. Azad, M. Barry, and G. Seitz, *On the structure of parabolic subgroups*, Com. in Algebra **18** (1990), 551–562.

[B] N. Bourbaki, *Groupes et algèbres de Lie, Chapitres 4,5 et 6*, Hermann, Paris, 1975.

[Ri1] R. Richardson, *Affine coset spaces of reductive algebraic groups*, Bull. London Math. Soc. **9** (1977), 38–41.

[Ri2] R. Richardson, *Finiteness Theorems for Orbits of Algebraic Groups*, Indag. Math. **88** (1985), 337–344.

[RRS] R. Richardson, G. Röhrle, and R. Steinberg, *Parabolic subgroups with abelian unipotent radical*, Invent. Math. **110** (1992), 649–671.

[R] G. Röhrle, *On the Structure of Parabolic Subgroups in Algebraic Groups*, (to appear in J. of Algebra).

[S] R. Steinberg, *Lectures on Chevalley Groups*, Yale University, 1967.

Mathematische Fakultät Universität Bielefeld 4800 Bielefeld, Germany

Contemporary Mathematics
Volume **153**, 1993

Bounds for Dimensions of Weight Spaces of Maximal Tori

GARY M. SEITZ

ABSTRACT. An upper bound is given for dimensions of weight spaces of arbitrary maximal tori of finite groups of Lie type, acting on irreducible modules in the natural characteristic.

Let G be a finite group of Lie type in characteristic p and V an irreducible module for G over an algebraically closed field of characteristic p. We establish an upper bound for dimensions of weight spaces of arbitrary maximal tori of G in their action on V, answering a question of R. Guralnick. The upper bound is roughly $\dim(V)/h$, where h is the Coxeter number of the associated simple algebraic group. There are two difficulties in establishing such a result. One is the open problem of determining dimensions of weight spaces of irreducible rational modules for simple algebraic groups. The other is the issue of when distinct weights at the level of algebraic groups can have the same restriction to a maximal torus of the finite group. This is complicated by the fact that G has several conjugacy classes of maximal tori and groups in distinct classes are typically nonisomorphic.

Given an irreducible representation of a simple algebraic group, rough information on dimensions of weight spaces relative to the dimension of the underlying space can be obtained by taking conjugates under the Weyl group. Of course, this does not work for the 0 - weight space, which is one of the most interesting cases. We get around this by using some curious arguments at the level of Lie algebras. To deal with the second issue we make use of a lemma from [1]. Our original proof of this lemma was complicated and was considerably improved by

---

1991 *Mathematics Subject Classification.* Primary 20G40; Secondary 20C20.
Key words and phrases. algebraic groups, weight spaces, maximal tori.
The author was supported in part by an NSF Grant.
This paper is in final form and no version of it will be submitted for publication elsewhere.

an elegant argument supplied by Steinberg. For completeness, we give the proof below (Lemma 2).

Let $\bar{G}$ be a simply connected simple algebraic group and σ a Frobenius morphism. Set $G = \bar{G}_\sigma$ and write $G = G(q)$ to indicate G is defined over the field of q elements. Suppose $\mathrm{rank}(\bar{G}) = n$ and h is the Coxeter number of $\bar{G}$. Let T be an arbitrary maximal torus of G. To state the main result we define an integer d, as follows. If $G \cong B_n(2^a)$ for $n \geq 2$, $G_2(3^a)$, or $F_4(2^a)$, let $d = -1, -2$, or -5, respectively. Otherwise, set $d = 1$.

THEOREM. *Let V be a nontrivial, irreducible, rational module for $\bar{G}$ and M a weight space for T on V. Assume $q > 4$ and $G \neq C_n(5)$, ${}^2B_2(8)$, or ${}^2F_4(8)$. Then*

$$\dim(M) \leq 1 + \dim(V) \cdot h/(h-1)(h+d).$$

If V is a twist of a restricted module, the bound can be improved to $1 + \dim(V)/(h+d)$. This will be clear from the proof of the theorem.

As an example, consider the Steinberg module for $SL_2(p)$. The 0 - weight space for the diagonal group has dimension 3. The bound of the last paragraph fails when $p = 5$, but holds for $p > 5$. So here the exclusion of $SL_2(5) = C_1(5)$ is essential.

Better bounds can probably be established by excluding certain modules. However, we note that if one considers the 0-weight space of a Frobenius twist of the adjoint module of $\bar{G}$, then the bound in the theorem is close to best possible.

The following is a consequence of the theorem and standard results concerning lifting irreducible representations of G to representations of $\bar{G}$.

COROLLARY. *Let $q > 8$ be a power of the prime p and let V a module for G over an algebraically closed field of characteristic p. Suppose G has no trivial composition factors on V. If T is any maximal torus of G and if M is any weight space of T on V, then*

$$\dim(M) \leq 1 + \dim(V)/n.$$

Let $\Sigma = \Sigma(\bar{G})$ be the root system of $\bar{G}$ with Weyl group W.

LEMMA 1. *Let $\alpha \in \Sigma$. The number of bases of Σ containing α is $|W|/h$.*

Proof. Let x be the number of bases containing α, y the number of conjugates of α under the Weyl group, and z the number of elements of a base with length equal to that of α. Count pairs (β, Π), such that β is a conjugate of α and Π is a base containing β. As W is regular on bases we get the equation $y \cdot x = |W| \cdot z$. Hence, $x = (|W| \cdot z)/y$.

If all roots in Σ are of the same length, then $y = |\Sigma|$ and $z = n$, which yields the result. In the other cases the number of long (short) roots in B_n; C_n; F_4; and G_2 is respectively $2n(n-1)$ $(2n)$; $2n$ $(2n(n-1))$; 24 (24); 6 (6). In each case we get the same number of bases containing α and this number is $|W|/h$. The result follows.

Fix a maximal torus T of G. Thus, there is a maximal torus $\bar{T}$ of $\bar{G}$ such that $T = \bar{T}_\sigma$.

LEMMA 2. *Let $q > 5$ and assume $G \neq Sz(8)$ or ${}^2F_4(8)$. If α and β are distinct roots in Σ, then $\alpha|T \neq \beta|T$. This holds for $q = 5$, unless $G = C_n(5)$ with $\beta = -\alpha$, a long root.*

Proof. The following argument is taken from (5.2) and (5.3) of [1]. Suppose $q \geq 5$ and $\alpha \neq \beta$ are roots with $\alpha|T = \beta|T$.

Let X be the character group of $\bar{T}$ and σ^* the action of σ on X. Then $X(\sigma^* - 1)$ is the annihilator of T (see the proof of II, 1.7 in [2]), so there is a weight ω in X with $\alpha - \beta = \omega(\sigma^* - 1)$.

We first establish the following claim. If γ is a root of Σ, then

$$(*) \qquad |\gamma| \leq 2|\omega|,$$

with equality only if $G = C_n, \gamma$ is long, and ω is conjugate to $\gamma/2$.

The Weyl group W is irreducible on $\mathbb{Q}\otimes X$ and preserves the form, so replacing γ by a conjugate we may assume $(\omega, \gamma) > 0$. Using the fact that ω is a weight and the Cauchy-Schwarz inequality we have

$$1 \leq 2(\omega, \gamma)/(\gamma, \gamma) \leq 2|\omega|/|\gamma|.$$

This gives the inequality in (*). If equality holds, then the second inequality implies $\omega = c\gamma$ for some c, while the first gives $c = 1/2$. But $\gamma/2$ can be a weight only when γ is a long root and $\bar{G} = C_n$. So this proves the claim.

We now return to the proof of the lemma. If G is a Suzuki or Ree group, set $q_1 = \sqrt{q}$. Otherwise, set $q_1 = q$. Then σ^* induces $q_1\tau$ on $\mathbb{R} \otimes X$, where τ is an isometry (see II, 1.5 in [2]). We then have

$$\begin{aligned}
|\omega(q_1\tau - 1)| &= |\alpha - \beta| \\
(1) \qquad &\leq |\alpha| + |\beta| && \text{(triangle inequality)} \\
(2) \qquad &\leq 4|\omega| && \text{(by (*))} \\
(3) \qquad &\leq (q_1 - 1)|\omega| && (q_1 \geq 5) \\
(4) \qquad &\leq |\omega q_1\tau| - |\omega| && (\tau \text{ is an isometry}) \\
(5) \qquad &\leq |\omega(q_1\tau - 1)| && \text{(triangle inequality)}
\end{aligned}$$

We must have equality at each stage. Equality in (1) forces α and β to be dependent, whence $\alpha = -\beta$. Equality in (2) and (*) imply $\bar{G} = C_n$. Equality in (3) implies $q_1 = q = 5$, while equality in (5) gives $\omega q\tau = d\omega$ with $d > 0$. As τ is an isometry, this gives $d = q$ and $\omega\tau = \omega$. Hence, $2\alpha = \alpha - \beta = \omega(q_1\tau - 1) = 4\omega$, so $\alpha = 2\omega$ and (*) shows α is long. This completes the lemma.

COROLLARY 3. *Let $q > 4$ and assume $G \neq C_n(5), Sz(8)$ or ${}^2F_4(8)$. If $\alpha \in \Sigma$, then T is not contained in* $\ker(\alpha)$.

Proof. Otherwise, α and $-\alpha$ would have the same restriction to T, contradicting Lemma 2.

At this point we assume the hypotheses of the theorem. Set $m = \dim(M)$.

The Lie algebra, $L(\bar{G})$, of $\bar{G}$ also acts on V. For $\alpha \in \Sigma$, let e_α denote a nonzero element of the $\bar{T}$ - weight space of $L(\bar{G})$ corresponding to α. So, e_α spans the Lie algebra of the $\bar{T}$ - root subgroup of $\bar{G}$ corresponding to α.

If Π is a base of Σ, let $M_\Pi = \cap_{\alpha\in\Pi} C_M(e_\alpha)$. Set $m_0 = \max(\dim(M_\Pi))$, the maximum over all bases of Σ.

LEMMA 4. $m - m_0 \leq \dim(V)/(h+1)$.

Proof. By Lemma 2, T has distinct weights on the spaces $e_\alpha M$ as α ranges over Σ. For Π a base of Σ, M/M_Π is isomorphic to a subspace of $V_\Pi = \sum_{\alpha\in\Pi} e_\alpha M$. Therefore, $\dim(V_\Pi) \geq m - m_0$.

Now consider $\sum_\Pi \sum_{\alpha\in\Pi} dim(e_\alpha M)$. By Lemma 1, we have included the term $\dim(e_\alpha M)$ with the same multiplicity for each root α and this multiplicity is $c = |W|/h$. We then have

$$(*) \quad c \cdot \sum_\alpha \dim(e_\alpha M) = \sum_\Pi (\sum_{\alpha\in\Pi} \dim(e_\alpha M) = \sum_\Pi \dim(V_\Pi) \geq |W|(m - m_0).$$

Moreover, $\sum_\alpha \dim(e_\alpha M) = \dim(\sum_\alpha (e_\alpha M) \leq \dim(V) - m$, since, by Lemma 2 and Corollary 3, the weights of T on $e_\alpha M$ are all distinct from each other and from that of T on M. Combining this with (*) we get $|W|(m - m_0) \leq c \cdot (\dim(V) - m) \leq c \cdot \dim(V) - c(m - m_0)$ and hence

$$(**) \qquad m - m_0 \leq (c/(c + |W|)) \cdot \dim(V).$$

The lemma follows as $c = |W|/h$.

We establish the theorem by induction on $\dim(V)$.

The Steinberg tensor product theorem [3] implies that V is a Frobenius twist of a module of the form $V_1 \otimes R$, where V_1 is a nontrivial restricted representation and R is the tensor product of twists of restricted representations, each with a different nontrivial twist. We may ignore the twist and assume $V = V_1 \otimes R$. Now $L(\bar{G})$ is trivial on R and it follows that $V|L(\bar{G})$ is homogeneous with irreducible summands isomorphic to $V_1|L(\bar{G})$.

Let Π be a base for which $m_0 = \dim(M_\Pi)$. For the moment, we exclude the cases $\bar{G} = B_n, C_n, F_4$, or G_2, with $p = 2, 2, 2$, or 3, respectively. Then the subalgebra of $L(\bar{G})$ generated by all e_α for $\alpha \in \Pi$ is the Lie algebra of the unipotent subgroup of the corresponding Borel subgroup of $\bar{G}$. As $L(\bar{G})$ is irreducible on V_1, we conclude that $M_\Pi \leq \langle v^+\rangle \otimes R$, where $\langle v^+\rangle$ is the fixed 1-space of $\langle e_\alpha : \alpha \in \Pi\rangle$ on V_1. Write $M_\Pi = \langle v^+\rangle \otimes D$.

If $m_0 \leq 1$, then Lemma 4 gives $m \leq 1 + \dim(V)/(h+1)$, which is better than the bound of the theorem. So now suppose $m_0 > 1$. Then R is not trivial. Now M is a T-weight space of V and e_α is trivial on $\langle v^+\rangle \otimes R$ for each $\alpha \in \Pi$. It follows that D is a T-weight space of R. Inductively, $\dim(D) - 1 \leq (h/(h^2 - 1))\dim(R)$. Counting conjugates of $\langle v^+\rangle$ under the Weyl group of $\bar{G}$, we see that $\dim(V_1) \geq h$

and so $\dim(V) \geq h \cdot \dim(R) \geq (h^2-1) \cdot (\dim(D)-1) = (h^2-1) \cdot (\dim(M_\Pi)-1) = (h^2-1) \cdot (m_0-1)$. This and Lemma 4 give

$$m - \dim(V)/(h^2-1) \leq m - (m_0 - 1) \leq 1 + \dim(V)/(h+1).$$

This yields,

$$m \leq 1 + (h/(h^2-1)) \cdot \dim(V),$$

as required.

We are left with the exceptional cases. The difficulty here is that the root vectors for fundamental roots do not generate the Lie algebra of the corresponding maximal unipotent group. We can omit the C_n case since $B_n(2^a) \cong C_n(2^a)$.

In the B_n and G_2 cases we can add an extra root vector, e_β, for β a long root in Σ and obtain a collection of root elements that generate the Lie algebra of the unipotent group. Indeed, if $\Pi = \{\alpha_1, \dots, \alpha_n\}$ with α_n short, we may take $\beta = \alpha_{n-1} + 2\alpha_n$ or $\alpha_1 + 3\alpha_2$, respectively. If $\bar{G} = F_4$, it is necessary to add two root elements. Here we use $\beta = \alpha_2 + 2\alpha_3$ and $\gamma = \alpha_2 + 2\alpha_3 + 2\alpha_4$. We now set $\tilde{\Pi} = \Pi \cup \{\beta\}$ or $\Pi \cup \{\beta, \gamma\}$ and call this an *augmented* base of Σ.

Repeat the above using augmented bases. Counting as in Lemma 1 we see that each long root is contained in c_ℓ augmented bases, where $c_\ell = |W|/2(n-1), |W|/6$, or $|W|/3$, according as $\bar{G} = B_n, F_4$, or G_2. Each short root is still contained in $c = |W|/h$ augmented bases. Proceed as in Lemma 4. We obtain an inequality like that in (*) only now the left side has two sums, one for short roots and one for long. The corresponding multiplicities are c and c_ℓ. Since $c_\ell > c$ we obtain (**) with c replaced by c_ℓ. We then obtain Lemma 4 with the inequality replaced by $m - m_0 \leq \dim(V)/(h+d)$. The rest of the proof proceeds as before.

References

[1] G. Seitz, *The root subgroups for maximal tori in finite groups of Lie type*, Pacific J. Math **106** (1983), 153-244.

[2] T. Springer and R. Steinberg, *Conjugacy classes*, Seminar on Algebraic Groups and Related Finite Groups, LNM 131, Springer-Verlag, Heidelberg, 1970, pp. E1-E99.

[3] R. Steinberg, *Representations of algebraic groups*, Nagoya Math. J. **22** (1963), 33-56.

Department of Mathematics, University of Oregon, Eugene, OR 97403, USA

E-mail address: seitz@bright.math.uoregon.edu

Contemporary Mathematics
Volume 153, 1993

Vector Bundles on Curves

C.S. SESHADRI

Dedicated to Prof. R. Steinberg on his 70th birthday

ABSTRACT. The moduli spaces of vector bundles on curves had been constructed using the Geometric Invariant Theory of Mumford. In an unpublished manuscript Faltings has sketched a method which avoids Geometric Invariant Theory. We give here an exposition of this method with essential improvements due to Madhav Nori.

1. Introduction

One of the most important applications of Mumford's Geometric Invariant Theory (GIT) is the construction of the moduli spaces of vector bundles on curves (cf. [7], [8], [13], [15]). In characteristic zero, it is possible to construct these moduli spaces (as projective varieties) when the degree of the bundles in question is coprime to the rank without the use of GIT (cf. Remark 5.5) but when the degree is not coprime to the rank (e.g. degree is zero) even to show that the required moduli space exists as a complex analytic space has not been possible so far without having recourse to GIT. When the characteristic of the ground field is $p > 0$, even the case when the degree is coprime to the rank (for construction as a projective variety) has been possible only by using GIT. Recall that GIT in char.$p > 0$, rests on what is known as "Mumford's Conjecture" whose proof by Haboush uses the celebrated Steinberg irreducibility theorem (cf. [3], [15], [19]).

Recently in an unpublished manuscript (cf. [2]) Faltings gives a construction of the moduli spaces of Higgs and parabolic bundles on smooth projective curves in characteristic zero without the use of GIT. Of course, one gets in particular, a

1991 Mathematics Subject Classification. 14H60.

construction of the usual moduli spaces of vector bundles on smooth projective curves without using GIT. Madhav Nori has pointed out an essential improvement of Faltings' proof (cf. Remark 4.3), which allows this construction to go through in arbitrary characteristic.

The main aim of this expository paper is to outline the above construction of Faltings (with the improvement due to Madhav Nori) for the case of the usual moduli spaces of vector bundles on smooth projective curves. The GIT approach is still much better (especially for the case when the degree of the bundles in question is not coprime to the rank); however it is instructive to have this alternate approach which may turn out to have some importance.

The results of Faltings give something new even if we admit GIT. It leads to a closer understanding of the homogeneous coordinate ring of the moduli space and raises interesting questions (cf. Remark 6.1). There is also a converse to the First main lemma so that we obtain a curious cohomological characterization of semi-stable bundles (cf. Theorem 6.2).

It seems that there should be no difficulty in extending the construction given here to the case of Higgs and parabolic bundles on smooth projective curves (in arbitrary characteristic). However, extending it to the higher dimensional case appears to present essential difficulties.

In retrospect one could say that the development of determinantal line bundles (cf. [4]) and Quillen's theorem on the positivity of these line bundles (cf. [12]) has made the present construction plausible.

One should also mention here that A. Hirschowitz has done work closely related to that of Faltings (cf. Remark 3.3 below).

After this paper was written, Madhav Nori has very recently communicated a more simplified approach which proves the two main lemmas of Faltings at the same time. This is given in Appendix II (§8).

It is a pleasure to thank Madhav Nori for all the help that he has given me in this work.

2. The determinantal line bundle

We fix a smooth projective curve X of genus $g \geq 2$ over an algebraically closed field k.

Let $\mathcal{F} = \{F_s\}_{s \in S}$ be a family of vector bundles on X parameterized by a scheme S i.e., $\mathcal{F}$ is a vector bundle on $X \times S$ and F_s is the restriction of $\mathcal{F}$ to $X \times s \approx X$. If $p : X \times S \longrightarrow S$ is the canonical projection, recall that $R^i p_*(\mathcal{F})$ $(i = 0, 1)$ is given locally on S as the cohomology of a complex $j : K^0 \longrightarrow K^1$ where K^i are free of finite rank (locally on S). Recall

DEFINITION 2.1. *We have a well defined line bundle $D(\mathcal{F})$ on S (called determinant of cohomology of $\mathcal{F}$) defined locally on S by* $(\det K^0)^{-1} \otimes (\det K^1)$, *where* "det" *denotes the usual determinant i.e., the highest exterior power.*

Suppose that $\chi(F_s) = 0$ for all $s \in S$ (suffices to assume for one $s \in S$). One sees that this is equivalent to assuming that rank $K^0 =$ rank K^1. Recall

DEFINITION 2.2. *Suppose that $\chi(F_s) = 0$ for all $s \in S$. Then we have a canonical section $\Theta(\mathcal{F}_1)$ of $D(\mathcal{F})$ (theta function) defined locally on S by the section* $(\det j)$ *of* $(\det K^0)^{-1} \otimes (\det K^1)$.

We have the following

LEMMA 2.3. *$\Theta(\mathcal{F})(s) \neq 0$ at $s \in S$ if and only if $h^0(F_s) = h^1(F_s) = 0$.*

The proof is immediate.

DEFINITION 2.4. *We say F_s is cohomologically trivial if $h^0(F_s) = h^1(F_s) = 0$.*

Thus $\Theta(\mathcal{F})(s) \neq 0$ if and only if F_s is cohomologically trivial.

Let E be a vector bundle on X. We denote by $E \otimes \mathcal{F}$ the family $\{E \otimes F_s\}_{s \in S}$ of vector bundles on X parameterized by S.

LEMMA 2.5. *Let E_1 and E_2 be two vector bundles on X such that* $\operatorname{rank} E_1 = \operatorname{rank} E_2$ *and* $\det E_1 = \det E_2$. *Then*

$$D(E_1 \otimes \mathcal{F}) = D(E_2 \otimes \mathcal{F}).$$

PROOF. Let $\mathcal{O}_X(1)$ be an ample line bundle on X. Then with the usual notations $E_1(m)$ and $E_2(m)$ are generated by global sections for some m. Then one knows that we have an exact sequence

$$0 \longrightarrow I \longrightarrow E_1(m) \longrightarrow \det(E_1(m)) \longrightarrow 0$$

where I is the trivial vector bundle of rank $(r-1)$ where $r = \operatorname{rank} E_1$. Hence we get

$$0 \longrightarrow I(-m) \longrightarrow E_1 \longrightarrow (\det E_1)((r-1)m) \longrightarrow 0$$

$$0 \longrightarrow I(-m) \otimes \mathcal{F} \longrightarrow E_1 \otimes \mathcal{F} \longrightarrow (\det E_1)((r-1)m) \otimes \mathcal{F} \longrightarrow 0.$$

Set $J_1 = I(-m) \otimes \mathcal{F}$ and $J_2 = (\det E_1)((r-1)m) \otimes \mathcal{F}$. By the properties of the functor $\mathcal{F} \longmapsto D(\mathcal{F})$ we have

$$D(E_1 \otimes \mathcal{F}) = D(J_1) \otimes D(J_2).$$

By hypothesis we have

$$0 \longrightarrow J_1 \longrightarrow E_2 \otimes \mathcal{F} \longrightarrow J_2 \longrightarrow 0.$$

Thus we conclude that $D(E_1 \otimes \mathcal{F}) = D(E_2 \otimes \mathcal{F})$.

LEMMA 2.6. *Suppose that for the family $\mathcal{F} = \{F_s\}_{s \in S}$, we have $\chi(F_s) = 0$. Let L be a line bundle on S. Then*

$$D(\mathcal{F}) = D(\mathcal{F} \otimes L) \quad (\textit{to be precise } D(\mathcal{F} \otimes p^*(L)).$$

PROOF. If $R^i p_*(\mathcal{F})$ is given locally on S as the cohomology of a complex $K^0 \longrightarrow K^1$, obviously $R^i p_*(\mathcal{F} \otimes L)$ is the cohomology of $K^0 \otimes L \longrightarrow K^1 \otimes \mathrm{L}$, The hypothesis $\chi(F_s) = 0$ implies that rank of K^0 = rank of K^1. Hence we have

$$(\det K^1) \otimes (\det K^0)^{-1} = (\det(K^1 \otimes L)) \otimes (\det(K^0 \otimes L))^{-1}$$

and then the above lemma follows.

LEMMA 2.7. *(Functoriality) Let $f : T \longrightarrow S$ be a morphism and $\mathcal{F}$ and $\mathcal{G}$ vector bundles on $X \times S$ and $X \times T$ respectively such that $(Id_X \times f)^*(\mathcal{F}) = \mathcal{G}$. Then we have*

(i) $f^*(D(\mathcal{F})) = D(\mathcal{G})$.

(ii) *If moreover* $\chi(F_s) = 0$, $f^*(\Theta(\mathcal{F})) = \Theta(\mathcal{G})$.

PROOF. This is immediate.

3. The first main lemma

LEMMA 3.1. *(First Main Lemma) Let F be a stable bundle on X. Then there exists a vector bundle E on X such that $E \otimes F$ is cohomologically trivial.*

PROOF. We first claim that there exists a vector bundle E_0 such that

(i) $\chi(E_0 \otimes F) = 0$.

(ii) E_0 is of arbitrary high rank.

(iii) $E_0 = \oplus_{i \in I} L_i$ - direct sum of line bundles L_i.

(iv) $\deg L_0$ is bounded by a constant independent of $r(E_0)$ (rank of E_0).

The above assertions are easily seen as follows. If V_i $(i = 1, 2)$ are vector bundles, we have

$$\frac{\chi(V_1 \otimes V_2)}{r(V_1)r(V_2)} = \mu(V_1) + \mu(V_2) - (g-1)$$

$\left(\text{recall } \mu(V_i) = \dfrac{\deg V_i}{r(V_i)}\right)$. Hence we have to choose E_0 as above such that

$$0 = \mu(E_0) + \mu(F) - (g-1).$$

If $E_0 = E_1 \otimes L$, $r(L) = 1$, we have

$$\mu(E_0) = \mu(E_1) + \deg L.$$

We can fix an L such that

$$-1 < \deg L + \mu(F) - (g-1) \leq 0.$$

Then we have to find E_1, as a direct sum of line bundles M_i, such that

$$\begin{cases} \mu(E_1) + x = 0, \quad \text{with} \quad -1 < x \leq 0 \\ x = \deg L + \mu(F) - (g-1). \end{cases}$$

We then see immediately that we can in fact find M_i such that $\deg M_i$ is either zero or one. Then

$$E_0 = \oplus L_i \quad \text{with} \quad L_i = M_i \otimes L$$

has the properties stated above.

Let $\{E_s\}_{s \in S}$ be a family of vector bundles on X parameterized by a scheme S such that

(a) $E_{s_0} = E_0$ for an $s_0 \in S$.

(b) The family has the versal property locally at s_0 i.e., we have a family of vector bundles $\{V_t\}_{t \in T}$ parameterized by a scheme T with $V_0 = E_0$ for $t_0 \in T$, then there is a morphism $f : T_0 \longrightarrow S$ where T_0 is a neighbourhood of t_0, $f(t_0) = s_0$ and the family $\{V_t\}_{t \in T_0}$ is the inverse of the family $\{E_s\}_{s \in S}$.

Suppose now that the above lemma is not true. Then $(E_s \otimes F)$ is *not* cohomologically trivial for every $s \in S$. Since $\chi(E_s \otimes F) = 0$ this means that

$$h^0(E_s \otimes F) = h^1(E_s \otimes F) \neq 0 \ \ \forall s \in S.$$

Further by the semi-continuity theorems, we can also suppose that $h^0(E_s \otimes F)$ (and hence $h^1(E_s \otimes F)$) is constant when s varies over a non-empty open subset of S. In fact we claim that we can suppose that we have a family of vector bundles $\mathcal{E} = \{E_s\}_{s \in S}$ parameterized by a scheme S such that $\forall s \in S$

(i) $h^0(E_s \otimes F) = h^1(E_s \otimes F) \neq 0$ and remains constant as s varies over S.

(ii) S is a versal family of simple bundles (dim End $E_s = 1$ $\forall s \in S$) with the correct dimension i.e., $\dim S = r^2(g-1)+1$ $(r = r(E_s))$. By local deformation theory note that since we are on a curve S is *smooth*.

(iii) There is an integer m_0 independent of the rank of E_s such that $E_s^*(m_0)$ is generated by global sections (E_s^* is the dual of E_s).

To see the above claim, note that since E_0 is a direct sum of line bundles whose degree is bounded by a constant independent of $r(E_0)$, there is an m_0 independent of $r(E_0)$ such that $E_0^*(m_0)$ is generated by global sections. Hence by the semi-continuity theorems, (i) and (iii) could be achieved by taking a suitable open subset U of S. We can also suppose E_s is *simple* when $s \in U$ (since we are working on a curve of genus ≥ 2, the set of simple bundles is *dense* in the versal family for E_0). Now to achieve (ii) take some $s \in U$ and a versal family S of the correct dimension for the simple bundle E_s. Again by restricting to suitable open subsets we can suppose that $\{E_s\}_{s \in U}$ is the inverse image by a morphism $g : U \longrightarrow S$ of the versal parameterized by S. Then denoting the versal family parameterized by U as $\{E_s\}_{s \in S}$, we see that it has the properties required above. Because of (1) and the fact that S is reduced, we have also the following:

(iv) if $p : X \times S \longrightarrow S$ is the canonical projection then $p_*(\mathcal{E} \otimes F)$ and $R^1 p_*(\mathcal{E} \otimes F)$ are *vector bundles* on S whose fibres at $s \in S$ are respectively isomorphic to $H^0(E_s \otimes F)$ and $H^1(E_s \otimes F)$ respectively.

Let $\mathcal{V} = \{V_a\}_{a \in A}$ be a family of vector bundles on X parameterized by a scheme A. Then recall that we have a canonical homomorphism

$$j : T_{A,a} \times H^0(V_a) \longrightarrow H^1(V_a)$$

where $T_{A,a}$ denotes the Zariski tangent space to A at a. This is easily seen as follows. By standard arguments we have only to show that if we take $A = \operatorname{Spec} D$, D the ring of dual numbers, then we have a canonical homomorphism

$$j_0 : H^0(V_a) \longrightarrow H^1(V_a),$$

V_a being the restriction of V to the closed subscheme X of $X_D = X \times \operatorname{Spec} D$. The kernel of the restriction homomorphism $\mathcal{V} \longrightarrow V_a$ which is a priori an $\mathcal{O}_{X_D}$-module has a canonical $\mathcal{O}_X$-module structure and this is easily seen to be isomorphic to V_a. We have therefore an exact sequence of $\mathcal{O}_{X_D}$-modules:

$$0 \longrightarrow V_a \longrightarrow \mathcal{V} \longrightarrow V_a \longrightarrow 0.$$

The connecting homomorphism $H^0(V_a) \longrightarrow H^1(V_a)$ is the homomorphism j_0 required above. Suppose now that $p_*(V)$ (p the canonical morphism $X_D \longrightarrow \operatorname{Spec} D$) is a vector bundle on Spec D whose fibre at 'a' is $H^0(V_a)$. Then the homomorphism $H^0(\mathcal{V}) \longrightarrow H^0(V_a)$ is surjective, which implies that the map j_0 is then zero.

The versal deformation space for V_a has tangent space $\approx H^1(V_a \otimes V_a^*)$ and we have a canonical homomorphism

$$\delta : T_{A,a} \longrightarrow H^1(V_a \otimes V_a^*) \quad \text{(Kodaira-Spencer map)}.$$

By functoriality etc., it is not difficult to see that we have a commutative diagram

$$\begin{array}{ccccc} T_{A,a} & \times & H^0(V_a) & \xrightarrow{j} & H^1(V_a) \\ \delta\downarrow & & id\downarrow & & id\downarrow \\ H^1(V_a \otimes V_a^*) & \times & H^0(V_a) & \longrightarrow & H^1(V_a) \end{array}$$

where the bottom row is the canonical map induced by the *contraction map*

$$(V_a \otimes V_a^*) \times V_a \longrightarrow V_a.$$

The bottom row should be viewed as the map j for the versal deformation space.

Let us now return to the family $\mathcal{E} \otimes F = \{E_s \otimes F\}_{s \in S}$, satisfying the properties (i) to (iv) above. By the preceding discussions, we have a canonical homomorphism

$$\text{(v)} \qquad T_{S,s} \times H^0(E_s \otimes F) \longrightarrow H^1(E_s \otimes F).$$

This homomorphism reduces to the zero map. Further we have the commutative diagram

$$\begin{array}{ccccc} T_{S,s} & \times & H^0(E_s \otimes F) & \longrightarrow & H^1(E_s \otimes F) \\ \delta\downarrow & & id\downarrow & & id\downarrow \\ H^1((E_s \otimes F) \otimes (E_s \otimes F)^*) & \times & H^0(E_s \otimes F) & \longrightarrow & H^1(E_s \otimes F) \end{array}$$

where the bottom row is the canonical map associated to the contraction

$$((E_s \otimes F) \otimes (E_s \otimes F)^*) \times (E_s \otimes F) \longrightarrow (E_s \otimes F)$$

and δ is the Kodaira-Spencer map. We have

$$T_{S,s} \approx H^1(E_s \otimes E_s)$$

and it is not difficult to see that δ identifies with the map (as F remains "constant" in the family)

$$H^1(E_s \otimes E_s^*) \longrightarrow H^1((E_s \otimes F) \otimes (E_s^* \otimes F^*))$$

induced from the canonical homomorphism

$$\begin{cases} k \longrightarrow F \otimes F^* \\ 1 \longmapsto (id). \end{cases}$$

From these considerations we deduce easily that the homomorphism (v) above identifies with the homomorphism

$$\text{(vi)} \qquad H^1(E_s \otimes E_s^*) \times H^0(E_s \otimes F) \longrightarrow H^1(E_s \otimes F)$$

induced by the canonical contraction map

$$(E_s \otimes E_s^*) \times (E_s \otimes F) \longrightarrow E_s \otimes F.$$

Since the map (v) is zero, it follows that (vi) is also the zero map. Using the duality theorem for cohomology we see that the map (vi) is equivalent to the canonical map

$$H^0(E_s \otimes F) \times H^0(E_s^* \otimes F^* \otimes K) \longrightarrow H^0(E_s^* \otimes E_s \otimes K)$$

(K being the canonical class on X)

or equivalently the canonical composition map

$$\text{(vii)} \qquad \mathrm{Hom}(E_s^*, F) \times \mathrm{Hom}(F, E_s^* \otimes K) \longrightarrow \mathrm{Hom}\,(E_s^*, E_s^* \otimes K)$$

(Hom denotes the space of global homomorphisms).

We see then that the canonical map (vii) is zero.

Let G_s be the sub-bundle of F generated generically by $\mathrm{Hom}(E_s^*, F)$. To be more precise, let $t_1, \cdots, t_\ell$ be a basis (over k) of $\mathrm{Hom}(E_s^*, F)$. Let t be the element of $\mathrm{Hom}(\oplus^\ell E_s^*, F)$ defined by $t = t_1 + \cdots + t_\ell$. Then G_s is the smallest sub-bundle of F containing the image of t. We have seen that $E_s^*(m_0)$ is generated by global sections (see property (iii) above). Now t induces a canonical homomorphism $t(m_0)$ of $\oplus^\ell E_s^*(m_0)$ into $F(m_0)$ and therefore a homomorphism of $H^0(\oplus^k E_s^*(m_0))$ into $H^0(F(m_0))$. We observe that $G_s(m_0)$ is the sub-bundle of $F(m_0)$ generated by the image of this map in $H^0(F(m_0))$. Note that $\deg G_s(m_0) \geq 0$, which implies that $\deg G_s$ is bounded below by a constant independent of the rank of E_s (since m_0 has this property by (iii)). Since G_s is a subbundle of F, $\deg G_s$ is also bounded above (for example by using the fact that F is stable, though this is not necessary). Thus $\deg G_s$ is bounded by an absolute constant independent of the rank of E_s. By cutting down S to a suitable open subset, we can assume without loss of generality that $\{G_s\}_{s \in S}$ is a family of subbundles of F. We see that the map (vii) factors as follows:

$$\begin{array}{ccccc} \mathrm{Hom}(E_s^*, F) & \times & \mathrm{Hom}(F, E_s^* \otimes K) & \longrightarrow & \mathrm{Hom}(E_s^*, E_s \otimes K) \\ \| & & \| & & \\ \mathrm{Hom}(E_s^*, G_s) & \times & \mathrm{Hom}(F/G_s, E_s^* \otimes K) & \longrightarrow & \mathrm{Hom}(E_s^*, E_s^* \otimes K) \end{array}$$

with

$$\begin{aligned} \mathrm{Hom}(E_s^*, F) &\simeq \mathrm{Hom}(E_s^*, G_s) \\ \mathrm{Hom}(F, E_s^* \otimes K) &\simeq \mathrm{Hom}(F/G_s, E_s^* \otimes K). \end{aligned}$$

We deduce that $G_s \neq 0$ and is a proper subbundle of F for otherwise we would have

$$\operatorname{Hom}(E_s^*, F) = H^0(E_s \otimes F) = 0$$
$$\operatorname{Hom}(F, E_s^* \otimes K) = H^1(E_s \otimes F) = 0$$

which leads to a contradiction.

Let Q be the Quot scheme of quotients of F with Hilbert polynomial equal to that of F/G_s. Then we see that $\dim Q$ is bounded by a constant independent of the rank of E_s, as a consequence of the fact that $\deg G_s$ has the same property. We have now a canonical morphism $f : S \longrightarrow Q$ defined by the family $\{F/G_s\}_{s \in S}$. Fix $s_1 \in S$ and let S_1 be the fibre of f through s_1. Then we see that

$$\text{(viii)} \qquad \begin{cases} \text{Codimension of } S_1 \text{ in } S \text{ is bounded by a constant} \\ \qquad \text{independent of the rank of } E_s. \end{cases}$$

Consider the family $\{E_s \otimes G_s\}_{s \in S}$. Then we have a canonical homomorphism (as in (v)).

$$\text{(v)}' \qquad T_{S,s} \times H^0(E_s \otimes G_s) \longrightarrow H^1(E_s \otimes G_s).$$

On the other hand we have also a canonical homomorphism (similar to (vi))

$$\text{(vi)}' \qquad H^1(E_s \otimes E_s^*) \times H^0(E_s \otimes F) \longrightarrow H^1(E_s \otimes F)$$

induced by the contraction map

$$(E_s \otimes E_s^*) \times (E_s \otimes F) \longrightarrow E_s \otimes F.$$

Unlike (v) and (vi), (v)′ and (vi)′ need not coincide but the *restriction* of (v)′ and (vi)′ to S_1 (i.e., to $T_{S_1,s}$) coincide (one understands easily what this restriction means since $T_{S_1,s}$ is a subspace of $T_{S,s} \approx H^1(E_s \otimes E_s^*)$) - because for the family $\{E_s \otimes G_{s_1}\}_{s \in S_1}$, G_s stays constant and then the considerations are as in the case of (v) and (vi) above. As in (v), the canonical map

$$\text{(v)}_1 \qquad T_{S_1,s} \times H^0(E_s \otimes G_{s_1}) \longrightarrow H^1(E_s \otimes G_{s_1}), \quad s \in S_1$$

is zero. Hence for the canonical map

$$H^0(E_s \otimes G_{s_1}) \times (H^1(E_s \otimes G_{s_1}))^* \longrightarrow T^*_{S,s_1}, \quad s \in S_1$$

dual to (v)′, the dimension of the image is bounded by a constant independent of the rank of E_s (since codim $T_{S_1,s}$ in $T_{S,s}$ is bounded by a constant independent of the rank of E_s). From our observations above that (v)′ and (vi)′ coincide when restricted to T_{S_1,s_1}, it follows that for the canonical composition map

$$\text{(vii)}_1 \quad \operatorname{Hom}(E_s^*, G_{s_1}) \times \operatorname{Hom}(G_{s_1}, E_s^* \otimes K) \longrightarrow \operatorname{Hom}(E_s^*, E_s^* \otimes K), \quad s \in S_1$$

the dimension of the image is bounded by a constant independent of the rank of E_s. We claim that this implies

$$\text{(ix)} \qquad \begin{cases} \qquad \dim \operatorname{Hom}(G_{s_1}, E_s^* \otimes K) \text{ (for } s \in S_1, \text{ say } s = s_1) \\ \text{is bounded by a constant independent of the rank of } E_s. \end{cases}$$

To prove this recall that G_{s_1} is the smallest submodule of F containing the image of $t : \oplus^\ell E^*_{s_1} \longrightarrow F$, where $t = t_1 + \cdots + t_\ell$ and $\{t_i\}$, $1 \leq i \leq k$, is a basis of $\mathrm{Hom}(E^*_{s_1}, F)$. We see that $t : \oplus^\ell E^*_{s_1} \longrightarrow G_{s_1}$ is generically surjective and hence surjective on a non-empty open subset U of X. Let $g_1, \cdots, g_b$ be a basis of $\mathrm{Hom}(G_{s_1}, E^*_{s_1} \otimes K)$. Then the composite maps $g_i \circ t_j$ $(1 \leq i \leq b, 1 \leq j \leq \ell)$ are in the image of the map (vii)$_1$. Hence $g_i \circ t = g_i \circ t_1 + \cdots + g_i \circ t_\ell$ are also in the image of (vii)$_1$. We see that $\{g_i \circ t\}$, $1 \leq i \leq b$, are linearly independent (by the surjectivity of t on U). Thus if $b \to \infty$, we would get a contradiction. This proves (ix). We have

$$\dim \mathrm{Hom}(G_{s_1}, E^*_{s_1} \otimes K) = h^1(E_{s_1} \otimes G_{s_1}).$$

Thus (ix) means that $h^1(E_{s_1} \otimes G_{s_1})$ is bounded by a constant independent of the rank of E_s. This implies that

$$\text{(x)} \quad \begin{cases} -\chi(E_{s_1} \otimes G_{s_1}) = h^1(E_{s_1} \otimes G_{s_1}) - h^0(E_{s_1} \otimes G_{s_1}) \text{ is bounded above by} \\ \qquad\qquad \text{a constant independent of the rank of } E_s. \end{cases}$$

On the other hand we have

$$\begin{aligned} -\frac{\chi(E_{s_1} \otimes G_{s_1})}{r(E_{s_1}) r(G_{s_1})} &= -\mu(E_{s_1}) - \mu(G_{s_1}) + (g-1) \\ &= (-\mu(E_{s_1}) - \mu(F) + (g-1)) + \mu(F) - \mu(G_{s_1}) \\ &= -\frac{\chi(E_{s_1} \otimes F)}{r(E_{s_1}) r(F)} + (\mu(F) - \mu(G_{s_1})) \\ &= \mu(F) - \mu(G_{s_1}) \quad (\text{since } \chi(E_{s_1} \otimes F) = 0) \\ &\geq \frac{1}{r(F)^2} \quad (\text{since } F \text{ is stable}). \end{aligned}$$

This implies that $-\chi(E_{s_1} \otimes G_{s_1}) \longrightarrow \infty$ as $r(E_s) \longrightarrow \infty$. This contradicts (x) above. This proves the first main lemma.

REMARK 3.2. (a) The proof of the first main lemma yields something stronger. Given a stable bundle F we found an E_0 such that $\chi(E_0 \otimes F) = 0$. Let $\{E_s\}_{s \in S}$ be a family of simple bundles parameterized by a smooth variety S which is versal at every point of S with $r(E_s) = r(E_0)$ (we can also suppose that $\dim S = r^2(g-1) + 1$, $r = r(E_0)$). Then the above proof shows that for generic s in S, $E_s \otimes F$ is cohomologically trivial, provided $r(E_s)$ is sufficiently large. Let S' be the closed subscheme of S consisting of $s \in S$ such that $\det E_s = \det E_0$. Then we have a stronger assertion, namely that for generic s in S', $E_s \otimes F$ is cohomologically trivial if $r(E_s)$ is sufficiently large. To see this we observe that $\mathrm{codim}\, S'$ in S is g and S' is also smooth. Then taking S'_1 similar to S_1 above, we observe that $\mathrm{codim}\, S'_1$ in S' is bounded by a constant independent of the rank of E_s. Then the above proof yields this stronger assertion, or we can say that we repeat the above argument for S' as for S.
(b) Let $F_1, \cdots, F_\ell$ be a finite number of stable bundles such that $\mu(F_1) = \mu(F_2) = \cdots = \mu(F_\ell)$. Then we observe that we can find E_0 such that $\chi(E_0 \otimes F) = 0$. Then if we choose $\{E_s\}_{s \in S'}$ as in (a), we see that for s generic in S', $(E_s \otimes F_j), 1 \leq j \leq \ell$, are cohomologically trivial. Suppose now that F is semi-stable and

$$F_1 \hookrightarrow F_2 \hookrightarrow \cdots \hookrightarrow F$$

a filtration such that $\mu(F_i) = \mu(F)$ and F_{i+1}/F_i is stable (it follows then that $\mu(F) = \mu(F_{i+1}/F_i)$), then if we choose E_s as above such that $E_s \otimes (F_{i+1}/F_i)$ are all cohomologically trivial, we deduce immediately that $E_s \otimes F$ is also cohomologically trivial. Thus more generally if $F_1, \cdots, F_\ell$ is a finite number of semi-stable bundles all with the same μ, we can find E_0 such that $\chi(E_0 \otimes F_i) = 0$, $1 \leq i \leq \ell$, and if we choose $\{E_s\}_{s \in S}$ as in (a) above, we see that for generic s in S', $(E_s \otimes F_j)$, $1 \leq j \leq \ell$, are all cohomologically trivial.

REMARK 3.3. In an unpublished paper A. Hirschowitz has proved the following result related to Lemma 3.1 (cf. §0.2.1, [1]). We are given integers r and d ($r > 0$). Let $n = \text{g.c.d.}(r, d)$. Then there exist a stable bundle E of rank r and degree d and a vector bundle F such that

(i) $\deg F = (-d + r(g-1))/n$, $r(F) = r/n$ (this implies that $\chi(E \otimes F) = 0$).

(ii) $h^0(E \otimes F) = h^1(E \otimes F) = 0$.

4. The second main lemma

Recall that if V is a semi-stable vector bundle on X, we have a filtration $V_1 \hookrightarrow V_2 \hookrightarrow \cdots \hookrightarrow V$ such that $\mu(V_i) = \mu(V)$ and V_{i+1}/V_i is stable (we have $\mu(V_{i+1}/V_i) = \mu(V)$). Then $\operatorname{gr} V = \oplus V_{i+1}/V_i$ is well-defined (i.e., independent of the filtration). If V and W are semi-stable with the same rank and μ, we say that V and W are gr-equivalent if $\operatorname{gr} V = \operatorname{gr} W$. If V is moreover stable then this equivalence is just isomorphism.

In this section, hereafter when we consider the family $\mathcal{F} = \{F_s\}_{s \in S}$ of vector bundles on X, S is a smooth projective curve, S.

LEMMA 4.1. *Let $\mathcal{F}$ be as above. Suppose that $\mathcal{F}\big|_{x \times S}$ is independent of $x \in X$ and hence equal to a vector bundle G on S. Then we have:*

(1) *$\mathcal{F}$ is obtained from G (or to be precise from $p^*(G)$, p being the canonical morphism $X \times S \longrightarrow S$) by a 1-cocycle $Z^1(X, \mathbf{Aut}\, G)$ ($\mathbf{Aut}\, G$ non-abelian sheaf on X with values in $\operatorname{Aut} G$) i.e., we have transition functions Θ_{ij} : $U_i \cap U_j \longrightarrow \operatorname{Aut} G$ with respect to an open covering $\{U_i\}$ of X defining a principal fibre space Θ on X with structure group $\operatorname{Aut} G$ and $\mathcal{F}$ is obtained by patching up the isomorphisms*

$$\Theta_{ij} : p_*(G)\big|_{U_j \times S} \longrightarrow p^*(G)\big|_{U_i \times S}.$$

Further F_s ($s \in S$) is the vector bundle obtained from Θ by extension of structure group associated to the canonical representation ρ_s of $\operatorname{Aut} G$ on the fibre G_s of G at s.

(2) *Let $G = \oplus^t H$ direct sum of t copies of a semi-stable vector bundle H. Then there is a vector bundle V of rank t on X such that $\mathcal{F} = V \otimes G$ (to be more precise equal to $q^*(V) \otimes p^*(G)$, q the canonical morphism $q : X \times S \longrightarrow X$).*

(3) *Let G be semi-stable. Then*

$$\deg(\mathcal{F}) = -\mu(G)\chi(F_s).$$

PROOF. We have $p^*(G)\big|_{x\times S} \simeq \mathcal{F}\big|_{x\times S}$. Consider the direct image sheaf $q_*(\mathrm{Hom}(p^*(G),\mathcal{F}))$. We see that it is a vector bundle on X whose fibre at $x \approx$ End G. From this it follows that given an isomorphism $G \longrightarrow \mathcal{F}\big|_{x\times S}$, this can be extended to an isomorphism over a neighbourhood U of x of $p^*(G)\big|_{V\times S}$ onto $\mathcal{F}\big|_{V\times S}$. From this it follows immediately that $\mathcal{F}$ is obtained by modifying $p^*(G)$ by an element of $Z^1(X, \mathrm{Aut}\, G)$. All the other assertions in (1) are now easy consequences.

Now to prove (2), if $G = \oplus^t H$, H stable, we observe that $\mathrm{Hom}(H,G) \simeq \oplus^t k$. From this it follows that $(q_*(p^*(H),\mathcal{F})$ is a vector bundle V on X of rank t. We see easily that $V\otimes H$ (or to be more precise $q^*(V)\otimes p^*(H)$) is isomorphic to $\mathcal{F}$ (in fact we have a bilinear map $V\otimes H$ to $\mathcal{F}$ etc.).

To prove (3) consider the following filtration of the semi-stable bundle G. Let H_1 be a stable subbundle of G such that $\mu(H_1) = \mu(G)$. Let G_1 be the largest subbundle of G of the form $\oplus^{r_1} H_1$ (r_1 copies of H_1). Repeating the process for G/G_1 we get a filtration

$$\begin{cases} G_1 \hookrightarrow G_2 \hookrightarrow \cdots \hookrightarrow G_\ell = G \\ G_i/G_{i-1} = \oplus^{r_i} H_i, \quad H_i \text{ stable, } \mu(H_i) = \mu(G) \\ \text{and } r_i \text{ is the largest with this property.} \end{cases}$$

Then we see that every automorphism of G leaves this filtration stable. From this it follows (by (1) and (2) of the Lemma) that we have a filtration of $\mathcal{F}$

$$\begin{cases} \mathcal{F}_1 \hookrightarrow \mathcal{F}_2 \hookrightarrow \cdots \hookrightarrow \mathcal{F}_\ell = \mathcal{F} \\ \mathcal{F}_i/\mathcal{F}_{i-1} = V_i \otimes H_i \\ V_i \text{ is a vector bundle on } X, \quad 1 \le i \le \ell, \quad V_0 = H_0 = \mathcal{F}_0 = (0). \end{cases}$$

We see that for $j = 0, 1$ and $1 \le i \le \ell$

$$\begin{cases} R^j p_*(V_i \otimes H_i) = (R^j p_*(V_i)) \otimes H_i, \\ R^j p_*(V_i) \text{ is a trivial vector bundle on } S \text{ of rank } h^j(V_i) \end{cases}$$

since V_i comes from X and H_i comes from S. From this it follows immediately that

$$D(V_i \otimes H_i) = (\det H_i)^{\chi(V_i)}$$

so that

$$\begin{aligned} \deg D(V_i \otimes H_i) &= -(\deg H_i)\chi(V_i) \\ &= -(\deg H_i)\chi(V_i) \\ &= -\frac{\deg H_i}{r(H_i)} \cdot r(H_i)\chi(V_i) \\ &= -\mu(H_i)\chi(\oplus^{r(H_i)} V_i) \\ &= -\mu(H_i)\chi((V_i \otimes H_i)\big|_{X\times s}), \quad s \in S. \end{aligned}$$

Since

$$D(\mathcal{F}) = \bigotimes_{i=1}^{\ell} (\det H_i)^{-\chi(V_i)}$$

it follows then easily that

$$\deg D(\mathcal{F}) = -\mu(G)\chi(F_s) \quad (\text{since } \mu(H_i) = \mu(G)).$$

LEMMA 4.2. *(Second main lemma) Let* $\mathcal{F} = \{F_s\}_{s\in S}$ *be a family of semi-stable vector bundles on* X*, parameterized by a smooth projective curve* S*. Let* E *be a vector bundle on* X *such that* $\chi(E \otimes F_s) = 0$*. Then*

$$\{\deg D(E \otimes \mathcal{F}) = 0\} \Longrightarrow \{\operatorname{gr} F_s\}_{s\in S} \text{ is constant as } s \text{ varies.}$$

PROOF. CASE I. Let us first consider the case when E is trivial of rank one. Then we have to prove:

$$\{\chi(F_s) = 0 \text{ and } \deg D(\mathcal{F}) = 0\} \Longrightarrow \{\operatorname{gr} F_s\}_{s\in S} \text{ is constant.} \tag{1}$$

We claim that we can suppose without loss of generality that $D(\mathcal{F})$ is in fact trivial. To see this, by Lemma 3.1 and Remark 3.2, we can find E_1 such that $\det E_1$ is trivial and $\Theta(E_1 \times \mathcal{F})$ does not vanish at an $s \in S$. We have $D(E_1 \otimes \mathcal{F})$ is of degree zero and has a non-trivial section. This implies that $D(E_1 \otimes \mathcal{F}) = D(E_0 \otimes \mathcal{F})$ (E_0 trivial) is trivial. Hence if the lemma had been proved for this case, one would conclude that

$$\{\operatorname{gr}(E_0 \otimes F_s) = \operatorname{gr}(\oplus^t F_s), t = r(E_0)\} \text{ is constant as } s \text{ varies.}$$

This implies that $\{\operatorname{gr} F_s\}_{s\in S}$ is constant, as is seen easily. Thus the above claim follows. Thus we have to show:

$$\{\chi(F_s) = 0 \text{ and } D(\mathcal{F}) \text{ trivial }\} \Longrightarrow \{\operatorname{gr} F_s\}_{s\in S} \text{ is constant.} \tag{2}$$

We assume then that the LHS of (1) holds. We observe that with the hypothesis $\chi(F_s) = 0$, $D(\mathcal{F})$ is trivial if and only if F_s is cohomologically trivial for every $s \in S$ (easy consequence using the fact that $\Theta(\mathcal{F})(s) \neq 0$ if and only if F_s is cohomologically trivial, see §2).

We claim that if L is a generic line bundle of degree zero, then $D(L \otimes \mathcal{F})$ is again trivial. For this we first show that $\deg D(L \otimes \mathcal{F}) = 0$ for any L of degree zero. First of all we have $\chi(L \otimes F_s) = 0$. Hence by Lemma 3.1 and Remark 3.2, we can find E_1 such that $\det E_1$ is trivial and $\Theta(E_1 \otimes L \otimes F_s) \neq 0$ at an $s \in S$. This implies that

$$\{\deg D(E_1 \otimes L \otimes \mathcal{F}) = \deg(D(L \otimes \mathcal{F})^{\otimes t}), \quad t = r(E_1)\} \text{ is } \geq 0$$

which implies that $\deg D(L \otimes \mathcal{F}) \geq 0$ (for any line bundle of degree zero). We have $\det(L \otimes L^{-1})$ is trivial. This implies that

$$D((L \oplus L^{-1} \otimes \mathcal{F}) = D(\mathcal{F})^{\otimes 2} = D(L \otimes \mathcal{F}) \cdot D(L^{-1} \otimes \mathcal{F}).$$

Hence we have

$$0 = \deg D(L \otimes \mathcal{F}) + \deg D(L^{-1} \otimes \mathcal{F})$$

and each term on the RHS is ≥ 0, which implies that $\deg D(L \otimes \mathcal{F})$ is zero. Now by the semi-continuity theorems, if we fix an $s_0 \in S$, for a generic L of degree zero $L \otimes F_{s_0}$ is cohomologically trivial (since when L is trivial this property holds). Then by the above discussion we get:

$$\{\text{for } L \text{ generic of degree zero, } \deg D(L \otimes \mathcal{F}) = 0 \text{ and } \Theta(L \otimes \mathcal{F})(s_0) \neq 0\}.$$

This implies the above assertion that for a generic L of degree zero, $D(L \otimes \mathcal{F})$ is trivial.

Let $x, y \in X$. Then by the foregoing discussions, for a generic line bundle L of degree one, we have $D(L(-x) \otimes \mathcal{F})$ and $D(L(-y) \otimes \mathcal{F})$ are trivial or equivalently $L(-x) \otimes F_s$ and $L(-y) \otimes F_s$ are cohomologically trivial for every $s \in S$ (the notation $L(-x)$ means L tensored by the ideal sheaf of x). We have

$$0 \longrightarrow L(-x) \longrightarrow L \longrightarrow L\big|_x \longrightarrow 0 \quad \text{exact (resp. for } y)$$

$$0 \longrightarrow (L \otimes F_s)(-x) \longrightarrow (L \otimes F_s) \longrightarrow (L \otimes F_s)\big|_x \longrightarrow 0 \quad \text{exact (resp. for } y).$$

Writing the cohomology exact sequence, we deduce that

$$\begin{aligned} H^0(L \otimes F_s) &\simeq (L \otimes F_s)\big|_x \\ &\simeq (L \otimes F_s)\big|_y \end{aligned}$$

which implies that

$$\begin{aligned} p_*(L \otimes \mathcal{F}) &\simeq (L \otimes \mathcal{F})\big|_{x \times S} \\ &\simeq (L \otimes \mathcal{F})\big|_{y \times S}. \end{aligned}$$

This gives that $(L \otimes \mathcal{F})\big|_{x \times S}$ is a constant family of vector bundles on S as x varies. Since L comes from X, it follows that $\{\mathcal{F}\big|_{x \times S}\}$ is a constant family of vector bundles on S as x varies. Denote by G the vector bundle $\mathcal{F}\big|_{x \times S}$ on S. Note that upto this point we have not used the hypothesis that F_s is semi-stable.

We shall show that we can assume without loss of generality that G is semi-stable i.e., the assertions (1) (or (2) above) would follow if we prove them for the case when G is semi-stable. Suppose then G is not semi-stable. Let

$$G_1 \hookrightarrow G_2 \hookrightarrow \cdots \hookrightarrow G_\ell = G \tag{3}$$

be the HN-filtration (Harder-Narasimhan filtration cf. [16]) for G with

$$\operatorname{gr}_{HN} G = \bigoplus_{i=1}^{\ell} H_i, \quad H_i = G_i / G_{i-1}, \quad 1 \leq i \leq \ell \quad (G_0 = (0)). \tag{4}$$

We have therefore the properties:

$$H_i \text{ is semi-stable and } \mu(H_1) > \mu(H_2) > \cdots > \mu(H_\ell). \tag{5}$$

Since the HN filtration is canonical, by (1) of Lemma 4.1 we obtain a canonical filtration of $\mathcal{F}$ as follows:

$$\begin{cases} \mathcal{F}_1 \hookrightarrow \mathcal{F}_2 \hookrightarrow \cdots \hookrightarrow \mathcal{F}_\ell = \mathcal{F}, \mathcal{F}_i\big|_{x \times S} \simeq G_i \\ \mathcal{F}_i / \mathcal{F}_{i-1} = \mathcal{G}_i \text{ and } \mathcal{G}_i\big|_{x \times S} \simeq H_i, \quad 1 \leq i \leq \ell. \end{cases} \tag{6}$$

This induces a filtration of F_s as follows:

$$(7)\qquad \begin{cases} F_{1,s} \hookrightarrow F_{2,s} \hookrightarrow \cdots \hookrightarrow F_{\ell,s} = F_s, \quad F_s = \mathcal{F}|X\times s \\ F_{i,s}/F_{i-1,s} = K_{i,s} = \mathcal{G}_i|X\times s, \quad 1 \leq i \leq \ell. \end{cases}$$

Then by (3) of Lemma 4.1, we get

$$\begin{aligned} \deg D(\mathcal{F}) &= -\sum_{i=1}^{\ell} \mu(H_i)\chi(K_i, s) \\ &= 0(\text{by } (1)). \end{aligned}$$

Using the simple computation

$$\begin{cases} \displaystyle\sum_{i=1}^{\ell} a_i b_i &= B_1(a_1 - a_2) + B_2(a_2 - a_3) + \cdots B_\ell(a_\ell) \\ \qquad B_j &= \displaystyle\sum_{i\leq j} b_i \end{cases}$$

and the fact that

$$\sum_{i\leq j} \chi(K_{i,s}) = \chi(F_{j,s})$$

we get

$$(8)\qquad \begin{cases} \deg D(\mathcal{F}) &= \displaystyle\sum_{1\leq i\leq \ell} -\chi(F_{i,s})(\mu(H_i) - \mu(H_{i+1})) = 0 \\ \mu(H_{\ell+1}) &= 0. \end{cases}$$

Since F_s is semi-stable and $\chi(F_s) = 0$, we have

$$(9)\qquad -\chi(F_{i,s}) \geq 0.$$

Besides we have

$$(10)\qquad \mu(H_i) - \mu(H_{i+1}) > 0, \quad 1 \leq i \leq \ell.$$

Using (8), (9) and (10), we deduce that

$$\begin{cases} \chi(F_{i,s}) = 0, \quad 1 \leq i \leq \ell \\ \text{i.e., } \mu(F_{i,s}) = \mu(F_s) = 0, \quad 1 \leq i \leq \ell. \end{cases}$$

This implies that $\{K_{i,s}\}_{s\in S}$ is a family of semi-stable bundles on X with $\mu(K_{i,s}) = \mu(F_s)$. Besides, by our hypothesis (i.e., the assertion (1) follows if G is semi-stable), we deduce immediately that $\{\text{gr } F_s\}$ is constant, as the filtration (7) has now the properties

$$\mu(F_{i,s}) = \mu(F_s) = \mu(K_{i,s}).$$

Thus we have completed the proof of the claim that it suffices to prove (1) (or (2)) when G is semi-stable.

Thus we have to prove (1) (or (2)) assuming that $G = \{\mathcal{F}|_{x\times S}\}$ which has been proved to be independent of x, is semi-stable. In this case, we claim that in fact we have $\{F_s\}_{s\in S}$ is constant (not merely $\{\mathrm{gr}\, F_s\}_{s\in S}$ is constant). For this we use the result of Madhav Nori that when s varies the $\mathrm{Aut}\, G$ modules G_s are all isomorphic (we shall indicate a proof of this in Appendix I). Then by (1) of Lemma 4.1, the claim follows. Thus we have proved Lemma 4.2 for the Case I as above.

CASE II (General case) Our hypotheses are $\chi(E\otimes F_s) = 0$ and $\deg D(E\otimes \mathcal{F}) = 0$ (of course F_s is semi-stable). Then by Case I, it follows that $(E\otimes\mathcal{F})|_{x\times S}$ is constant as x varies. Let then G be the bundle on S, $G = \mathcal{F}|_{x\times S}$. We have

$$(E\otimes\mathcal{F})|_{x\times S} = \oplus^{r(E)} G.$$

Again if G is semi-stable, we are through. Let then G be not semi-stable and

$$(3)' \qquad G_1 \hookrightarrow G_2 \hookrightarrow \cdots \hookrightarrow G_\ell$$

be the HN filtration for G with

$$(4)' \text{ and } (5)' \qquad \begin{cases} \mathrm{gr}_{HN} G = \bigoplus_{i=1}^{\ell} H_i, \quad H_i = G_i/G_{i-1}, \quad 1\le i\le \ell \\ \mu(H_1) > \mu(H_2) > \cdots > \mu(H_i). \end{cases}$$

We obtain again a filtration of $\mathcal{F}$:

$$(6)' \qquad \begin{cases} \mathcal{F}_1 \hookrightarrow \mathcal{F}_2 \hookrightarrow \cdots \hookrightarrow \mathcal{F}_\ell = \mathcal{F}, \quad \mathcal{F}_i|_{x\times S} \simeq G_i \\ \mathcal{F}_i/\mathcal{F}_{i-1} = \mathcal{G}_i \quad \text{and} \quad \mathcal{G}_i|_{x\times S} \simeq H_i, \quad 1\le i\le \ell. \end{cases}$$

This induces a filtration of F_s:

$$(7)' \qquad \begin{cases} F_{1,s} \hookrightarrow F_{2,s} \hookrightarrow \cdots \hookrightarrow F_{\ell,s} = F_s, \quad F_s = \mathcal{F}|_{x\times S} \\ F_{i,s}/F_{i-1,s} = K_{i,s} = \mathcal{G}_i|_{x\times S}, \quad 1\le i\le \ell. \end{cases}$$

Since H_i is semi-stable, as we did in Case I, using the result of Madhav Nori, it follows that $\{K_{i,s}\}_{s\in S}$ is constant. We have only to show

$$(11) \qquad \mu(F_s) = \mu(F_{i,s}) = \mu(K_{i,s})$$

to complete the proof of Lemma 4.2. For this we observe that the HN filtration of $(E\otimes\mathcal{F})|_{x\times S}$ is:

$$(3)'' \qquad \oplus^{r(E)} G_1 \hookrightarrow \oplus^{r(E)} G_2 \hookrightarrow \cdots \hookrightarrow \oplus^{r(E)} G_\ell, \quad G_i \text{ as in } (3)'.$$

Then we see easily that the canonical filtration of $E\otimes\mathcal{F}$ induced by this filtration is

$$(6)'' \qquad \begin{cases} E\otimes\mathcal{F}_1 \hookrightarrow E\otimes\mathcal{F}_2 \hookrightarrow \cdots \hookrightarrow E\otimes\mathcal{F}_\ell, \ (E\otimes\mathcal{F}_i)|_{x\times S} \simeq \oplus^{r(E)} G_i \\ (E\otimes\mathcal{F}_i)/(E\otimes\mathcal{F}_{i-1}) = (E\otimes\mathcal{G}_i), \\ \qquad\qquad (E\otimes\mathcal{G}_i)|_{x\times S} \simeq \oplus^{r(E)} H_i, \quad 1\le i\le \ell \end{cases}$$

where G_i, $\mathcal{F}_i$ are $\mathcal{G}_i$ are as in $(3)'$ and $(6)'$ above. The filtration $(6)''$ induces a filtration:

$$(7)'' \qquad \begin{cases} E\otimes F_{1,s} \hookrightarrow E\otimes F_{2,s} \hookrightarrow \cdots \hookrightarrow E\otimes F_{\ell,s} = E\otimes F_s \\ E\otimes F_{i,s}/E\otimes F_{i-1,s} = E\otimes K_{i,s} = (E\otimes\mathcal{G}_i)|_{X\times s}, \quad 1\le i\le \ell. \end{cases}$$

Then as in the case of Case I, it follows that

$$\chi(E \otimes F_{i,s}) = 0 \text{ i.e., } \mu(E \otimes F_{i,s}) = \mu(E \otimes F_s), \quad 1 \leq i \leq \ell.$$

This implies that

$$\begin{cases} \mu(F_i, s) & = \mu(F_s) \text{ and consequently} \\ \mu(K_i, s) & = \mu(F_s), \quad 1 \leq i \leq \ell. \end{cases}$$

This proves (11) above. This completes the proof of Lemma 4.2.

REMARK 4.3. The proof of Lemma 3.1 (First main lemma) is as in Faltings (cf. [2]). The proof of Lemma 4.2 (Second main lemma) is as in Faltings (cf [2]) upto the reduction to the case when G is semi-stable. The fact that this implies Lemma 4.2 without any hypothesis on the characteristic of the ground field is due to Madhav Nori.

5. Construction of the moduli space

We shall now outline a proof of the following result, proved in ([13], [16]) using GIT.

THEOREM 5.1. *Let $M(n, d)$ be the set of equivalence classes of semi-stable vector bundles of rank n and degree d on X under the* gr*-equivalence relation (cf §4). Then there is a natural structure of a normal projective variety of dimension $n^2(g-1)+1$ (recall genus of $X = g \geq 2$) on $M(n, d)$. Further if $\{V_t\}_{t \in T}$ is a family of vector bundles on X of rank n and degree d, then*

$$T^{ss}(\textit{resp. } T^s) = \{t \in T | V_t \textit{ is semi-stable (resp. stable)}\}$$

is open in T and the canonical map $T^{ss} \longrightarrow M(n, d)$ (resp. $T^s \longrightarrow M(n, d)$) is a morphism. In fact $M(n, d)$ is a coarse moduli space.

PROOF. The proof is by induction on the rank and may be divided into the following steps:

I. Given n, d there is a smooth quasi-projective variety $R(n, d) = R$ and a family of vector bundles $\mathcal{F} = \{F_s\}_{s \in R}$ of rank n and degree d on X parameterized by R such that

(a) given a *semi-stable* vector bundle V on X of rank n and degree d there is an $s \in R$ such that $V \simeq F_s$.

(b) R has the *local universal property* with respect to families of semi-stable bundles i.e., if $\mathcal{V} = \{V_t\}_{t \in T}$ is a family of *semi-stable* vector bundles on X of rank n and degree d parameterized by a scheme T, then given $t_0 \in T$, there is a neighbourhood T_0 of t_0 and a morphism $f : T_0 \longrightarrow R$ such that the family $\mathcal{V}$ is the inverse image by f of the family $\mathcal{F}$.

(c) There is a group $G = G(n)$ isomorphic to a PGL operating on R such that

$$F_s \simeq F_t \Longleftrightarrow t = gs; \quad g \in G \text{ and } s, t \in R.$$

The existence of R with $\mathcal{F}$, G etc. as above can be found in ([13], [16]). One shows that the set of semi-stable bundles of a given rank and degree is bounded so that there is an m sufficiently large such that if V is semi-stable bundle of rank n and degree d, $V(m)$ is the quotient of the trivial bundle E of rank equal to $h^0(V(m))$. One takes the quot scheme $Q = Q(E/P)$ of quotients of E with Hilbert polynomial P equal to that of $V(m)$. The group $G = PGL(E)$ operates canonically on Q and R is a suitable G stable open subset of Q. If $\{E \longrightarrow V_q(m)\}$, $q \in Q$ is the canonical family of quotients of E associated to Q, $F_s = V_s$ for $s \in R$.

We denote by R^{ss} (resp. R^s) the subsets of R:

$$R^{ss}(\text{resp. } R^s) = \{s \in R | F_s \text{ is semi-stable (resp. stable)}\}.$$

Obviously R^{ss} and R^s are G-stable subsets of R. They are usually shown to be open in R using GIT (cf. [13], [16]). We shall prove this below without using GIT.

II. *Semi-stable reduction.* Let $\mathcal{V} = \{V_t\}_{t \in T}$ be a family of vector bundles on X parameterized by a smooth curve T such that if $t \neq t_0$ ($t_0 \in T$), V_t is semi-stable. Then there is a morphism $f : T_1 \longrightarrow T$ which is finite over a neighbourhood of t_0 and a family $\mathcal{W} = \{W_t\}_{t \in T_1}$ of vector bundles on X parameterized by T_1 such that W_t is semi-stable for all $t \in T_1$ and $\mathcal{W}$ is the inverse image of the family $\mathcal{V}$ by f over $T - t_0$.

The above property can also be called as the *properness of the moduli functor* associated to $M(n,d)$ and would, à posteriori, imply the properness of $M(n,d)$ and can be seen to be equivalent to it. One way of proving the properness of the moduli functor for $M(n,d)$, which has now become standard is to prove an if and only if criterion between semi-simple vector bundles and semi-stable points in the GIT sense and then use GIT (cf. [14], [16], the best proof now is in [18]). The properness of the moduli functor for $M(n,d)$ without using GIT was first given by Langton (cf. [5]), in fact even for the higher dimensional case. We admit this proof (cf. [5], [6]).

III. *Openness of R^{ss} and R^s.* Because of our induction hypothesis, we can assume that $R^{ss}(m,e)$ and $R^s(m,e)$, $m < n$, are open in $R(m,e)$. We have a group $G(m)$ operating on $R(m,e)$ (see I above). Let Z be a scheme and $G(m)$ act trivially on Z. Consider the diagonal action of $G(m)$ on $R(m,e) \times Z$. Let Γ be a closed $G(m)$ stable subset of $R(m,e)^{ss} \times Z$. Then we claim that

$$(*) \qquad \begin{cases} p(\Gamma) \text{ is closed in } Z,\ p \text{ being the canonical projection} \\ p : R(m,e)^{ss} \times Z \longrightarrow Z. \end{cases}$$

By standard arguments, to prove $(*)$ we can assume without loss of generality that Z is a smooth curve and $p(\Gamma)$ is *dense* in Z, so that under these hypotheses one has only to show that $p : \Gamma \longrightarrow Z$ is surjective. We see easily that there is a closed subvariety $\Gamma_1 \hookrightarrow \Gamma$ such that $\dim \Gamma_1 = 1$ and $p(\Gamma_1)$ is dense in Z. By performing base change by $\Gamma_1 \longrightarrow Z$ etc., we can assume without loss of generality that $p : \Gamma_1 \longrightarrow Z$ is birational so that it is in fact an *open immersion*. If p is surjective, we are through. Let then $z_0 \in Z$, $z_0 \notin p(\Gamma_1)$. If $q : \Gamma_1 \longrightarrow R(m,e)^{ss}$ is the morphism induced by the projection on $R(m,e)^{ss}$, then the inverse image by q of the family $\{F_s\}_{s \in R(m,e)^{ss}}$, gives a family of semi-stable bundles parameterized by $p(\Gamma_1)$ (identifying Γ_1 with $p(\Gamma_1)$). Then by using II, we can "extend" this family to a family of semi-stable bundles parameterized by a variety which is finite over a neighbourhood of z_0. Using the local universal property of $R(m,e)^{ss}$, we conclude easily that if Γ_2 is the closure of the $G(m)$ saturation of Γ_1, then $z_0 \in p(\Gamma_2)$. But we have $\Gamma_2 \hookrightarrow \Gamma$ which implies that $z_0 \in p(\Gamma)$. From this we conclude easily that $Z = p(\Gamma)$. Thus the assertion $(*)$ is proved. We have used II, but strictly speaking, since we have used it only for lower rank, this could follow by our induction hypotheses.

We see easily that the following two statements are equivalent:

(i) Vector bundle V is *not* semi-stable.

(ii) There exists a vector bundle W such that

$$r(W) < r(V) \quad \text{and} \quad \mu(W) > \mu(V)$$

and there is a non-zero homomorphism

$$f : W \longrightarrow V.$$

Let $\{V_t\}_{t \in T}$ be a family of vector bundles of rank n and degree d parameterized by a scheme T (in particular $T = R$) and

$$T^{ss}(\text{resp. } T^s) = \{t | V_t \text{ is semi-stable (resp. stable)}\}.$$

We see also easily the following:

(iii) Let W be any semi-stable with $r(W) < n$. Then there is an m_0 such that if $\deg W > m_0$, the only homomorphism of W into V_t (any $t \in T$) is the zero homomorphism.

Now combining (i), (ii) and (iii), we see that

(iv) There exist e_1 and e_2 such that $t \notin T^{ss}$ if and only if there is a semi-stable vector bundle W such that

$$r(W) < n, \quad \mu(W) > \mu(V_t), \quad e_1 \leq \deg W \leq e_2$$

and there is a non-zero homomorphism

$$f : W \longrightarrow V_t.$$

Let now e be such that $e_1 \leq e \leq e_2$. Consider

$$\begin{cases} \Gamma_e \hookrightarrow R(m,e)^{ss} \times T \text{ defined by} \\ \Gamma_e = \{(s,t) | s \in R(m,e)^{ss}, \; t \in T \text{ and there is a} \\ \qquad \text{non-zero homomorphism } F_s \longrightarrow V_t\}. \end{cases}$$

By the semi-continuity theorems, Γ_e is closed in $R(m,e)^{ss} \times T$ and is obviously $G(m)$ stable. Hence by $(*)$, $p(\Gamma_e)$ is closed in T. By (iv)

$$T - T^{ss} = \bigcup_{e_1 \leq e \leq e_2} p(\Gamma_e)$$

so that $T - T^{ss}$ is closed and hence T^{ss} is open in T.

Now to prove that T_s is open, observe that $t \in T^{ss} - T^s$ if and only if there is a semi-stable bundle W such that $r(W) < n$, $\mu(W) = \mu(V_t)$ and there is a non-zero homomorphism $f : W \longrightarrow V_t$. Then a similar argument as above shows that $T^{ss} - T^s$ is closed in T^{ss}.

Note that the above argument for openness is almost the same as in [9] where unitary representations of certain Fuchsian groups play the same role as II here.

IV. *Construction of* $M(n,d)$. Consider the family $\mathcal{F} = \{F_s\}_{s \in R^{ss}}$ of semi-stable vector bundles of rank n and degree d parameterized by $R^{ss} = R^{ss}(n,d)$ having the properties in I above.

DEFINITION 5.2. *Let E be a vector bundle on X such that $\chi(E \otimes F_s) = 0$. Let $r = r(E)$ and $L = \det E$. We denote the line bundle $D(E \otimes \mathcal{F})$ on R^{ss} by $D(r, L, \mathcal{F})$ as $D(E \otimes \mathcal{F})$ depends only on r and L (cf. Lemma 2.5). As E varies with given r and L recall we have canonical sections $\Theta(E \otimes \mathcal{F})$ (theta functions - denoted also as $\Theta(E, \mathcal{F})$) of $D(r, L, \mathcal{F})$.*

Recall that we have a canonical action of $G = G(n)$ on R^{ss}. We claim that

$$(*) \qquad D(r, L, \mathcal{F}) \ \text{and} \ \Theta(E, \mathcal{F}) \ \text{are} \ G\text{-invariant}$$

To prove this, by I we have

$$F_s \simeq F_{g \cdot s}, \quad s \in R^{ss}, \quad g \in G.$$

From this by applying the semi-continuity theorems and using the fact that F_s is stable for $s \in R^s$ (which implies that $\operatorname{Aut} F_s \simeq k^*$, k base field), we deduce that

$$g^*(\mathcal{F}) \simeq \mathcal{F} \otimes p^*(M) \quad \text{on} \quad X \times R^s,$$

where M is a line bundle on R^s and g is considered as an automorphism of $X \times R^{ss}$ leaving $X \times R^s$ stable. Now R^{ss} is smooth. It can also be shown without much difficulty that

$$\operatorname{codim}(R^{ss} - R^s) \ \text{in} \ R^{ss} \geq 2.$$

Hence we deduce that

$$g^*(\mathcal{F}) \simeq \mathcal{F} \otimes p^*(M) \quad \text{on} \quad X \times R^{ss}.$$

Since $g^*(E \otimes \mathcal{F}) \simeq E \otimes g^*(\mathcal{F})$, we get

$$g^*(E \otimes \mathcal{F}) \simeq (E \otimes \mathcal{F}) \otimes p^*(M).$$

Now by Lemma 2.6, we deduce that

$$D(E \otimes \mathcal{F}) \simeq D(g^*(E \otimes \mathcal{F})), \quad g \in G.$$

In fact, it is not difficult to see that we have an action of G lifting the action of G on R^{ss}. We leave the details. Then by functoriality (Lemma 2.7), it follows that $g^*(\Theta(E, \mathcal{F})) = \Theta(E, \mathcal{F})$. Thus the claim $(*)$ is proved.

Now by Lemma 3.1 (First main lemma), we can find $E_0, \cdots, E_N$ such that the sections $\{\Theta(E_i, \mathcal{F})\}$, $0 \le i \le N$, have no base points on R^{ss}. Let

$$\theta : R^{ss} \longrightarrow \mathbb{P}^N$$

be the morphism into the projective space defined by $\{\Theta(E_i, \mathcal{F})\}$. More generally, it $\mathcal{V} = \{V_t\}_{t \in T}$ is a family of semi-stable vector bundles of rank n and d, parameterized by T, we get a morphism

$$\theta_T : T \longrightarrow \mathbb{P}^n$$

defined by $\{\Theta(E_i, \mathcal{V})\}$. Then the above $\theta = \theta_{R^{ss}}$. We refer to θ_T as *the theta morphism*, (we have of course fixed the E_i, $0 \le i \le N$). The theta morphisms are functorial in T. Now by II (semi-stable reduction) and $(*)$ of III, we see that $M_1 = \theta(R^{ss})$ is closed in $\mathbb{P}^N$ and hence it is a projective variety. Let $\phi : R^{ss} \longrightarrow M(n, d)$ be the set theoretic mapping such that if $s \in R^{ss}$, then $\phi(s) = \text{gr}\, F_s$. Then we claim that we have the following:

(1) a factorisation of θ

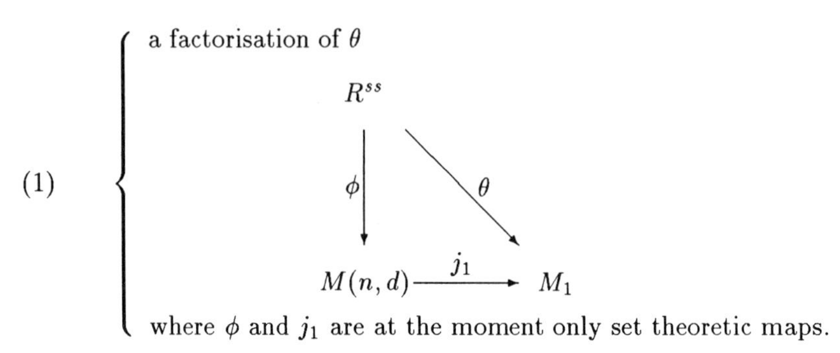

where ϕ and j_1 are at the moment only set theoretic maps.

To prove (1) one has to show that if $s_1, s_2 \in R^{ss}$ are such that $\text{gr}\, F_{s_1} = \text{gr}\, F_{s_2}$, then $\phi(s_1) = \phi(s_2)$. If F_{s_1} is stable, then we see immediately that the G-invariance of θ implies what we want. Suppose then F_{s_1} is not stable. Suppose for simplicity that

$$(2) \quad \begin{cases} 0 \longrightarrow F_1 \longrightarrow F \longrightarrow F_2 \longrightarrow 0, \quad F = F_{s_1} \\ F_1, F_2 \text{ stable}, \quad \mu(F_i) = \mu(F), \quad i = 1, 2 \\ \text{and the extension is not trivial.} \end{cases}$$

Then (2) defines a non-zero element a in $\text{Ext}\,(F_2, F_1)$. Then the set $\{ta\}_{t \in A^1}$ $(t \in k)$ defines a family of vector bundles $\mathcal{V} = \{V_t\}_{t \in \mathbb{A}^1}$ parameterized by the affine line $\mathbb{A}^1$ such that $V_1 = F$, $V_0 = F_1 \oplus F_2$ and

$$(3) \quad \begin{cases} 0 \longrightarrow q^*(F_1) \longrightarrow \mathcal{V} \longrightarrow q^*(F_2) \longrightarrow 0 \\ q \text{ canonical map } X \times \mathbb{A}^1 \longrightarrow X. \end{cases}$$

We see that all the vector bundles $\{V_t\}$, $t \neq 0$, are mutually isomorphic, in fact the canonical action of the multiplicative group $\mathbb{G}_m$ on $\mathbb{A}^1$ lifts to an action on the family $\mathcal{V}$. By the local universal property of R^{ss}, we see that there is a neighbourhood U of (0) in $\mathbb{A}^1$ and a morphism $\varepsilon : U \longrightarrow R^{ss}$ such that $\{V_t\}_{t \in U}$ is the inverse image of $\{F_s\}_{s \in R^{ss}}$ by ε. We see that $\varepsilon(U - (0))$ is contained in a

G-orbit in R^{ss}, since all V_t, $t \neq 0$, are mutually isomorphic. By the G-invariance of θ, it follows that θ is constant on $\varepsilon(U - (0))$ and hence it constant on $\varepsilon(U)$. From this we deduce easily that

$$\theta(s_1) = \theta(s_2), \quad \text{where} \quad F_{s_1} = F \quad \text{and} \quad F_{s_2} = F_1 \oplus F_2. \tag{4}$$

By an easy generalisation we deduce that if $F_{s_1} = F$ and $F_{s_2} =$ direct sum of stable bundles with $\operatorname{gr} F_{s_1} = \operatorname{gr} F_{s_2}$, then $\theta(s_1) = \theta(s_2)$. Thus the factorisation (1) follows. In fact, we have shown that this is just a consequence of the G-invariance of θ.

Consider the normalisation $i : M \to M_1$ of M_1 in the function field of R^{ss}. Then M is a normal projective variety and if $\delta : R^{ss} \longrightarrow M$ is the canonical morphism, we see that δ is also G invariant. Hence, as in (1) above, δ factorises through $M(n,d)$ and we have the following diagram (with the obvious commutativity properties):

(5)

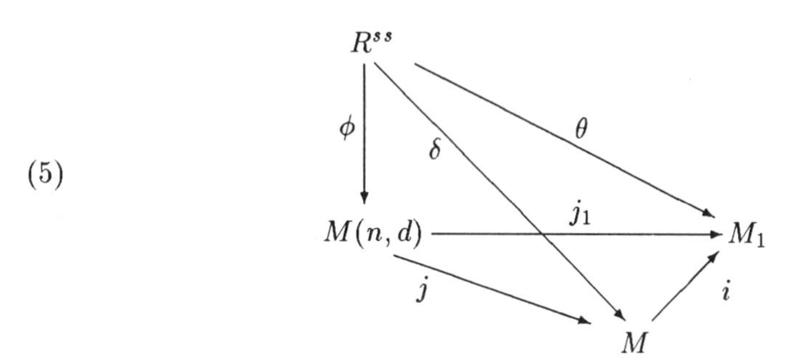

The map j is for the moment only set theoretic. What is to be shown is that j is bijective. Then $M(n,d)$ acquires the structure of a normal projective variety through j and it is not difficult to show that it satisfies the properties of Theorem 5.1.

To prove the bijectivity of j we shall use the properties of parabolic moduli (cf. [6],[16]) which can be proved without using GIT. Fix a point $p \in X$. For a vector bundle V of rank n and degree d, we consider the following parabolic structure:

(a) a flag of V at p:

$$V_p = F^1 V_p \supset F^2 V_p \supset \cdots \supset F^r V_p$$

such that $\dim F^i V_p$ is fixed i.e., the "flag type" is fixed.

(b) Weight α_i to $F^i V_p : 0 \leq \alpha_1 < \cdots < \alpha_r < 1$.

It is easily seen that the weights α_i and the flag type can be chosen so that we have (in fact we can take the flag type to be the "complete flag type" i.e., $r = n$ or the flag type with $r = 2$ and $\dim F^2 V_p = n - 1$):

(i) V parabolic semi-stable $\Longrightarrow$ V parabolic stable.

(ii) V parabolic semi-stable (hence parabolic stable by (i)) $\Longrightarrow$ underlying bundle is semi-stable.

(iii) V parabolic bundle with underlying bundle stable $\Longrightarrow V$ parabolic stable. It can be shown shown without much difficulty that we have then the following property:

(iv) V semi-stable (of course of rank n and degree d). Then there exists V' semi-stable such that $\operatorname{gr} V = \operatorname{gr} V'$ and V' has a parabolic structure such that V' is parabolic stable.

Then we can construct a smooth variety PR^s and a smooth morphism $g: PR^s \longrightarrow R^{ss}$ such that PR^s parameterizes all stable parabolic structures on the family $\{F_s\}_{s\in R^{ss}}$. Now g is not surjective but the image of g is an open subset R' of R^{ss} and given F_s, $s \in R^{ss}$, there is an F_t, $t \in R'$ such that $\operatorname{gr} F_t = \operatorname{gr} F_s$ (by (iv) above). We have $R^s \subset R'$. The fibre of g over $t \in R^s$is the flag variety B of flags of the vector space V_p of the type fixed above. More generally for $t \in R'$, the fibre of g over t is an open subset of B. The action of G on R^{ss} lifts to a proper free action of G on PR^s and g is a G-morphism. Now the orbit space

$$PM(n,d) = PR^s \bmod G$$

exists as an *algebraic space* (in the sense of M. Artin), further it is *complete* since the semi-stable reduction theorem holds also for parabolic semi-stable bundles (cf. [6],[16]). Then we have the following diagram (with the obvious commutativity properties):

(6)

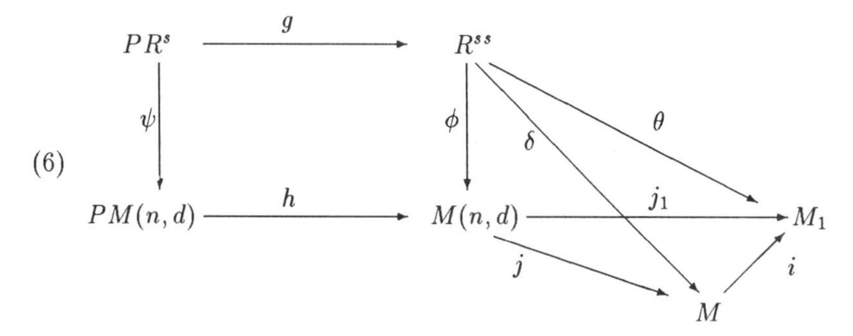

Here ψ is the canonical morphism $PR^s \longrightarrow PR^s \bmod G = PM(n,d)$. The map h is for the moment only set theoretic and induced by the G-invariant morphism g. However, the maps $j\circ h : PM(n,d) \longrightarrow M$ and $j_1 \circ h : PM(n,d) \longrightarrow M_1$ are morphisms. We note that h is surjective (by (iv) above). Of course j, j_1, θ and δ are surjective and i is a finite morphism.

We shall now show that the fibres of j_1 are finite. Let $m \in M_1$. Then $Z = (j_1 \circ h)^{-1}(m)$ is a closed subset of $PM(n,d)$. To show that $j_1^{-1}(m)$ is finite, it suffices to show that if W is an irreducible component of Z, then $h(W)$ reduces to the point of $M(n,d)$. Then we see easily that it suffices to show that if C is a closed irreducible *curve* in $PM(n,d)$ such that $(j_1 \circ h)(C) = m$, then $h(C)$ reduces to a point in $M(n,d)$. Let $D = \psi^{-1}(C)$. Then $\psi : D \longrightarrow C$ is a principal G-bundle which is the restriction to C of the principal G-bundle $\psi : PR^s \longrightarrow PM(n,d)$. We have a canonical family $\{W_t\}_{t\in PR^s}$ of parabolic stable bundles parameterized by PR^s and if we forget the parabolic structure, this family is simply the inverse image by g of the family $\{F_s\}_{s\in R^{ss}}$ parameterized by R^{ss}. Restrict this family to D and write it as $\mathcal{W} = \{W_t\}_{t\in D}$. Forgetting the parabolic structure we write this family as $\mathcal{W}' = \{W'_t\}_{t\in D}$. We have only to show that $\operatorname{gr} W'_t$ is independent of t. Observe that the morphism $(\theta \circ g) : D \longrightarrow M_1$ is the "theta morphism" into the projective $\mathbb{P}_N$ (i.e., it is defined by the functions $\Theta(E_i, \mathcal{W}')$). We have of course $(\theta \circ g)(D) = m$.

We can suppose without loss of generality that C is a closed *smooth* curve in $PM(n,d)$, otherwise we will have to work with its normalisation. Then the principal fibre space $\psi : D \longrightarrow C$ has local sections(since our structure group G is the projective group, this is a consequence of Tsen's theorem. The general case is due to Steinberg) $s_i : U_i \longrightarrow D$ where $\{U_i\}$ is an open covering of C. If $\mathcal{V}_i$ is the inverse image of $\mathcal{W}$ by s_i, then $\mathcal{V}_i$ defines a family of parabolic stable bundles on X parameterized by U_i. It is an easy exercise to show that $\{\mathcal{V}_i\}$ patch up to define a family of parabolic stable bundles $\mathcal{V} = \{V_t\}_{t\in C}$ on X parameterized by C. Forgetting the parabolic structure, we obtain a family $\mathcal{V}' = \{V'_t\}_{t\in C}$ of semi-stable vector bundles on X parameterized by C. Consider the "theta morphism" $\alpha : C \longrightarrow \mathbb{P}^N$ defined by the theta functions $\Theta(E_i, \mathcal{V}')$. By functoriality of "theta morphisms" if α_i is the restriction of α to U_i, we have

$$\alpha_i = \theta \circ g \circ s_i. \tag{7}$$

This implies that $\alpha_i(U_i) = m$ i.e., the "theta morphism" α maps C onto the single point $m \in M_1$. This implies that $\deg D(E_i \otimes \mathcal{V}) = 0$. Then by Lemma 4.2 (Second Main Lemma) it follows that $\operatorname{gr} V'_t$ is constant as t varies over C or that $\operatorname{gr} W'_t$ is constant as t varies over D. As remarked above, this shows that the fibres of j_1 are finite.

Let θ_s (resp. δ_s) denote the restriction of θ (resp. δ) to R^s. The action of G on R^s is a proper free action so that the orbit space $R^s \bmod G$ exists as a smooth algebraic space in the sense of M. Artin. Since θ and δ are G-invariant, θ_s and δ_s induce the following commutative diagram:

(8)

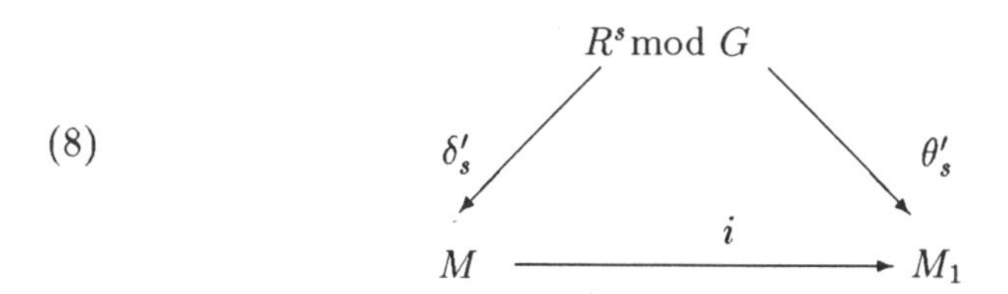

We see that θ'_s are quasi-finite since the fibres of f_1 are finite. By definition M is the normalisation of M_1 in the function R^{ss} but we see that in fact M can be taken to be the normalisation of M_1 in the G-invariant subfield of the function field of R^{ss}. But the G-invariant subfield of R^{ss} is the function field of $R^s \bmod G$. Hence M is the normalisation of M_1 in the function field of $R^s \bmod G$ and hence since $R^s \bmod G$ is smooth, we conclude that δ'_s is an open immersion. Thus $M^s = \delta'_s(R^s \bmod G) = \delta'_s(R^s)$ is open in M. It is not a priori clear that $\delta^{-1}(M^s) \subset R^s$ (for a point of R^{ss} and R^s may have the same image in M under δ). But we see that there is a non-empty open subset U of M^s such that $\delta^{-1}(U) \subset R^s$ (for one sees easily that $\delta(R^{ss} - R^s)$ is a proper closed subset of M).

Consider the morphism

$$\tau = (j \circ h) : PM(n,d) \longrightarrow M. \tag{9}$$

We claim that for every $u \in U$, $\tau^{-1}(u) \approx$ the flag variety B of the type fixed above. This is an immediate consequence of the fact that $\delta^{-1}(U) \subset R^s$ and $g : PR^s \longrightarrow R^{ss}$ is a fibre space over R^s with fibre B. Now by the connectedness theorem, all the fibres of τ are connected, since we have seen that the generic fibre of τ is connected. We shall now show that this implies the bijectivity of the map j.

The map j is of course surjective and the fibres of j are finite (since the fibres of j_1 are finite and i is a finite morphism, see (6) above). Suppose then that for $m \in M$, $j^{-1}(m) = \{m_i\}$, $i = 1, \cdots, r$, ≥ 2 and m_i are distinct points in $M(n,d)$. Each m_i represents an associated graded equivalence class of semi-stable vector bundles of rank n and degree d. Now if $\{V_t\}_{t \in T}$ is a family of semi-stable vector bundles of rank n and degree d on X parameterized by a scheme T, we see that the subset T_i defined by

$$T_i = \{t \in T | \operatorname{gr} V_t = m_i\}$$

is closed in T (the argument is as in II above, in fact simpler). Applying this to the family $\mathcal{W}' = \{W_t'\}_{t \in PR^s}$ of vector bundles on X parameterized by PR^s obtained by forgetting the parabolic structure of the canonical family $\mathcal{W} = \{W_t\}_{t \in PR^s}$ as parabolic vector bundles parameterized by PR^s, we obtain closed subsets H_i of PR^s:

$$H_i = \{t \in PR^s | \operatorname{gr} W_t' = m_i\}. \tag{10}$$

We see that H_i are G-invariant and mutually disjoint. Hence we have mutually disjoint (non-empty) closed subsets, $Z_i, (1 \leq i \leq r, r \geq 2)$ in $PM(n,d)$ such that $H_i = \psi^{-1}(Z_i)$. We have

$$\tau^{-1}(m) = \bigcup_{1 \leq i \leq r} Z_i \tag{11}$$

so that the fibre of τ over m is *not* connected. This leads to a contradiction. Thus j is bijective and the proof of Theorem 5.1 is complete.

REMARK 5.3. To prove that the fibres of j_1 are finite, we have used properties of parabolic moduli. This could be avoided if one uses the following (and a similar argument as above):

Let $\mathcal{V} = \{V_t\}_{t \in T_0}$ be a family of semi-stable vector bundles parameterized by a smooth (irreducible) curve T_0. Then we can find a finite cover $h : T_1 \longrightarrow T_0$ and a family of semi-stable vector bundles $\mathcal{W} = \{W_t\}_{t \in T}$, parameterized by the projective completion T of T_1 such that

$$\operatorname{gr} W_{h(t)} = \operatorname{gr} V_{h(t)}, \quad t \in T_1.$$

This assertion, which is a priori stronger than semi-stable reduction (cf. II above), could probably be proved in the same manner as II above. In any case the above assertion could be deduced by using parabolic moduli by arguments similar to those employed above.

However, it is not clear whether one can prove the bijectivity of j in a direct manner.

REMARK 5.4. In the above proof if char. $k = 0$ say $k = \mathbb{C}$, the use of parabolic moduli could be avoided as follows. The set $M(n,d)$ is a compact topological space being identified with the space of certain unitary representations of a Fuchsian group (cf. [9]). Then in the diagram (1) above, ϕ and j_1 are continuous maps. Now j_1 is a finite topological covering of M_1 (or we can use the diagram (9) where j is a finite topological covering of M). Then using a result of Grauert and Remmert, $M(n,d)$ gets a canonical complex structure such that j_1 (resp. j) is complex analytic. Then by standard results, $M(n,d)$ is projective, ϕ is a morphism etc.

REMARK 5.5. Let $k = \mathbb{C}$ and $(n,d) = 1$. Then Theorem 5.1 could be proved in one of the following two ways:

(a) The structure of a compact algebraic space on $M(n,d)$ is easily seen to be a Moishezon space (the field of meromorphic functions on $M(n,d)$ has transcendence degree equal to the dimension of $M(n,d)$). Now one can show that $M(n,d)$ has a canonical Kähler structure (cf. [10]). Then under these conditions, it is known that $M(n,d)$ is projective.

(b) Using [9] and local deformation theory, one sees that $M(n,d)$ has the structure of a compact complex analytic manifold. It is probably an easy consequence of Quillen's work (cf. [12]) that the line bundle $D(E \otimes \mathcal{F})$ considered in the proof of Theorem 5.1 has a canonical "Quillen metric" which is positive.

6. Some consequences

REMARK 6.1. The line bundle $D(r, L, \mathcal{F})$ on R^{ss} considered in the proof of Theorem 5.1 goes down to an ample line bundle on $M(n,d)$. We denote this by $D(r,L)$. It is not difficult to see that this does not depend upon the choice of R^{ss} but only on r and L. Let B be the graded k-algebra

$$B = \bigoplus_{t \geq 0} \Gamma(M(n,d), D(r,L)^t)$$

so that $M(n,d) = \operatorname{Proj} B$. Let $A(r,L)$ be the graded subalgebra of B generated by $\Theta(E,\mathcal{F})$ as E varies with $r(E) = r$ and $\det E = L$. By the proof of Theorem 5.1, it follows that B is finite over $A(r,L)$. We can ask how close $A(r,L)$ is to B.

Using the First and Second main lemmas (Lemma 3.1 and Lemma 4.2) and GIT, we get the following characterization of semi-stable bundles:

THEOREM 6.2. *Let F be a vector bundle on X. Then F is semi-stable if and only if there exists a vector bundle E on X such that $E \otimes F$ is cohomologically trivial i.e., $h^0(E \otimes F) = h^1(E \otimes F) = 0$.*

PROOF*. By Lemma 3.1 and Remark 3.2 (b), we see that if F is semi-stable, then there exists a vector bundle E such that $E \otimes F$ is cohomologically trivial.

*See later (Lemma 8.3) for a much simpler proof.

We have only to prove the converse. Suppose then that F is a vector bundle of rank n and degree d and there exists E such that $E \otimes F$ is cohomologically trivial. Let $R = R(n,d)$ be the smooth variety as in I in the proof of Theorem 5.1, parameterizing a family of vector bundles $\mathcal{F} = \{F_s\}_{s \in R}$. We can suppose that the $F \simeq F_{s_0}$ for some $s_0 \in R$. Now if $E^1 = \oplus^\ell E$ (direct sum of E ℓ times), we see that $E^1 \otimes F$ is also cohomologically trivial. Hence we can suppose without loss of generality that $r(E)$ is sufficiently large. Hence if $\phi : R^{ss} \longrightarrow M(n,d)$ is the canonical morphism, $D(E \otimes \mathcal{F})$ goes down to a very ample line bundle D on $M(n,d)$ (cf. Remark 6.1). Let $\xi = \det F$. Set

$$R_\xi = \{s \in R | \det F_s = \xi\}.$$

Then R_ξ is a closed G-stable subvariety of R and $s_0 \in R_\xi$. Set

$$\phi_\xi : R_\xi^{ss} \longrightarrow M(n,d)_\xi, \phi_\xi = \phi\big|_{R_\xi^{ss}}.$$

Here $M(n,d)_\xi$ is a closed subvariety of $M(n,d)$ and is the moduli space of semi-stable vector bundles V or rank n with $\det V = \xi$ under the gr-equivalence relation. Let $D(E \otimes \mathcal{F})_\xi$ denote the restriction of $D(E \otimes \mathcal{F})$ to R_ξ^{ss}. Then if D_ξ is the restriction of D to $M(n,d)_\xi$, we see that $D(E \otimes \mathcal{F})_\xi$ goes down to D_ξ.

Recall that R is an open G-stable subset of a Quot scheme Q. We have a good choice of a G-linearized very ample line bundle L on Q. Let Q_1 denote the closure of R in Q. It has been shown by C. Simpson (cf. [18]) that:

$$q \in Q_1 \text{ is GIT semi-stable } \Longleftrightarrow q \in R^{ss}. \tag{1}$$

Recall that $q \in Q_1$ is GIT semi-stable if and only if there exists a G-invariant section f of L^t $(t > 0)$ on Q_1 such that $f(q) \neq 0$. Let Q_ξ be the closure of R_ξ^{ss} in Q_1. Then we get

$$q \in Q_\xi \text{ is GIT semi-stable } \Longleftrightarrow q \in R_\xi^{ss}. \tag{2}$$

In (2) we take of course semi-stable points for the G linearized ample line bundle which is the restriction of L to Q_ξ. We denote this by L_ξ. By GIT the restriction of a suitable power of L_ξ to R_ξ^{ss} goes down to an ample line bundle on $M(n,d)_\xi$. We have the following result (cf [1]).

$$\text{Pic } M(n,d)_\xi \simeq \mathbb{Z}, \tag{3}$$

we deduce that there exist integers $a, b > 0$ such that the restrictions of L_ξ^a and $D(E \otimes \mathcal{F})^b$ to R_ξ^{ss} go down to the same line bundle on $M(n,d)$. Hence we can assume without loss of generality that

$$L_\xi \big|_{R_\xi^{ss}} = D(E \otimes \mathcal{F})_\xi. \tag{4}$$

Our hypothesis implies that there exists a G-variant section (namely $\Theta(E, \mathcal{F})$) which does not vanish at s_0. Hence we deduce that there is a G-invariant section δ of L_ξ on R_ξ^{ss} which does not vanish at s_0. By GIT we have

$$\begin{cases} M(n,d)_\xi = \text{Proj } B = \text{Proj } B', \quad \text{where} \\ B = \oplus_{m \geq 0} B_m, \quad B_m = G\text{-variant sections of } L_\xi^m \text{ on } Q_\xi \\ B' = \oplus_{m \geq 0} B'_m, \quad B'_m = \Gamma(M(n,d)_\xi, D_m^m). \end{cases} \tag{5}$$

We get

$$B_m = B'_m, \quad m \gg 0. \tag{6}$$

Hence we see that there is an ℓ such that δ^ℓ extends to a section (of course G-invariant) of L_ξ^ℓ to the whole of Q_ξ (in fact by taking the normalisation of Q_ξ which does not affect semi-stable points, we can take $\ell = 1$, (cf. Theorem 4.1, [17])). Thus without loss of generality δ can be taken as a G-invariant section of L_ξ on the whole of Q_ξ. Besides $\delta(s_0) \neq 0$, $s_0 \in R_\xi$. Then by (2) above we conclude that $s_0 \in R_\xi^{ss}$ i.e., $F_{s_0} = F$ is semi-stable. This proves the theorem.

7. Appendix I

Though not explicitly stated the following is essentially in [11].

THEOREM 7.1. *(Nori) Let V be a semi-stable vector bundle on X. Then as $x \in X$ varies the End V modules V_x are all mutually isomorphic. Hence the same fact holds for the* Aut V *modules V_x.*

PROOF. Consider the category $\mathcal{C}_\mu$ of vector bundles W on X such that $\mu(W) = \mu$. Consider the functor

$$\begin{cases} T_x : \mathcal{C}_\mu \longrightarrow (\text{Finite dimension } k\text{-modules}) \\ T_x(W) = W_x \end{cases}$$

is faithful i.e., the canonical map

$$\operatorname{Hom}(W_1, W_2) \longrightarrow \operatorname{Hom}(T_x W_1, T_x W_2)$$

is injective (cf [13]).

Let $\mu(V) = \mu$ so that $V \in \mathcal{C}_\mu$. Let $\mathcal{C}$ be the abelian subcategory of $\mathcal{C}_\mu$ generated by V i.e., the objects of $\mathcal{C}$ are subquotients of direct sums of V. The objects of $\mathcal{C}$ can be supposed to form a set. Let T_x be the functor

$$T_x : \mathcal{C} \longrightarrow (k\text{-mod}) \text{ - assumed finite dimensional}$$

$$T_x(W) = W_x, \quad x \in X.$$

The proof of the theorem could be divided into the following steps:

I. There is a uniquely determined finite dimensional k-algebra A such that we have the following factorisation for T_x:

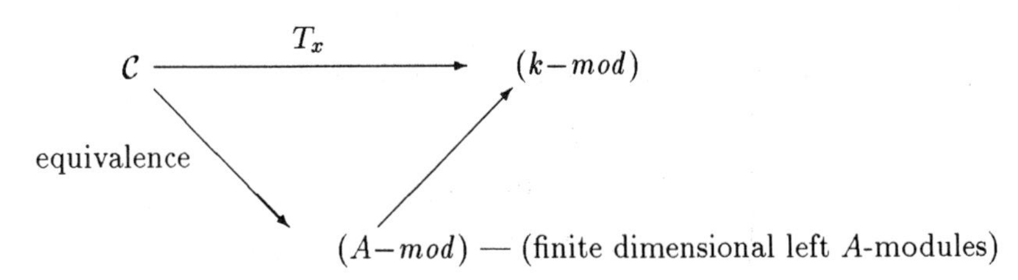

PROOF OF I. The crucial point is to define A and find the object in $\mathcal{C}$ (vector bundle) which would correspond to A by the above equivalence.

Let us now define A as follows:

$$A \hookrightarrow \prod_{W \in \mathrm{Obj}\, \mathcal{C}} \mathrm{End}(T_x W),$$

$$A = (a_W), \quad a_W \in \mathrm{End}\ W_x \quad (T_x W = W_x)$$

such that (a_W) satisfies the following condition

$$\begin{array}{ccc} T_x W_1 & \xrightarrow{T_x f} & T_x W_2 \\ a_{W_1} \uparrow & \curvearrowright & \uparrow a_{W_2} \\ T_x W_1 & \xrightarrow{T_x f} & T_x W_2 \end{array} \qquad \forall f \in Hom_{\mathcal{C}}(W_1, W_2) \tag{1}$$

We have a canonical homomorphism $A \longrightarrow \mathrm{End}(V_x)$ and we check easily that it is injective so that A is finite dimensional. Now if $S \hookrightarrow \mathrm{Obj}\ \mathcal{C}$ is a *finite* subset, we can define

$$A_S \hookrightarrow \prod_{W \in S} \mathrm{End}(W_x)$$

similar to $(*)$ above. We have $A = \varprojlim_S A_S$ and since A is finite dimensional, we have

$$A = A_S \quad \text{for some } S. \text{ We fix this S.} \tag{2}$$

Note that every object of $\mathcal{C}$ gets a canonical A module structure and this gives a canonical morphism

$$\mathcal{C} \longrightarrow (A\text{-mod}). \tag{3}$$

If W_1, W_2 are any objects of $\mathcal{C}$ we define

$$\begin{cases} (T_x W_1)^* \otimes W_2 = \oplus^{r(W_1)} W_2 \text{ with the identification} \\ ((T_x W_1)^* \otimes W_2)_x = W_{1,x}^* \otimes W_{2,x}. \end{cases} \tag{4}$$

The reason for this notation will become clear soon. Now given $f \in \mathrm{Hom}_{\mathcal{C}}(W_1, W_2)$, define

$$\begin{cases} f_1 : (T_x W_1)^* \otimes W_1 \longrightarrow (T_x W_1)^* \otimes W_2 \\ f_1 = Id \otimes f \\ f_2 : T_x W_2^* \otimes W_2 \longrightarrow (T_x W_1)^* \otimes W_2 \\ f_2 = (T_x f)^* \otimes Id. \end{cases} \tag{5}$$

We see that f_1 and f_2 are morphisms in $\mathcal{C}$. We now set

$$(6)\quad \begin{cases} g : (T_xW_1)^* \otimes W_1 \oplus (T_x^* W_2 \otimes W_2) \longrightarrow (T_xW_1)^* \otimes W_2 \\ g = f_1 - f_2 \quad \text{(the notation understood in the obvious sense)} \\ \oplus_{W \in S}(T_xW)^* \otimes W \xrightarrow{\alpha(f)} (T_xW_1)^* \otimes W_2 \\ \text{canonical projection} \searrow \qquad \swarrow g \\ (T_xW_1)^* \otimes W_1 \oplus (T_xW_2)^* \otimes W_2 \end{cases}$$

Note that $\alpha(f)$ defined by the above commutative diagram is again a morphism in $\mathcal{C}$. We observe that

$$(7)\qquad B(S) = \bigcap_{f \in \mathrm{Hom}_{\mathcal{C}}(W_1, W_2),\ W_i \in S} \ker \alpha(f), \quad B(S) \in \mathrm{Obj}\ \mathcal{C}.$$

We see that writing the kernel of $\alpha(f)$ at the fibre level at x is essentially writing the commutativity condition in (1) and thus without much difficulty we find that

$$\begin{cases} B(S)_x = A(S) = A. \\ B(S)_x \hookrightarrow \oplus_{W \in S}(T_X W)^* \otimes T_x W \end{cases}$$

Now $B(S)_x$ is a ring and we check that right multiplication is induced by an endomorphism of $\oplus_{W \in S}(T_xW)^* \otimes W$ which is a bundle containing $B(S)$ (all in $\mathcal{C}$). Hence right multiplication by elements of $B(S)_x$ induce endomorphisms of $B(S)$ (for if V_1 is a subbundle of V_2 with V_1, V_2 in $\mathcal{C}_\mu$ and $\beta \in \mathrm{End}\ V_2$ such that $T_x\beta(V_1, x) \subset V_{1,x}$, then one sees easily that $\beta(V_1) \subset V_1$). Then one sees easily that we have a canonical morphism (A-mod)$\longrightarrow \mathcal{C}$. On the other hand by (3) we have a canonical morphism $\mathcal{C} \longrightarrow$ (A-mod) and with these we get the required equivalence and the proof of I.

II. Let us now identify $\mathcal{C}$ with (A-mod) through T_x i.e., we identify W with the A module W_x obtained as above. Let y be another point of X. Consider the functor

$$T_y : \ \mathcal{C} \longrightarrow (k\text{-mod})$$

$$T_y(W_x) = W_y, \quad W \in \mathcal{C}.$$

Then we claim that there exists a right projective A module M such that

$$T_y(W_x) = M \otimes_A W_x.$$

This is an immediate consequence of the following:

Lemma 7.2. *Let S : (A-mod) $\longrightarrow$ (B-mod) be an exact functor (all finite dimensional). Then there exists ${}_BM_A$ which is right A projective such that*

$$S(W) = M \otimes_A W.$$

Proof. We set $M = S(A)$ and we observe that M has canonically left B and right A module structures. If $W \in$ (A-mod), writing a presentation

$A^{n_1} \longrightarrow A^{n_2} \longrightarrow W \longrightarrow 0$ and applying the functor S, we see easily that $S(W) = M \otimes_A W$. Because of exactness of S, M is A-flat which implies M is A-projective in our case.

III. Let us write the right projective A module M as M_y as it depends on y. Now M_y is the pull-back of a projective module M'_y over $A/\text{rad } A$. Now $A/\text{rad } A$ is semi-simple and finite dimensional and hence there exist only a finite number (upto isomorphism) of projective modules over $A/\text{rad } A$. Since y runs over the *connected* curve X, it follows that $M'_y \approx M'_x$. We have $M_x = A$ so that we can take $M_y = A$. Then we see that

$$\begin{aligned} T_y(W) &= W_y \\ &= A \otimes_A W_x \approx W_x. \end{aligned}$$

Hence the functor T_y can be *identified* with T_x in this manner. The theorem follows since End W (in $\mathcal{C}$) is the commutatant of A in End W_x.

8. Appendix II (Nori's method of proving Faltings' results)

A family $\mathcal{E} = \{E_s\}_{s \in S}$ of vector bundles on X, parameterized by a scheme S, is said to be *versal* if

(a) S is smooth, connected and

(b) the infinitesimal deformation map

$$\rho_s : T_{s,s} \longrightarrow H^1(\text{End } E_s) \quad (T_{S,s} = \text{tangent space at } s)$$

is surjective for every $s \in S$.

We say that the family $\mathcal{E} = \{E_s\}_{s \in S}$ is *versal with fixed determinant* if the family of line bundles $\{\det E_s\}_{s \in S}$ is constant and moreover in addition to (a) above, we have

(b)′ the infinitesimal deformation map

$$\rho'_s : T_{S,s} \longrightarrow H^1(\text{End}^0 E_s)$$

is surjective, where $\text{End}^0 E_s$ is the subbundle of End E_s consisting of endomorphisms of trace zero.

Since $g \geq 2$, one knows that versal families (resp. versal families with fixed determinant) exist in every rank and that

$$\begin{cases} \dim Z \geq m^2(g-1) + 1 (\text{resp. } \geq (m^2-1)(g-1)) \\ m = \text{rank } E_s. \end{cases}$$

Let

$$(*) \qquad 0 \longrightarrow \mathcal{E}' \longrightarrow \mathcal{E} \longrightarrow \mathcal{E}'' \longrightarrow 0$$

be an exact sequence of vector bundles on $X \times S$, S being a scheme. Following our practice, we write $\mathcal{E} = \{E_s\}_{s \in S}$, $\mathcal{E}' = \{E_s\}_{s \in S}$ and $\mathcal{E}'' = \{E''_s\}_{s \in S}$. Set

$$R = \ker\{\text{End } \mathcal{E} \longrightarrow \text{Hom}(\mathcal{E}', \mathcal{E}'')\}$$

where End and Hom represent locally free sheaves or vector bundles on $X \times S$. Now $R = \{R_s\}_{s \in S}$ defines a family of subbundles of End $\mathcal{E} = \{\text{End } E_s\}_{s \in S}$. Further R is a sheaf of Lie subalgebras of End $\mathcal{E}$, associated to a parabolic subgroup which is canonically associated to the exact sequence $(*)$ above. Then the infinitesimal deformation map $\rho_s = \rho(\mathcal{E})_s$ factors as follows:

$$\begin{array}{ccc} & T_{S,s} & \\ \rho(*)_s \swarrow & & \searrow \rho_s \\ H^1(X, R_s) & \xrightarrow{\ i\ } & H^1(X, \text{End } E_s) \end{array} \tag{1}$$

where i is the homomorphism induced by the canonical inclusion $R_s \longrightarrow \text{End } E_s$. Note that we have also canonical homomorphisms from R to End $\mathcal{E}'$ and End $\mathcal{E}''$ respectively and these induce homomorphisms:

$$\begin{aligned} i' &: H^1(X, R_s) \longrightarrow H^1(X, \text{End } E'_s) \\ i'' &: H^1(X, R_s) \longrightarrow H^1(X, \text{End } E''_s). \end{aligned} \tag{2}$$

We see that similar to (1) above, we have:

$$i' \circ \rho(*)_s = \rho(\mathcal{E}')_s, \quad i'' \circ \rho(*)_s = \rho(\mathcal{E}'')_s \tag{3}$$

Suppose now that

$$0 \longrightarrow \mathcal{E}' \longrightarrow \mathcal{E} \longrightarrow \mathcal{E}'' \longrightarrow 0 \tag{$(*)^0$}$$

is an exact sequence of vector bundles on $X \times S$ such that the family $\mathcal{E} = \{E_s\}_{s \in S}$ has fixed determinant. Then we set

$$R^0 = \ker\{\text{End}^0 \mathcal{E} \longrightarrow \text{Hom}(\mathcal{E}', \mathcal{E}'')\}$$

and the same considerations, as above, go through and we have $(1)^0$, $(2)^0$, $(3)^0$ similar to (1), (2), (3) above.

LEMMA 8.1. *Let $\mathcal{E} = \{E_s\}_{s \in S}$ be a versal family (resp. a versal family with fixed determinant) of vector bundles on X parameterized by S. Let S' be a closed (or more generally a locally closed) subscheme of S such that Codim $S' = \ell$. Suppose that we have an exact sequence*

$$0 \longrightarrow \mathcal{E}' \longrightarrow \mathcal{E}\big|_{X \times S'} \longrightarrow \mathcal{E}'' \longrightarrow 0 \tag{$*$}$$

of vector bundles on $X \times S'$. Then we have:

(a) $h^1(X, \text{Hom}(E'_s, E''_s)) \leq \ell, \quad s \in S'$

(b) *if the family $\mathcal{E} = \{E''_s\}_{s \in S}$ is constant, then*

$$h^1(X, \text{Hom}(E_s, E''_s)) \leq \ell, \quad s \in S'.$$

PROOF. For every $s \in S'$, we have now the following commutative diagram:

(4)

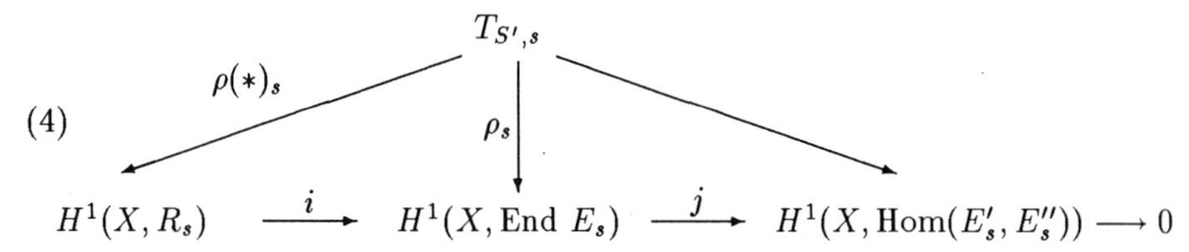

By our hypothesis (versality and the fact that codim S' in $S \leq \ell$) we see that $\text{codim}\, \theta \leq \ell$. Further j is surjective since $\text{Hom}(E'_s, E''_s)$ is a quotient bundle of End E_s and we are on a curve. It follows then that $\text{codim}\, j \circ \rho_s \leq \ell$. But the bottom row of the above diagram is exact. Then using the commutativity of the above diagram we see that $j \circ \rho_s = 0$. This implies the assertion (a) above.

To prove (b) note that we have the following commutative diagram:

$$\begin{array}{ccc} R_s & \hookrightarrow & \text{End } E_s \\ \downarrow & & \downarrow \\ \text{End } E''_s & \hookrightarrow & \text{Hom}(E_s, E''_s) \end{array}$$

where the vertical arrows are surjective. We have now the following commutative diagram:

(5)

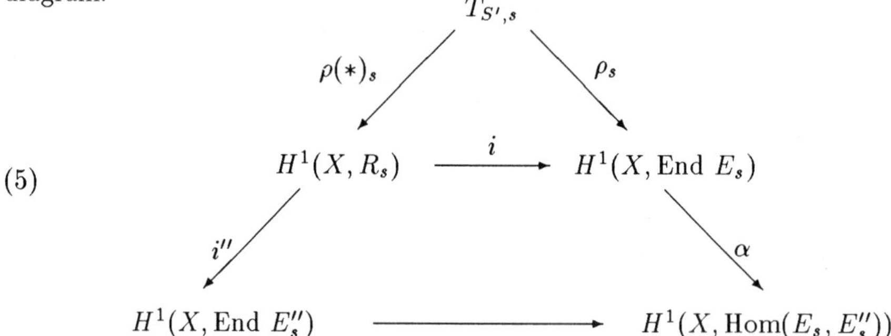

We have obvious exact sequences on the sides of the triangle. As we saw for (4) above, again α is surjective so that $\text{codim}\, \alpha \circ \rho_s \leq \ell$. Since the family $\mathcal{E}'' = \{E''_s\}_{s \in S}$ is constant, it follows that its infinitesimal deformation map $\rho(\mathcal{E}'')_s = i'' \circ \rho(*)_s$ is zero. Hence again by the commutativity of the above diagram, it follows that $\alpha \circ \rho_s = 0$. This proves (b).

To deal with the case of fixed determinant, we see that the above proof goes through replacing R by R^0 etc.

LEMMA 8.2. *Let $\mathcal{E}, \mathcal{E}', \mathcal{E}'', S$ and S' be as in Lemma 8.1. Suppose moreover that*

(i) $\chi(X, \text{Hom}(E_s, F)) = 0$ *(for all s or equivalently for one $s \in S$) where F is a vector bundle of rank r.*

(ii) rank $E'' \leq r$ *(or more generally $\leq t$). Then there is a constant $C = C(\ell, r)$ which depends only on ℓ and r (or more generally t) such that if* rank $\mathcal{F} >$ *C, we have*

$$\mu(E''_s) \geq \mu(F) \quad \forall \quad s \in S'$$

where μ denotes the slope of the vector bundle (i.e., equal to deg /rank *).*

PROOF. We have the following exact sequence

$$0 \longrightarrow \operatorname{End} \mathcal{E}'' \longrightarrow \operatorname{Hom}(\mathcal{E}, \mathcal{E}'') \longrightarrow \operatorname{Hom}(\mathcal{E}', \mathcal{E}'') \longrightarrow 0$$

of vector bundles on $X \times S$. Then by the property (a) of Lemma, we get

$$(6) \quad \begin{cases} -\ell & \leq \chi(X, \operatorname{Hom}(E'_s, E''_s)) \\ & = \chi(X, \operatorname{Hom}(E_s, E''_s)) - \chi(X, \operatorname{End} E''_s), \quad s \in S' \\ & = r(E_s)r(E''_s)(1 - g + \mu(E''_s) - \mu(E_s)) - r(E''_s)^2(1-g) \\ & \text{Here } r(E_s) \text{ denote the rank of } E_s \text{ etc.} \end{cases}$$

Now we have

$$\chi(X, \operatorname{Hom}(E_s, F)) = 0 \Longleftrightarrow 1 - g + \mu(F) - \mu(E_s) = 0.$$

Hence the R.H.S. of the inequality (6) is equal to

$$-r(E_s)r(E''_s)(\mu(F) - \mu(E''_s)) - r(E''_s)^2(1-g).$$

Then (6) takes the form

$$(7) \qquad r(E_s)r(E''_s)(\mu(F) - \mu(E''_s)) \leq \ell + r(E''_s)^2(g-1), \quad s \in S'.$$

Suppose that $\mu(F) - \mu(E''_s) > 0$, then we see easily that

$$r(F)r(E''_s)(\mu(F) - \mu(E''_s)) \geq 1.$$

Hence multiplying (7) by $r(F)$ we get

$$r(E_s) \leq \ell r(F) + r(F)r(E''_s)^2(g-1), \quad s \in S'.$$

Hence if we set $C = \ell r + rt^2(g-1)$, we see that if $r(F_s) > C$, we have

$$\mu(E''_s) \geq \mu(F) \ \forall \ s \in S'.$$

This completes the proof of the lemma.

LEMMA 8.3. *Let E, F be vector bundles on X such that* $\operatorname{Hom}(F, E) = F^* \otimes E$ *is cohomologically trivial. Then E is semi-stable (by symmetry it follows that F^* and hence F is semi-stable).*

PROOF. Suppose that E is not semi-stable. Then there is a subbundle G of E such that $\mu(G) > \mu(E)$. Since $\operatorname{Hom}(F, E)$ is cohomologically trivial, in particular we have $\chi(X, \operatorname{Hom}(F, E)) = 0$. This means that

$$1 - g + \mu(E) - \mu(F) = 0.$$

Hence we get

$$1 - g + \mu(G) - \mu(F) > 0.$$

Hence by Riemann-Roch, we deduce that $\chi(X, \operatorname{Hom}(F, G)) \neq 0$, which implies that there is a non-zero homomorphism of F into G and hence into E. This leads to a contradiction since $h^0(X, \operatorname{Hom}(F, E)) = 0$ and the lemma follows.

REMARK 8.4. Note that the above proof is quite elementary and far simpler than the one given for Theorem 6.2 above using GIT.

LEMMA 8.5. *Let $F_0, \cdots, F_\ell$ be stable bundles on X which are mutually distinct and $\mu(F_0) = \cdots = \mu(F_\ell)$. Let $F = F_0 \oplus \cdots \oplus F_\ell$ and $r = r(F)$. Let $\mathcal{E} = \{E_s\}_{s \in S}$ be a versal family (resp. versal family with fixed determinant) of vector bundles on X. We suppose also that there is an $s_0 \in S$ such that E_{s_0} is unstable i.e., E_{s_0} is not semi-stable. We suppose that $\chi(X, \mathrm{Hom}(E_s, F_0)) = 0$ (hence $\chi(X, \mathrm{Hom}(E_s, F_i)) = 0$, $0 \le i \le \ell$) and that $r(E_s) > C = C(\ell, r)$ (C as in Lemma 8.2). We see that such families exist. Set*

$$D_i = \{s \in S | h^0(X, \mathrm{Hom}(E_s, F_i)) \ne 0\}, \quad 0 \le i \le \ell.$$

Then $D_0 \cap \cdots \cap D_\ell$ is a closed subset of S of pure codimension $(\ell+1)$ (i.e., every irreducible component is of codimension $(\ell+1)$).

PROOF. We see that D_i is the zero locus of the section $\Theta(\mathcal{F}^*, E_i)$ of the line bundle $D(\mathcal{E}^* \otimes F_i)$. Hence if D_i is neither empty nor equal to S, then D_i is closed and of pure codimension one in S. Now by Lemma 8.3, we see that $s_0 \in D_i$ for all i, $0 \le i \le \ell$. Hence D_i is not empty. Now the above Proposition for $\ell = 0$ means that D_i is of pure codimension one. Hence to show this it suffices to show that D_i is not the whole of S. We shall first prove this even though it follows from the case of general ℓ. Suppose then say that $D_0 = S$. Then by replacing S by a non-empty open subscheme, we can suppose without loss of generality that we have a non-trivial homomorphism

$$\begin{cases} \alpha_0 : \mathcal{E} \longrightarrow p_X^*(F_0) \text{ on } X \times S \\ \mathcal{E}'' = \alpha_0(\mathcal{E}) \text{ is a subbundle of } p - X^*(F_0) \\ p_X : X \times S \longrightarrow X \text{ canonical projection.} \end{cases}$$

We have the following exact sequence of vector bundles

$$0 \longrightarrow \mathcal{E}' \longrightarrow \mathcal{E} \longrightarrow \mathcal{E}'' \longrightarrow 0 \quad (\mathcal{F}' = \ker \alpha_0).$$

Then if $\mathcal{E}'' = \{E_s''\}_{s \in S}$, by Lemma 8.2, we have

$$\mu(E_s'') \ge \mu(F_0), \quad s \in S.$$

But since F_0 is stable, this means that $E_s'' = F_0$ for all $s \in S$. Hence $\mathcal{E}'' = p_X^*(F_0)$ i.e., $\{E_s''\}_{s \in S}$ is a constant family. Hence by the property (b) of Lemma 8.1, we get

$$h^1(X, \mathrm{Hom}(E_s, F_0)) = 0, \quad s \in S.$$

Since $\chi(X, \mathrm{Hom}(E_s, F_0)) = 0$, we deduce that

$$h^0(X, \mathrm{Hom}(E_s, F_0)) = 0 \text{ for all } s \in S.$$

This leads to a contradiction, for $D_0 = S$ means that

$$h^0(X, \mathrm{Hom}(E_s, F_0)) \ne 0 \text{ for every } s \in S.$$

This proves the Proposition for the case $\ell = 0$.

The above argument extends easily to the case of a general ℓ as follows: We have seen that $D_0 \cap \cdots \cap D_\ell$ is non-empty. Since D_i is defined locally by one equation, (being the section of a line bundle), we see that

$$\operatorname{codim}(D_0 \cap \cdots \cap D_\ell) \leq \ell + 1.$$

Suppose that $(D_0 \cap \cdots \cap D_\ell)$ is not of pure codim ℓ. Then there exists a closed irreducible subscheme S' of S such that

$$S' \subset D_0 \cap \cdots \cap D_\ell, \quad \operatorname{codim} S' = \ell.$$

To show that this leads to a contradiction.

Replacing S' by a suitable open subset, we can assume without loss of generality that we have the following:

(i) We have homomorphisms $\alpha_i : \mathcal{E} \longrightarrow p_X^* F_i$, $0 \leq i \leq \ell$, such that $(\alpha_i)_s : E_s \longrightarrow F_i$ is non-zero $\forall\ i$ and $s \in S'$.

(ii) Set

$$\alpha = \bigoplus_{i=0}^{\ell} \alpha_i : \mathcal{E}\big|_{X \times S'} \longrightarrow p_X^*(F).$$

Then $\alpha(\mathcal{E}) = \mathcal{E}''$ is a subbundle of $p_X^*(E)$ so that if we write $\mathcal{E}'' = \{E_s''\}_{s \in S}$, E_s'' is a subbundle of F for all $s \in S'$:

$$0 \longrightarrow \mathcal{E}' \longrightarrow \mathcal{E} \longrightarrow \mathcal{E}'' \longrightarrow 0 \quad (\mathcal{E}' = \ker \alpha).$$

By Lemma 8.2, we have $\mu(E_s'') \geq \mu(F)$. Then by (ii) and the semi-stability of E, we see that

$$\mu(E_s'') = \mu(F) \quad \text{for all } s \in S'.$$

Now F is a direct sum of *distinct* stable bundles $\{F_i\}$ all with the same μ. Further E_s'' $(s \in S')$ is a subbundle of F with the same μ meets every F because of (1) above. We then conclude easily (using the fact that semi-stable bundles with the same μ form an abeilan category and F is a direct sum of distinct simple objects in this category) that in fact

$$E_s'' = F \quad \text{for all} \quad s \in S'.$$

Now by the property of Lemma 8.1, we have

$$h^1(X, \operatorname{Hom}(E_s, F)) \leq \ell, s \in S'.$$

But $\chi(X, \operatorname{Hom}(E_s, F)) = 0$, since $\chi(X, \operatorname{Hom}(E_s, F_i)) = 0$ for $0 \leq i \leq \ell$ and $s \in S'$. Hence we see that

$$h^0(X, \operatorname{Hom}(E_s, F)) \leq \ell, \quad s \in S'.$$

On the other hand since $(\alpha_i)_s \neq 0$, $0 \leq i \leq \ell$, $s \in S'$, we see that

$$h^0(X, \operatorname{Hom}(E_s, F)) = \sum_{i=0}^{\ell} h^0(X, \operatorname{Hom}(E_s, F_i)) \geq (\ell + 1).$$

This leads to a contradiction and the proposition follows.

PROPOSITION 8.6. *Let* $\mathcal{V} = \{V_t\}_{t\in T}$ *be a family of vector bundles on* X *parameterized by a scheme. Set*

$$T^{ss} = \{t \in T | V_t \text{ is semi-stable}\}.$$

Then T^{ss} *is open.*

PROOF. Because of the local nature of the assertion, we can suppose without loss of generality that $\mu(V_t)$ is constant as t varies. Let $t_0 \in T^{ss}$. It suffices to show that there is a closed subset D of T such that $T - T^{ss} \subset D$ and $t_0 \notin D$. By Lemma 8.1 (for the case $\ell \ominus 0$), we see that there is a vector bundle E such that $\operatorname{Hom}(E, V_{t_0})$ is cohomologically trivial. Since $\mu(V_t)$ is constant, we see that $\chi(X, \operatorname{Hom}(E, V_t)) = 0$ for all $t \in T$. Then if $t \in T - T^{ss}$, by Lemma 8.3, we see that $\operatorname{Hom}(E, V_t)$ is not cohomologically trivial. Hence if D is the zero set of the section $\Theta(E^*, \mathcal{V})$ of the line bundle $D(E^* \otimes \mathcal{V})$ we see that

$$T - T^{ss} \subset D, \quad t_0 \notin D.$$

REMARK 8.7. Note that the above proposition has been proved in a different manner in Theorem 5.1 above.

PROPOSITION 8.8. *Fix integers* d, $n(\geq 1)$ *and an integer* k *which is sufficiently large i.e.,* $k > C(1, n)$ *(the constant as in Lemma 8.2). Fix also a line bundle such that*

$$(*) \qquad 1 - g + \frac{d}{n} - \frac{\deg L}{k} = 0.$$

Suppose that we are given two semi-stable vector bundles V_1, V_2 *such that*

(a) $r(V_1) = r(V_2) = n$ *and* $\mu(V_1) = \mu(V_2) = d/n$.

(b) *There is a stable bundle in the decomposition of* $\operatorname{gr} V_1$ *into stable bundles which does not figure in that of* $\operatorname{gr} V_2$.

Then there exists a vector bundle E *such that*

(i) $\det E = L$ *and* $r(E) = k$, *which implies that* $1 - g + \mu(V_1) - \mu(E) = 0$, *and*

(ii) $\operatorname{Hom}(E, \operatorname{gr} V_1)$ *is not cohomologically trivial and* $\operatorname{Hom}(E, \operatorname{gr} V_2)$ *is cohomologically trivial.*

PROOF. Let $F_1, \cdots, F_t$ be the distinct stable bundles which figure in the decomposition of $\operatorname{gr} V_2$ and F_0 a stable bundle in that of $\operatorname{gr} V_1$ such that $F_0, F_1, \cdots, F_t$ are distinct. Let D_i, $0 \leq i \leq t$ be as in Lemma 8.5. Then applying Lemma 8.5(for $\ell = 0, 1$) we see that D_0 is of pure codimension one and $D_0 \cap D_i (1 \leq i \leq t)$ is of pure codimension two in S (S as in Lemma 8.5). Hence there exists an $s \in D_0$ such that $s \notin D_i$, $1 \leq i \leq t$. This implies that there exists E as desired such that $\operatorname{Hom}(E, F_0)$ is not cohomologically trivial and $\operatorname{Hom}(E, F_i)$ is cohomologically trivial $(1 \leq i \leq t)$. This implies immediately the assertion (ii) and the Proposition follows.

The construction of the moduli space $M(n, d)$ given in Theorem 5.1 can be simplified in many aspects using the above results. We shall now briefly indicate how this can be done.

As in the proof of Theorem 5.1, we take $R = R(n,d)$, the group G operating on R and the family $\mathcal{F} = \{F_s\}_{s\in R}$ of vector bundles of rank n and degree d, parameterized by R. Then by Proposition 8.6, it follows that R^{ss} is a G-stable open subset of R. We consider a versal family $\{E_s\}_{s\in S}$ of vector bundles of rank k with fixed determinant L (i.e., $\det F_s = L$) as in Proposition 8.5 and Lemma 8.5 above. Now the determinantal line bundle $D(E_s^* \otimes \mathcal{F})$ remains constant as s varies. Let us denote this line bundle on R^{ss} by $\mathcal{D}$. We claim that the theta functions $\{\Theta(E_s^*, \mathcal{F})\}$ which are sections of $\mathcal{D}$ are all contained in a finite dimensional vector space as s varies over S. One way of seeing this is to use the parabolic moduli used in the proof of Theorem 5.1. Using the functoriality of determinantal line bundles and theta functions as well as the G-invariance of theta functions (as in the proof of Theorem 5.1 above), we see that the theta functions $\Theta(E_s^*, \mathcal{F})$ can be identified with sections of a line bundle on the complete algebraic space $PM(n,d)$ and hence they are contained in a finite dimensional vector space. Let $\{\Theta(E_i^*, \mathcal{F})\}$, $0 \le i \le N$, be a basis of the linear space of sections of $\mathcal{D}$ generated by the theta functions $\{\Theta(E_s^*, \mathcal{F})\}$, $s \in S$. Then we have a morphism (theta morphism) and a factorisation as in the above proof of Theorem 5.1.

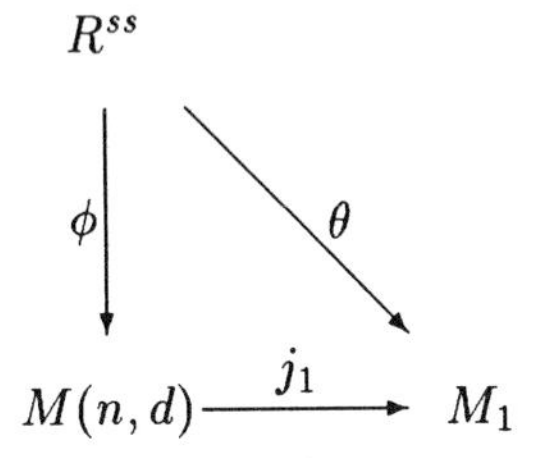

Of course j_1 is surjective. We see now that the fibres of j_1 are finite. This is an immediate consequence of Proposition 8.8. For by this Proposition, if $m_1, m_2 \in M(n,d)$ are represented by $\operatorname{gr} V_1$ and $\operatorname{gr} V_2$ and $j_1(m_1) = j_1(m_2)$, then the stable bundles figuring in the decomposition of $\operatorname{gr} V_1$ and $\operatorname{gr} V_2$ are identical but may occur with different multiplicities. Now it is immediate that the fibres of j_1 are finite. We show then that R^s is open and non-empty. We then take the normalization M of M_1 in the function field of R^{ss} and construct the moduli space as in the above proof of Theorem 5.1.

References

1. J.M. Drezet and M.S. Narasimhan, Groupe de Picard des variétés de modules de fibrés semi-stables sur les Courbes algébriques, Invent. Math. **97** (1989), 53-94.

2. G. Faltings, Unpublished manuscript.

3. W.J. Haboush, Reductive groups are geometrically reductive, Ann. of Math. **102** (1975), 67-83.

4. F. Knudsen and D. Mumford, The projectivity of the moduli space of stable curves I: Preliminaries on "det" and "Div", Math. Scand. **39** (1976), 19-55.

5. S. Langton, Valuative criteria for families of vector bundles on algebraic varieties, Ann. of Math. **101** (1975), 88-110.

6. V. Mehta and C.S. Seshadri, Moduli of vector bundles on curves with parabolic structures, Math. Ann. (1980), 205-239.

7. D. Mumford, Projective invariants of projective structures and applications, Proc. Intern. Cong. Math. Stockholm, (1962), 526-530.

8. D. Mumford and J. Fogarty, Geometric invariant theory, Second enlarged edition, Springer Verlag (1982).

9. M.S. Narasimhan and C.S. Seshadri, Stable and unitary vector bundles on a compact Riemann surface, Ann. of Math. **82** (1965), 540-567.

10. M.S. Narasimhan, Elliptic operators and differential geometry of moduli spaces of vector bundles on compact Riemann surfaces, Proc. Intern. Conf. on Functional Analysis and related topics, Tokyo, (1969), 68-71.

11. Madhav Nori, The fundamental group scheme, Proc. Indian Acad. Sci. (Math. Sci.) **91** (1982), 73-122.

12. D. Quillen, Determinants of Cauchy-Riemann operators over a Riemann surface, Funct. Anl. Appl. **19** (1985), 31-34.

13. C.S. Seshadri, Space of unitary vector bundles on a compact Riemann surface, Ann. of Math. **85** (1967), 303-336.

14. C.S. Seshadri, Mumford's conjecture for $GL(2)$ and applications, Proc. Bombay Colloq. on alg. geometry, Oxford Univ. Press, (1969), 347-371.

15. C.S. Seshadri, Theory of moduli, Proc. Symposia in Pure Math. Vol. XX1X, Algebraic Geometry, Arcata (1974).

16. C.S. Seshadri, Fibrés vectoriels sur les Courbes algébriques (rédigé par J.M. Drezet), Astérisque 96, (1982).

17. C.S. Seshadri, Quotient spaces modulo reductive algebraic groups, Ann. of Math. **95** (1972), 511-556.

18. Carlos T. Simpson, Moduli of representations of the Fundamental group of a smooth projective variety, To appear.

19. R. Steinberg, Representations of algebraic groups, Nagoya Math. J. 22, (1963), 33-56.

School of Mathematics, SPIC Science Foundation, 92, G.N. Chetty Road,
T. Nagar, Madras-600 017, Tamil Nadu, India.
E-mail address: css@ssf.ernet.in

Recent Titles in This Series

(Continued from the front of this publication)

(See the AMS catalog for earlier titles)

47